BIO-INSPIRED TECHNOLOGIES FOR THE MODERN WORLD

Emerging Trends and Applications

BIO-INSPIRED TECHNOLOGIES FOR THE MODERN WORLD

Emerging Trends and Applications

R. Ramakrishna Reddy, PhD
T. Pullaiah, PhD

AAP | APPLE ACADEMIC PRESS

First edition published 2025

Apple Academic Press Inc.
1265 Goldenrod Circle, NE,
Palm Bay, FL 32905 USA
760 Laurentian Drive, Unit 19,
Burlington, ON L7N 0A4, CANADA

CRC Press
2385 NW Executive Center Drive,
Suite 320, Boca Raton FL 33431
4 Park Square, Milton Park,
Abingdon, Oxon, OX14 4RN UK

Library and Archives Canada Cataloguing in Publication

Title: Bio-inspired technologies for the modern world : emerging trends and applications / R. Ramakrishna Reddy, PhD, T. Pullaiah, PhD.
Names: Reddy, R. Ramakrishna (Rajuru Ramakrishna), 1952- author | Pullaiah, T., author.
Description: First edition. | Includes bibliographical references and index.
Identifiers: Canadiana (print) 20240306880 | Canadiana (ebook) 20240306902 | ISBN 9781774914823 (hardcover) | ISBN 9781774914830 (softcover) | ISBN 9781003455233 (ebook)
Subjects: LCSH: Biotechnology.
Classification: LCC TP248.2 .R43 2024 | DDC 660.6—dc23

Library of Congress Cataloging-in-Publication Data

Names: Reddy, R. Ramakrishna (Rajuru Ramakrishna), 1952- author. | Pullaiah, T., author.

Title: Bio-inspired technologies for the modern world : emerging trends and applications / R. Ramakrishna Reddy, T. Pullaiah.

Description: First edition. | Palm Bay, FL : Apple Academic Press, 2024. | Includes bibliographical references and index. | Summary: "Technology will continue to drive biology, and biology will continue to drive technology. Mother Nature gives us ample opportunity to understand and observe her preserved secrets relating to the science and engineering fields. Scientists and inventors study the characteristics of things in nature to come up amazing and astonishing technologies and products invented as a result. This new volume delves into a sampling of these bio-inspired technologies. It looks at technological issues that have been tackled with the help of biologically inspired engineering, by such things in nature as bionic plants, the lotus leaf, insects and beetles, geckos, bats, spiders, and butterflies. It considers bio-inspired technologies applied in water purification, for business lessons, in healthcare and medicine, and more. This unique volume is an inspiring resource for professionals, researchers, scholars, engineers, and businessmen and businesswomen interested in the latest developments by studying the wonders of natural science"-- Provided by publisher.

Identifiers: LCCN 2024004237 (print) | LCCN 2024004238 (ebook) | ISBN 9781774914823 (hardback) | ISBN 9781774914830 (paperback) | ISBN 9781003455233 (ebook)

Subjects: MESH: Biotechnology

Classification: LCC TP248.2 (print) | LCC TP248.2 (ebook) | NLM TP 248.2 | DDC 660.6--dc23/eng/20240228

LC record available at https://lccn.loc.gov/2024004237

LC ebook record available at https://lccn.loc.gov/2024004238

ISBN: 978-1-77491-482-3 (hbk)
ISBN: 978-1-77491-483-0 (pbk)
ISBN: 978-1-00345-523-3 (ebk)

About the Authors

R. Ramakrishna Reddy, PhD
Department of Physics, Sri Krishnadevaraya University, Anantapur, Andhra Pradesh, India

R. Ramakrishna Reddy, PhD, is a former Professor in the Department of Physics at Sri Krishnadevaraya University in Andhra Pradesh, India. He has over 30 years of teaching and research experience and received his PhD in spectroscopy (Physics) in 1981 from Sri Venkateswara University, Tirupati. He served as a Registrar of the Sri Krishnadevaraya University, Anantapur, in 2006, Yogi Vemana University, Kadapa, in 2009, and served as a Principal to S.K. University College of Engineering and Technology, Anantapur from Sep 2009–June 2012. His research includes spectroscopy and molecular structure, solid-state and materials chemistry, and atmospheric science. Prof. Reddy has received many prestigious awards. To cite a few, he received the UGC research award for 1999, by UGC, New Delhi; Andhra Pradesh Best Teacher Award for the year 2005; American Biographical Institute Inc., USA "Man of the Year 1995;" Bharat Jyothi Award for the year 2006 by Indian International Friendship Society, New Delhi; Best Scientist and Fellowship Awards (FNESA) for 2006 and 2007 by the National Environmental Science Academy, New Delhi, etc. He served as a UGC-BSR faculty fellow and UGC emeritus professorship for 2012–2015 and 2017–2019, respectively. He was chosen as one of the visiting scientists to IUCAA (Astronomy and Astrophysics), Pune, continuously for 15 years, and IISc, Bangalore. He has carried out 12 research projects [ISRO, UGC, DST] and published 250 research articles in reputed national and international journals. Under his guidance, 21 students earned their doctoral degrees. He is a reviewer for various international scientific journals and a life member of scientific societies. He has authored four books related to atmospheric science published in Germany, two for graduate students, and 12 popular science books. He has published around 152 popular science articles in Telugu dailies/weeklies.

T. Pullaiah, PhD
Former Professor, Department of Botany, Sri Krishnadevaraya University, Andhra Pradesh, India

T. Pullaiah, PhD, is a former Professor in the Department of Botany at Sri Krishnadevaraya University in Andhra Pradesh, India, where he has taught for more than 35 years. He has held several positions at the university, including Dean, Faculty of Biosciences, Head of the Department of Botany, Head of the Department of Biotechnology, and Member of Academic Senate. He was the President of Indian Botanical Society (2014), President of the Indian Association for Angiosperm Taxonomy (2013), and Fellow of Andhra Pradesh Akademi of Sciences. Under his guidance, 54 students have obtained their doctoral degrees. He has authored 70 books, edited 40 books, and published over 340 research papers. His books include *Redsanders: Silviculture and Conservation* (Springer), *Genetically Modified Crops* (Springer), *Sandalwood: Silviculture, Conservation and Applications* (Springer), *Advances in Cell and Molecular Diagnostics* (Elsevier), *Camptothecin and Camptothecin Producing Plants* (Elsevier), *Paclitaxel* (Elsevier), *Monograph on Brachystelma and Ceropegia in India* (CRC Press), *Ethnobotany of India* (five volumes, Apple Academic Press), *Global Biodiversity* (four volumes, Apple Academic Press), and *Invasive Alien Species* (four volumes, Wiley Blackwell). He was also a member of Species Survival Commission of the International Union for Conservation of Nature (IUCN). Professor Pullaiah received his PhD from Andhra University, India, attended Moscow State University, Russia, and worked as postdoctoral fellow from 1976–1978.

Contents

Abbreviations

ADHD	attention deficit/hyperactivity disorder
AFM	atomic force microscope
AI	artificial intelligence
ANU	Australian National University
BA	bat algorithm
BCI	brain-computer interface
CITE	comprehensive initiative on technology evaluation
CMBR	Center for Micro-Bio-Robotics
COVID-19	coronavirus disease 19
DARPA	Defense Advanced Research Projects Agency
DOPA	3,4-dihydroxy-L-alanine
EAER	European Aerospace Research and Development
ECOM	electric endodontic motor
EMTs	emergency medication technicians
EPS	extracellular polymetric substance
FM	frequency-modulated
GFPs	green fluorescence proteins
GIS	geographic information systems
HSR	high spatial resolution
HTLT	high throughput laser texturing
IoT	internet of things
ISE	Institute for Solar Energy Systems
ISS	International Space Station
IT	information technology
IUCN	International Union for Conservation of Nature
IWCC	in-water cleaning and capture
JPL	Jet Propulsion Laboratory
LAL	limulus amebocyte lysate
LINV	International Laboratory of Plant Neurobiology
LMICs	low and middle-income countries
MAVs	micro-air vehicles
MIT	Massachusetts Institute of Technology
MMP-2	matrix metalloproteinase-2
MPI-IS	Max Planck Institute for Intelligent Systems

MT	metric tons
NDE	non-destructive assessment
NHS	N-hydroxysuccinimide
NIR	near-infrared
NPs	nanoparticles
NWT	nanotechnology-enabled water treatment
PDMS	polydimethylsiloxane
PLA	polylactic acid
PS	polystyrene
RO	reverse osmosis
RPA	robotics in precision agriculture
RPM	remote patient monitoring
SARS-CoV-2	severe acute respiratory syndrome coronavirus 2
SC2r-CoVs	SARS-CoV-2-related coronaviruses
SDI	silt density index
SEM	scanning electron microscope
SLPs	silk-like polymers
SMI	soil moisture index
UAVs	unmanned aerial vehicles
UF	ultrafiltration
USTC	University of Science and Technology in China
UV	ultraviolet
VOCs	volatile organic compounds
VPNs	virtual private networks
VSS	variable stiffness structure
WHO	World Health Organization

Preface

Technology will continue to drive biology, and biology will continue to drive technology. Taking inspiration from nature to solve human problems is the idea behind "biomimicry," or "biomimetics." Bioinspired technologies can be applied to solve every day problems in a low cost and sustainable way. In order to harness the most from nature's inventions, it is critical to bridge between the fields of biology and engineering. "Bio-inspired solutions usually represent not just an incremental change to an existing technology but also a total rethink of how to solve a problem."

Many technological issues have been tackled with the help of biologically inspired engineering, such as materials research; nano-bionic optical sensors; needle-free injections; fire extinguishers; self-cleaning surfaces; eco-friendly house paint, fabrics, coatings; adhesive-equipped robots to catch space junk; solar cell applications; adhesive materials based on the nanostructures; bulletproof material and medicine; NanoSeptic self-cleaning surfaces; strong adhesive materials; and solar cell applications. The microstructures, qualities, and powers of organisms like the lotus leaf, the Bombardier beetle, geckos, bats, butterflies, spiders, and more inspire technological advances. There is a clear explanation of these ideas in this book. Many aspects of healthcare delivery will have to adapt to new technologies. Scientists from a variety of disciplines are looking into animal leadership styles in an effort to better understand what makes a good human leader. This book will comprise a number of unique chapters that are rarely collected in one volume.

The first author gave several Radio talks and lectures on topics of science popularization in the local vernacular language. He also wrote many newspaper articles in the local language on such topics. We thought this popular information should also be brought into the English language for international audiences. The result is this book.

This publication is an ideal reference source for professionals, researchers, scholars, engineers, and businessmen/businesswomen interested in the latest developments in Nature Science. We are grateful to various scientists for their support in permitting their publications to improve the book's quality. We don't claim this work as exhaustive. Readers are requested to give suggestions, errors, and omissions so that the same can be rectified in future editions.

We wish to thank Dr. Bharat Bhushan, Academy Professor, The Ohio State University (San Jose, CA), Elsevier Masson, Legal Department, Ronald J. Lewarchik, Springer Nature, Lisa S. Baik, and John R. Carlson, Alyssa Y. Stark, Hartmut Wisch, Shutter stock and Getty Images for their help for the permission of the figures. We thank all the journal editors, publishers, and authors for permitting us to use their figures in our book, and we acknowledged them at appropriate places. We thank the team of Apple Academic Press for their support.

The first author would like to express his warmest appreciation to his children, R. Sree Vani and R. Vamsydhar Reddy, and especially to his wife, C. Ramathulasamma for their endurance. He is also thankful to his beloved students, Dr. K. Ramagopal, and Dr. Kanike Raghavendra Kumar, Associate Professors, for their continuous support and encouragement.

—R. Rama Krishna Reddy, PhD
T. Pullaiah, PhD

CHAPTER 1

Bio-Inspired Sustainable and Durable Superhydrophobic Materials–The Lotus Effect

ABSTRACT

Lotus is a perennial and sacred water plant. The scientific community is becoming increasingly interested in the sacred lotus. It relates to the small family of Nelumbonaceae. Ultra-hydrophobicity shown by lotus leaves is also known as the "Lotus effect." As its unique utility, a wide spectrum of biomimetic superhydrophobic surfaces is manufactured. Scientists began to imitate these natural templates inspired by nature. The micro- and nanostructure of the leaves is the key to the lotus plants' intrinsic super-hydrophobicity, as described in this chapter. Today's self-cleaning superhydrophobic products engineer is designed inspired by nature and become essential. It supports the industry in developing self-cleaning products. Non-wettable rainwear and sheets for ships are the other applications of the lotus effect. Other home and vehicular applications are sanitary goods in bathrooms, toilets, and motor vehicle shields to reduce dirt matter sticking and easier cleaning. This chapter also provides insights into technical breakthroughs in manufacturing lotus leaf-inspired self-cleaning surfaces, eco-friendly house paint, fabrics, coatings, solar cell applications, biodegradable water-repellent materials, nanotechnology, building materials, and industrial applications. Scientists have developed a nanotextured dust-reduction coating with hydrophobic properties inspired by the lotus leaf. Scientists at Texas A&M University developed various Nanomaterials inspired by the Lotus flower. They have numerous applications in the field of biomedicine. Different scientists have attempted to create biodegradable, water-resistant polymers inspired by lotus leaves.

Bio-Inspired Technologies for the Modern World: Emerging Trends and Applications.
R. Ramakrishna Reddy & T. Pullaiah (Authors)

1.1 INTRODUCTION

The lotus, often known as the Indian lotus, is one of the most prized flowers in the world. It is scientifically referred to as *Nelumbo nucifera*. The family of lotus-lily plants, Nelumbonaceae, is a delicious aquatic plant indigenous to tropical and subtropical Asia. There are 150 varieties of lotus on the globe, but only a handful are commonly available. In Hinduism, lotus flowers are revered as a sign of the divine or immortality in humanity. The lotus blossom was significant in ancient Egyptian culture (Figure 1.1). It appeared in numerous works of art during that time period (Barthlott & Neinhuis, 1997; Bhushan, 2012; Wong, 1987).

FIGURE 1.1 Lotus leaf and blossom.

Source: Reprinted with permission from Lian et al. (2020).

The stunning and unique nature of the Lotus plant's symmetry and colors are noteworthy. The lotus flowers will be kept along with the side of deity figures. In many Eastern cultures, the flower carries tremendous symbolic importance. Each Lotus flower color also has its unique symbolism. A white lotus symbolizes purity for Buddhists, and a yellow lotus is related to spiritual ascension. The wax protective coating on its leaves restores its beauty. Even though it's called a flower, this exquisite beauty is so much more! Lotuses have a unique life cycle compared to other flowers. As a result of the lotus's extensive root system, the lotus blossom is submerged in water at night. It reblooms the following day and looks to be completely free of blemishes.

Because of its water-repelling properties, the lotus is well-known. In nature, a wide variety of plant leaves have water-repellent qualities. During the rainy season, lotus leaves are drenched with rainwater. There is evidence that the leaves have waxed and are accumulating dirt and debris as they go. Waterproof and self-cleaning surfaces are known as "Lotus Effects." The typical size of a lotus structure is between 1 and 20 nanometers (nm). Using a scanning electron microscope (SEM), German scientists Wilhelm Barthlott & Neinhuis (1997) examined the surfaces of lotus leaves. Wax crystals of nanoscopic diameters between 10 and 100 nanometers in diameter cover leaf surfaces (Figure 1.2(a)–(f)). Rain, fog, or dew can displace a water droplet, for example. Frequently, a more noticeable drop forms around the leaf's center, which falls off when the leaf sways (Figure 1.2(b) and (c)). The superhydrophobic surface must have a micro- and nanostructured hierarchical feature. It's made of secreted crystalloid epicuticular wax, which has a low surface energy. There are contact angles of over 150° on the structures of Lotus leaf structures. Skin features such as papillae and hairs enhance this impression (Sanjay et al., 2014).

Chemical products of technological advancement will be used to duplicate the lotus effect on a variety of object surfaces, including glass, metal, stone, wood, plastic, absorbents, and fabrics, for example. In addition, the ability to protect wood doors, furniture, and carpets from staining and destruction is an advantage for businesses, households, and individuals. It extends the life of the car and the windshield and reduces fogging on rainy days, all while saving money on routine maintenance and cleaning. In the market, the Lotus effect makes nanotechnology one of the most potent technologies available. Researchers who were inspired by the lotus leaf in their research were able to develop and produce useful compounds for a variety of technological uses. Paints, tiles, fabrics, and floors are just a few of the many technological concerns being studied by these researchers.

1.2 WHAT IS LOTUS-EFFECT? HOW LOTUS PLANTS CLEAN THE DIRT?

The term "Lotus effect" was coined due to its ability to self-clean. The well-known lotus effect is one example of how bionics creatively employs natural phenomena. Self-cleaning surfaces have been designed based on Lotus leaves' water-repellent and self-cleaning properties. The same self-cleaning qualities can be achieved by mimicking these traits. The lotus effect can be applied to a wide variety of surfaces. There are typical features such as self-cleaning windows and solar cells (nanografi.com).

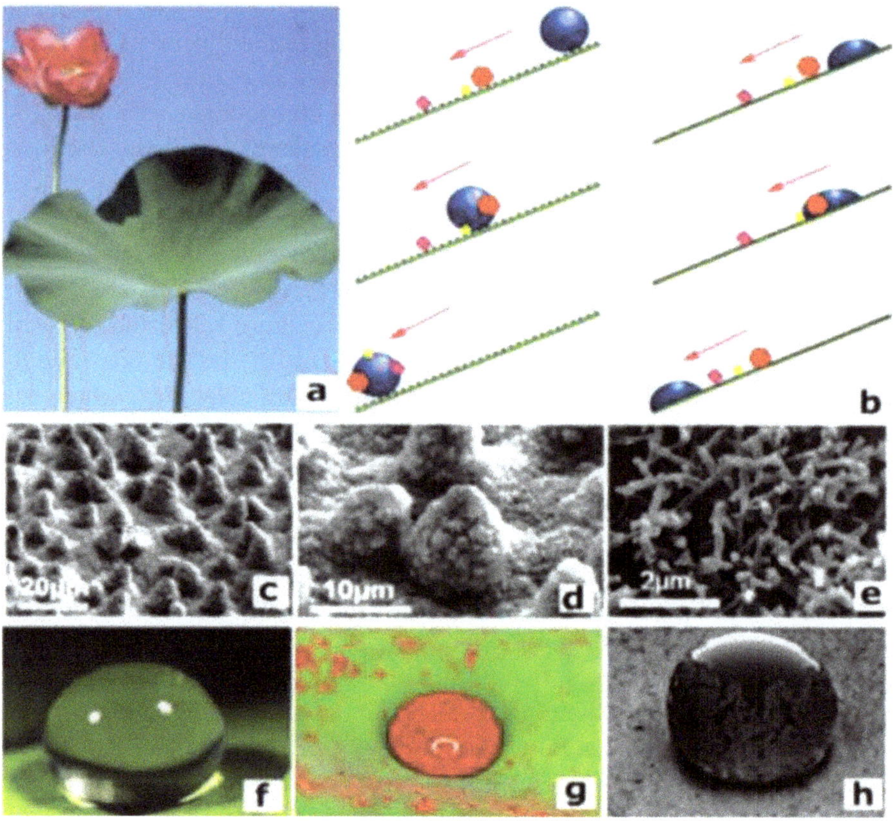

FIGURE 1.2 The water-repellent surface of a lotus leaf. Macroscopic appearance of the surface (a); self-cleaning property (b) and (c); and several multiples of SEM images of the lotus leaf surface (d–f).

Source: Reprinted with permission from Lian et al. (2020).

Using a high-resolution electron microscope, the microstructure of a lotus leaf was examined. The electron microscope displays on Lotus leaves tiny bumps measuring 5–10 μm in height and 15–30 μm apart (Figure 1.2(d)–(f)). Waxy crystals measuring one nanometer cover the surface of the lumpy and uneven Lotus leaf (Figures 1.2 and 1.3). With the Lotus appearance, waxy substance and small rough surfaces are crucial. About 1 nm-diameter wax crystals cover the lumpy, uneven surface of the Lotus leaf (Figures 1.2 and 1.3).

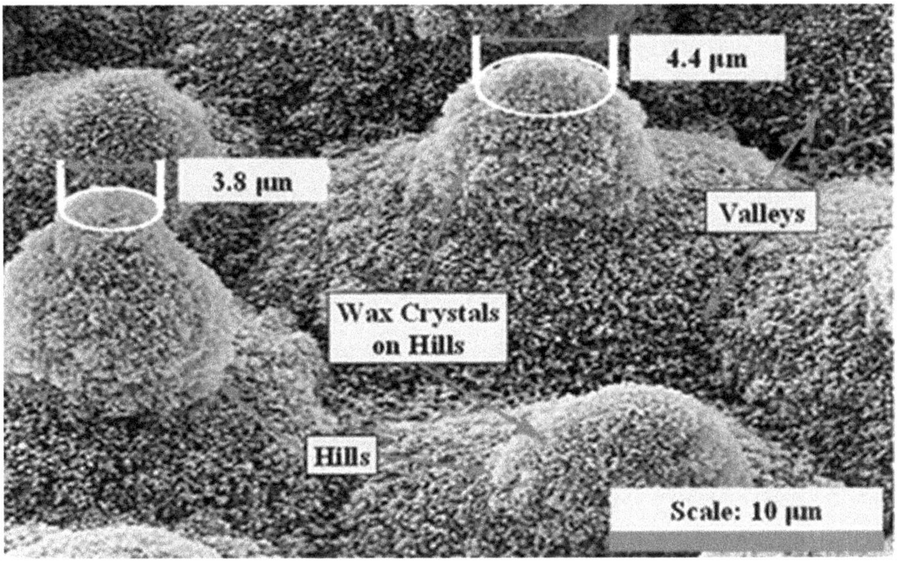

FIGURE 1.3 An electron microscope reveals surface structures of a paint that replicates the Lotus effect. This image depicts the microscale structure comprising hills with the valleys among them, as well as the nanoscale wax crystals.

Source: Reprinted with permission from Ensikat, et al., (2011). Copyright © 2022 by the authors. Licensee MDPI, Basel, Switzerland. https://creativecommons.org/licenses/by/4.0/

Lotus leaf has a waxy layer and dislikes water (hydrophobic). The Greek words Hydro (water) and Phobos (aversion) both signify water-repellent nature. Hydrophobic surfaces can be made hydrophobic by using this feature (Figures 1.1 and 1.2). Leaves become very hydrophobic if the area in which water and dirt come into touch is reduced. It is the surface that determines the droplet's shape, and water can then spread out and moisten the surface. Hydrophilic, or "water-loving," describes the surface attractant to water. Surface-to-tangent water angle must be less than 80° for this to work. When two things come into contact with one another, a "contact angle" is formed. In this circumstance, the contact angle is more significant than 80° (Figure 1.4).

Wax on the lotus leaf's surface prevents glue from adhering. The wax layer's rough surface will cause the creation of air pockets on its surface. These gaps inhibit soil adhesion to the leaf. They leave them accessible for the water to wash off. On a smooth surface, the dirt particles shift with the water drop. The dirt particles adhered on a rough surface at the dripping while unreeling the page and washed off in such a way that it was called the "Lotus effect." Waterfowl feathers and the wings of butterflies are examples of superhydrophobic surfaces found in nature. This category includes surfaces whose water contact angles are more than 150°. Lotus leaf non-wetting surface is recognized for its unique micro/nanoscale hierarchical patterns. Scientists have utilized this concept to create water-resistant and self-cleaning materials and textiles (Cheng et al., 2005; Onda et al., 1996).

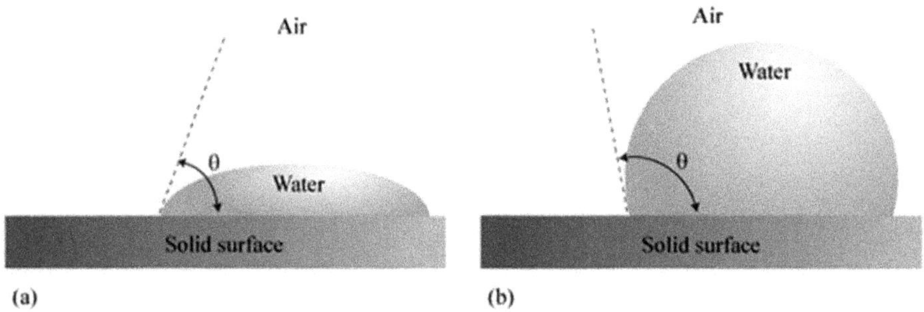

FIGURE 1.4 (a) The movement of water across a hydrophilic material. (b) Formation of water droplets on hydrophobic surfaces.

Source: Reprinted with permission from Papierowska et al. (2018) http://creativecommons.org/licenses/by/4.0/

Table 1.1 uses the contact angles of surfaces to categorize them. Water droplets rolling down the leaf's surface pick up the dirt and carry it away (Figure 1.5). In general, the surfaces depicted above have low drag during fluid flow. In scientific literature, superhydrophobic surfaces on the leaves and stems of Lotus plants are well-documented (Boreyko & Chen, 2010; Cheng, 2005; Onda et al., 1996). Lotus leaves are covered in nanostructures. Hydrophobic wax crystals with a diameter of 1 nm coat the surfaces. It creates a rough surface on the nanoscale.

TABLE 1.1 Surface Classification Depending on Their Contact Angle

Incidence Angle (Theta)	Surface Variety	Material
~0	Super-hydrophilic surface	UV irradiated TiO_2
> 30	Hydrophilic nature	Glass
30–90	Intermediate*	Aluminum
90–140	Hydrophobic nature	Plastic
140+	Superhydrophobic nature	Lotus leaf

FIGURE 1.5 Lotus leaves with a pleasant green color and the plant has "self-cleaning" (superhydrophobic) leaves. Water does not wet the lotus leaves because the leaf's surface is extremely hydrophobic because of the waxy coating it has.

Source: Reprinted with permission from Koch et al. (2009)

The lotus leaf surface is classified as super-hydrophobic or hydrophobic based on the contact angle (Figure 1.6). A hydrophobic surface is one with a slope greater than 90°. In order to be referred to as "super-hydrophobic," the contact angle must be greater than 150°.

Superhydrophobicity in lotus plants is astounding (Figure 1.7). The micro- and nanostructure of the leaves hold the key. Its surface is covered in tiny bumps. A phenomenon known as the lotus effect occurs when water comes into touch with wax-covered bumps.

Because of the openings between the papillae, a layer of air is kept in place. Instead of spreading out across the material, it causes the water's surface tension to form droplets. An interesting occurrence is observed whenever liquid falls upon one of these non-wetting surfaces and condenses into little spheres.

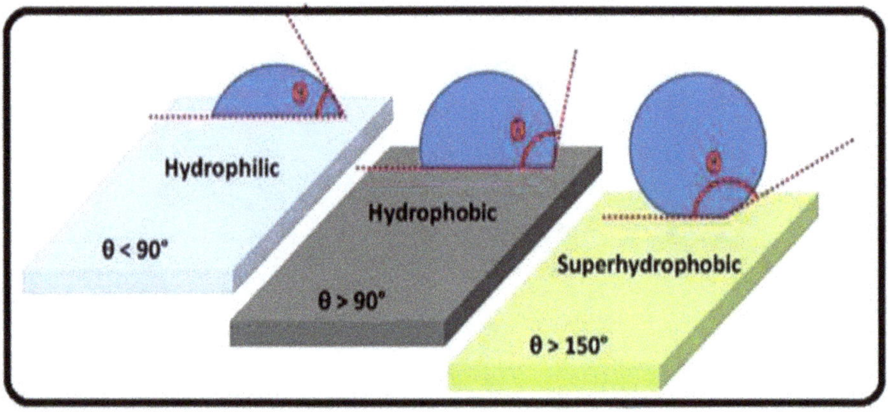

FIGURE 1.6 Different surfaces are based on the contact angle. Nurhan Onar Camlibel and Buket.

Source: Reprinted with permission from Lian et al. (2020).

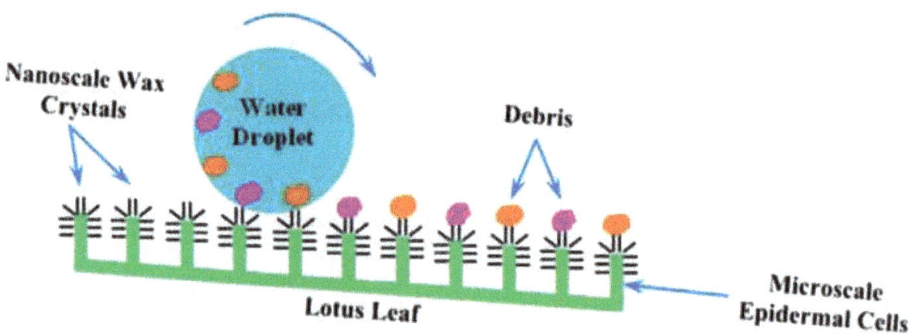

FIGURE 1.7 Lotus leaves with self-cleaning characteristics (*Nelumbo nucifera*) are displayed.

Source: Reprinted with permission from Collins & Safiuddin (2022).

1.3 THE CONCEPT OF THE LOTUS EFFECT

The Lotus leaves do not get wet and have self-cleaning properties. Superhydrophobic surfaces on the lotus leaf have existed for hundreds of years and have gained importance because of Nanotechnology and the Lotus Effect (Darmanin & Guittard, 2015; Koch et al., 2009; Neinhuis & Barthlott, 1997). Scientists recognized that Lotus leaves, butterfly wings, gecko feet, and shark skins were notifying as superhydrophobic surfaces and self-cleaning ones. Water repellency and self-cleaning are the focus of significant research at the moment. To put it another way, the lotus leaf surface is made up of primary and secondary alcohols, long-chain hydrocarbons, aldehydes, and triterpenes (Neinhuis & Barthlott, 1997) and consists of wax crystals that are hydrophobic (1 nm in diameter). The behavior of superhydrophobic, chemical composition and topography of Lotus leaves were investigated by biologists. The results of the study indicate that the leaves have two levels of structure: microstructure (measured in microns or thousands of millimeters) and Nanostructure. Both systems consist of a waxy covering that stops water from accessing the lotus leaf's surface.

1.4 SIGNIFICANCE OF LOTUS EFFECT

Many researchers investigated the Lotus Effect and found possible technological applications (Barthlott & Neinhuis, 1997). The lotus effect gains significance in creating self-cleaning materials. Under the name "Lotus-Effect," Neinhius & Barthlott (1997) have patented it. Two different properties are identified in this effect: Hydrophobicity (water repulsion) and the other superhydrophobicity. The first is physical, whereas the second is chemical. The contact angle is a significant differentiating factor between these two. If the contact angle is more than 150° (Figure 1.8), hydrophobic properties occur.

FIGURE 1.8 A hydrophilic surface and a hydrophobic surface are both affected by water's ability to remove dust particles.

Source: Reprinted with permission from Alhilo et al., 2021. https://creativecommons.org/licenses/by/3.0/legalcode.en.

Various Scientists (Johnson & Dettre, 1964; Marmur, 2006; Latthe et al., 2014) examined the hydrophobic surfaces. The leaves of the *Nelumbo* show ultra-hydrophobicity. Inspired by nature, engineers today have developed self-cleaning and superhydrophobic goods that can be easily cleaned. Water-repellent coatings and materials for boats, equipment, and surfaces exposed to fresh or saltwater, as well as materials for construction that require minimum upkeep, are all vital. They are impervious to the weather, stain-resistant, and dry even when submerged. In our daily lives, they have become vital and well-known novel items.

The lotus effect was used in some important goods. Self-cleaning super-hydrophobic materials can be used in a variety of ways. As a result, they've earned a place in our lives as useful and interesting items. Coatings derived from superhydrophobic surfaces were cited by Latthe et al. (2014a). There is a wide range of uses for textiles that are resistant to stains and remain dry even when submerged in water. According to Cheng et al. (2006), a water-repellent property for self-cleaning has been developed from lotus leaves and is now being used in paint on building walls as well as fabrics, speedboats, and swimming or diving costumes.

1.4.1 ARTIFICIAL LOTUS EFFECT

The "Lotus Effect" is named after the Lotus plant because of its self-cleaning capabilities. If this concept is created artificially, it is considered an artificial lotus effect. The lotus plant has given its name to a natural self-cleaning mechanism. A research team led by Ajayaghosh (2008) has successfully produced a superhydrophobic, self-cleaning surface. They achieved this by developing carbon nanotubes having a nanometer-thick paraffin coating. It is similar to the Artificial Lotus effect. Repellency explains that water cannot coat the leaf and rolls off (Ajayaghosh, 2008). This research explains the carbon nanotubes. They bound organic molecules to the cover of the tubes using a self-assembly method. This research explains that aromatic rings stick themselves firmly to the honeycomb structure of the annotates. Another finding one can identify is that the hydrocarbon chains act like a paraffin-like coating. A group of scientists pointed out the dispersion of these adducts applied to glass, metal, and mica surfaces by a team of scientists. They yield a water-repellent character, once dry, and exhibit self-cleaning properties. After SEM image observation, researchers concluded that the coating does not have a stable structure as the lotus leaves. They have nanoscale roughness.

A tilt angle of 2° is enough to cause water droplets to roll off. Here, water droplets remove any dust from the surface.

1.5 LOTUS FLOWER SELF-CLEANING HOUSE PAINT FOR THE ECO-FRIENDLY HOME

Lotus leaves contain tiny bumps and hairs. These permit them to entrap water. Due to gravity, all raindrops roll off the leaves. These raindrops take all the dirt with them and appear very clean. Lotus leaf surface (superhydrophobic surfaces) is one of the natural self-cleaning. A house paint called "Lotusan" was synthesized (http://ecosalon.com/eco-friendly-house-paint-self-cleaning-lotus-flower/) by a German company called ISPO. It is an eco-friendly house paint (Buczynski, 2013; Carbonea & Mangialardi, 2005) and uses a microstructure-modeled nature. This method will be used in textiles, windows, sprays, other products, and paint industries. Scanning electron microscopy (SEM) images and microstructures are given in Figure 1.9.

The microstructures of the Lotus leaf and Lotusan are depicted in Figure 1.9. Figure 1.10 is a 7,000-es magnification of these cells! Exactly what the Sto Corporation desired for its Lotusan sealant and paint.

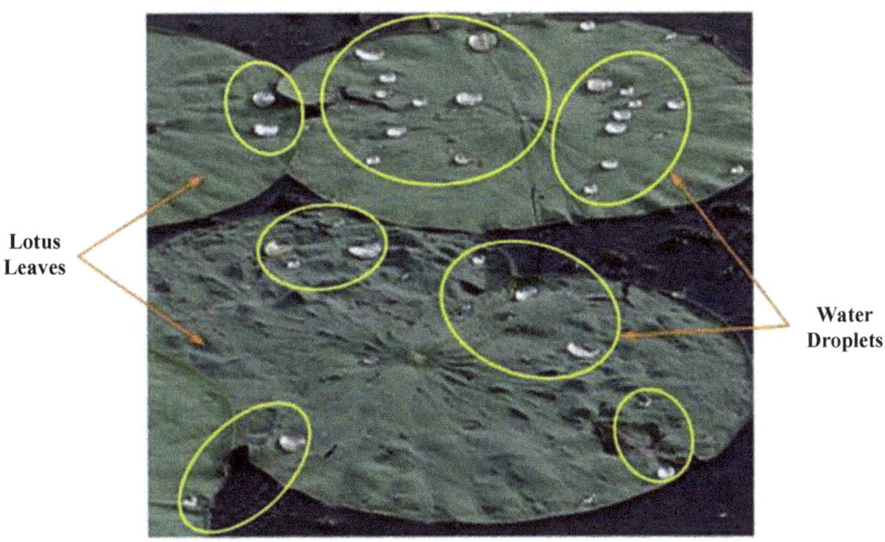

FIGURE 1.9a Lotus leaf and water drops;

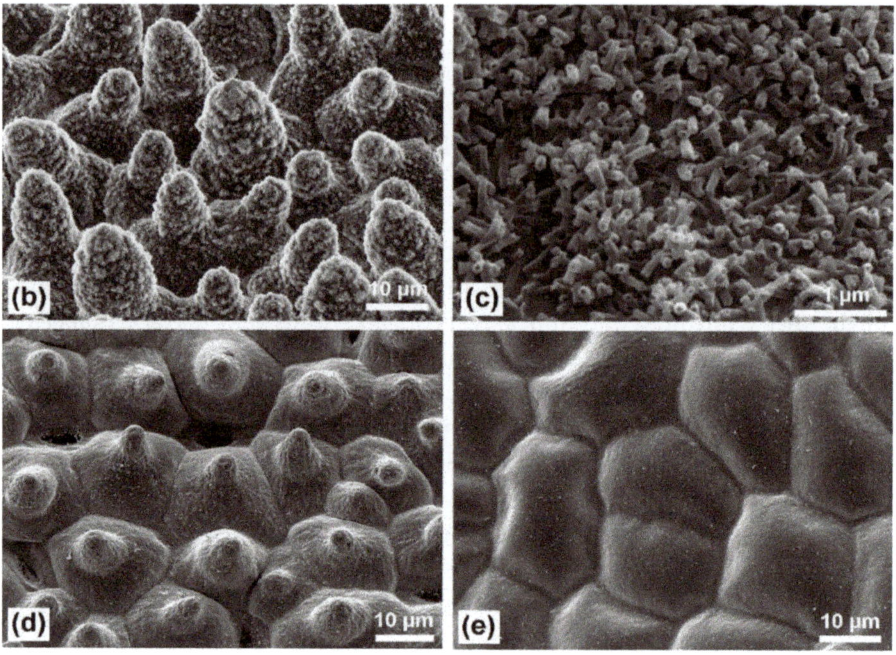

FIGURE 1.9b (b) Scanning electron microscopy (SEM) image of the upper leaf side prepared by 'glycerol substitution' shows the hierarchical surface structure consisting of papillae, wax clusters and wax tubules. (c) Wax tubules on the upper leaf side. (d) Upper leaf side after critical-point (CP) drying. The wax tubules are dissolved, thus the stomata are more visible. Tilt angle 15°. (e) Leaf underside (CP dried) shows convex cells without stomata.

Source: Reprinted with permission from Ensikat et al., 2011. http://creativecommons.org/licenses/by/2.0.

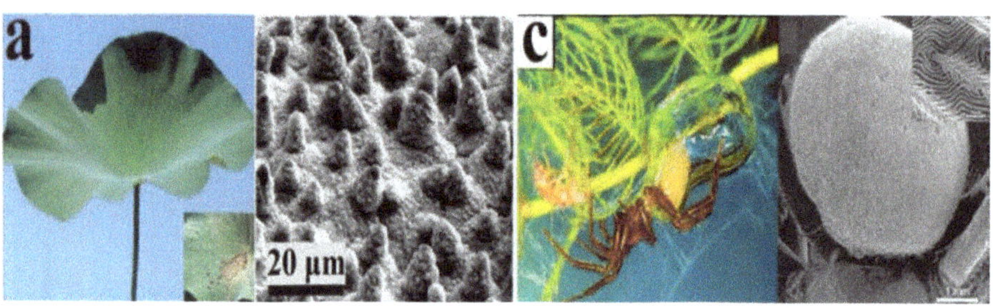

FIGURE 1.10 (a) Images of a superhydrophobic lotus leaves with self-cleaning properties and scanning electron microscopy (SEM) images of micro-/nano- surface structure.

Source: Reprinted with permission from Cai et al. (2021). Copyright: © 2021 by the authors. Licensee MDPI, Basel, Switzerland. http://creativecommons.org/licenses/by/4.0/.

Lovely facades, according to the Lotus Effect, remain beautiful in perpetuity. StoColorLotusan® is based on the idea that dirt falls off when it rains. Inspired by the lotus petal, Lotus Technology developed a paint that is highly water resistant and has a unique microtextured surface. As a result, water and dirt adhesion is noticeably reduced as a result of this combination (Figures 1.11 and 1.12). Unlike other paints, Lotusan does not enable dirt to adhere to its surface. Water's inherently non-sealing nature is the reason behind this. The dirt is carried away by the water when it exits a Lotusan-sealed building. Lotusan, like the lotus leaf, is a self-cleaning material.

FIGURE 1.11 Lotusan paint.

Source: Adapted from https://stemazing.org/wp-content/uploads/2017/07/lotusPaint.pdf

1.6 ADVANTAGE OF THE PRODUCT

Labor work is involved in the process of washing the exterior of a building. The outer part of the buildings often collects dirt and microorganisms developed on the exterior side of the building because of the weather exposure on the unprotected, weather-exposed sides of the building, soil deposits, and moisture provide an ideal environment for the growth of microorganisms (https://www.sto-sea.com/en/about-sto/sto-innovations/sto-innovations.html). The painted surfaces of a structure must be cleaned frequently with chemicals that are harmful to the environment. The surface of the normal coating is slightly hydrophobic and considerably more wettable.

FIGURE 1.12 Dirt was removed with the raindrops.

Source: Reprinted with permission from Koch et al., 2009.

In the StoColor® Lotusan® (Figure 1.13) paint, unique properties like water-repellent nature exist (Figure 1.13), and it is self-cleaning. The paint (https://www.stocorp.com/lotusan/) causes a series of hills and valleys and follows the Lotus effect principle. It operates against algae, bacteria, and fungus to build up. The specialty of the paint is keeping walls clean and dry for longer.

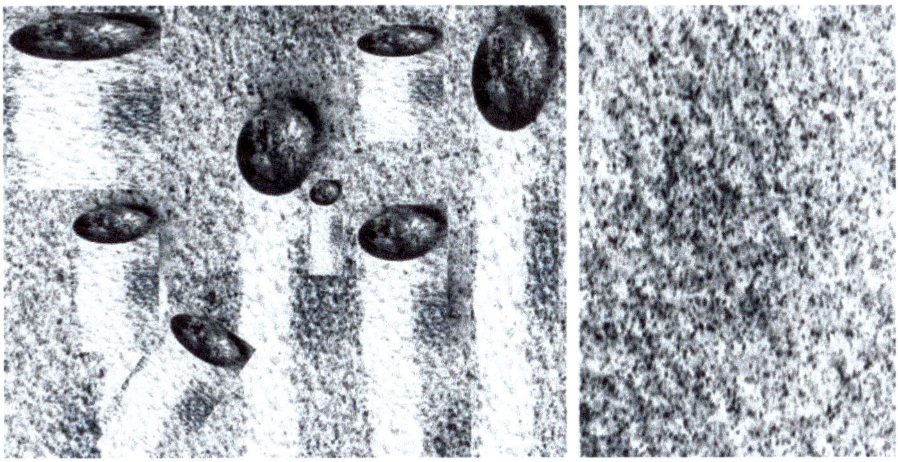

FIGURE 1.13 The "StoLotusan color" facade painted to the wall and its action (lotus effect).
Source: Adapted from https://w\ww.sto.com/biomimetics/en/biomimetics/lotusan/index.html.

1.7 ADVANTAGES OF STOLOTUSAN PAINTS

This paint is highly resistant to dirt and dust and self-cleaning when water hits. Resists the growth of microorganisms in its tendency and faces humid conditions. Considered to be UV-resistant—retains color very well (http://www.p3.co.uk/collateral/StoLotusan_Colour.pdf).

1.8 HOW DOES THE LOTUS-EFFECT® TECHNOLOGY WORK TECHNICALLY?

Lotus-Effect® Technology facade is considered as high-water repellency. It contains mineral rock flour and metal oxides in a microstructure nature. This lotus leaf principle is now being applied from biology to technology. It is one impressive example of successful bionics to date. It is challenging to apply to textiles. The material repels everything from sticking with the fabric (https://www.technicaltextile.net/articles/acquiring-natures-inspiration-lotus-effect-on-textiles-5174). The force of surface tension is the basic principle involved in it. When water falls over a regular fabric, it leaks and damages the beauty. Water drops will not be absorbed by the fabric and will remain as such and roll down when tilted. The treated fabric can withstand water and other liquids.

In addition to suits, coats, pants, shower curtains, and tablecloths, the Lotus effect may be found in a variety of other textile applications. Fluorocarbon-based fabrics are utilized in the fabrication of athletic and industrial clothing to restore their repellency. These garments pave the way for future intelligent fabric breakthroughs.

1.9 FUNCTIONAL SURFACES

Semi-aquatic, water-repellent *Nelumbo nucifera*, popularly known as the sacred lotus, is a plant. As the name implies, this plant's superhydrophobic surfaces display what is known as the "Lotus Effect." Researchers believe that superhydrophobic materials could be utilized in a range of applications. Self-cleaning coatings can be applied to windshields, optical equipment, window glasses, solar panels, anti-corrosion coatings, and interior fabrics. The selection of biological materials with the ideal combination of minute surface textures and chemical properties is an important stage. The waxy ridges on the leaves of lotus plants allow water to roll off and carry away dirt and other detritus. Auto-scrubbing paint StoCoat Lotusan was inspired by this "lotus effect," which wicks water into microscopic ridges on surfaces, such as the pitcher plant's surface, to create extremely slippery surfaces. These concepts led to the creation of SLIPS (Slippery Liquid-Infused Porous Surfaces) and superwicking surfaces for indirect cooling. Sharklets utilize the bacteria-resistant scales of a shark's skin.

1.10 ALL-WEATHER 'LOTUSHIRT' MADE BY NANOTECHNOLOGY

Sandle Tim (2018) stated that the Fiber Secret Company had founded two significant products to bring tech-industry innovation to the world of men's clothing. Lotushirt is its first kind of product. By applying Nanotechnology, a new shirt was produced. The resultant nano-fabrics have been fabricated from nanoscopic fibers. The polyethylene-based superhydrophobic fabric was introduced into the market, which is self-cleaning. By using nanofiber technology and the Lotus Effect, this fabric has been fabricated. This shirt fiber has the ability to resist water and stains. This technology has attracted the attention of numerous designers because of its resemblance to the self-cleaning lotus petal (the "lotus effect"). As the wax crystals become larger, they generate three-dimensional structures that are less than a few nanometers in size on the surface, creating a unique roughness.

The garment's nanotech fabric, which mimics the hydrophobic structure of the lotus leaf, is called the Lotus shirt (Sachidanandham & Thamima, 2019). The material's distinct characteristics are plainly seen. A clothing that has water trapped on the outside resists stains and stays dry inside because of this. This clothing does not require ironing, as it is wrinkle-free (Figure 1.14). It all begins with the 'Flying Fish' underarm netting, a breathable, lightweight fabric. The shirt's makers say that it is fast drying, wrinkle-free, flexible, breathable, and comfortable (https://www.kickstarter.com/projects/454615361/lotushirt-the-stainless-wrinkle-free-shirt?ref=8obb3z).

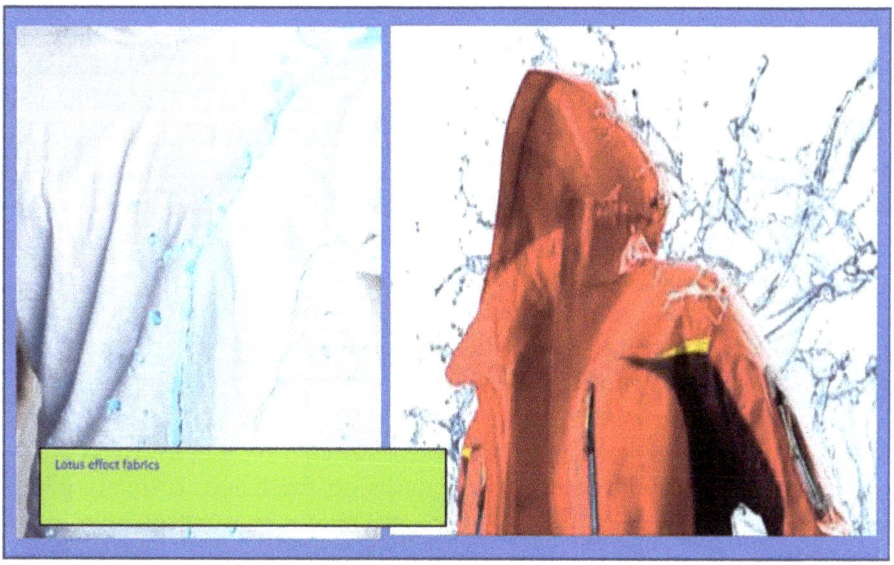

FIGURE 1.14 The lotus effect is utilized to make fabrics with incredible qualities because of the fibers produced from this flower.

Source: Adapted from https://www.fiber2fashion.com/industry-article/6589/the-legacy-of-the-celestial-flower-lotus-fiber-fabrics.html.

Researchers mentioned that the Lotushirt research details were first sent to Digital Journal (01-Apr-2018—*Digital Journal*, http://www.digitaljournal.com/56 | Science Reporter | August 2018) for review. They tested and were impressed with the shirt, appreciating its unique advantages like "its lightweight; a very easy to wear shirt. It feels very comfortable." The shirt was fabricated from a type of polyester (80%) together with "memory yarn," which imparts the stretchability of having more than a million micro-holes within the fabric. In the past, numerous garment firms began using Nanotechnology as a foundation. These Silver nanoparticles (NPs) have antibacterial properties, which means they kill the microorganisms that generate the odor. NPs are incorporated into fabrics to reduce the frequency with which they need to be washed while also preventing the development of unpleasant odors. From socks to t-shirts, things with nanotechnology have garnered attention and remain popular today.

1.11 LOTUS LEAF-INSPIRED SELF-CLEANING SURFACES

Water-repelling or hydrophobic materials can also be created using nanotechnology. Because of nano-scale features on the leaves, the foliage of many plants is hydrophobic. Researchers harnessed the lotus effect to manufacture artificial superhydrophobic materials, which can be used to make umbrellas. In addition to being naturally water-resistant, silicon is enhanced by the use of nano-sized patterns. This material's surface does not absorb water when a droplet touches it; instead, it rolls up and slides off.

When water falls on a typical fabric, it spreads across the fabric's surface, making it wet. A waterfall drop on the treated superhydrophobic cloth surface will remain in place and roll down. Additionally, nanotechnology can be employed to create hydrophobic or water-repelling materials. For example, nanotechnology can also be employed to create hydrophobic materials. Nanoscale features on the leaves of many plants cause their foliage to be hydrophobic. Artificial superhydrophobic materials that can be used to construct umbrellas were created using the lotus effect.

The usage of nano-sized patterns enhances silicone's inherent water resistance. Rather than being absorbed, a water droplet rolls up and slides off the surface of these materials. Nanotechnology can also be used to create water-resistant or hydrophobic materials. Many plants' foliage is hydrophobic because of the nanoscale characteristics of the leaves. Artificial superhydrophobic materials have been created by scientists using the lotus effect. The inclusion of nanoscale patterns improves silicone's natural water resistance. When it's tipped, the water drops fall through the fabric. Rather than retaining moisture, the superhydrophobic property-treated fabric beads up. It doesn't matter what liquids fall on this fabric; they don't leave any stains or blemishes.

1.12 SELF-CLEANING FINISHES INTO TEXTILE

In this busy life, nowadays, people do not have enough time to clean their daily clothes. The concept of self-cleaning will be understood by the surface of Lotus leaves. Nanotechnology has opened a new door for scientists to develop a self-cleaning surface that can clean itself without laundering. Using nanotechnology, applying self-cleaning properties on textile surfaces helps develop new products (Sachidanandham & Thamima, 2019; Tomar & Yadav, 2019). This research is an interdisciplinary one and will cover many areas. It includes different methods of self-cleaning processes and types of chemicals used. Self-cleaning fabrics and products have a wide range of uses in the healthcare and athletic apparel industries.

1.13 SOLAR CELL APPLICATIONS

Self-cleaning superhydrophobic surfaces have been described by Heckenthaler et al. (2019) for use in solar cell applications. Solar cell maintenance is a time-consuming and challenging process. The culprits are reflection suppression and transmittance. Due to their water- and dirt-repellent properties, organic solar cells benefit from transparency and antireflection. Utilizing a multifunctional surface improves the photovoltaic power conversion efficiency. Rugged surface textures and hydrophobic wax coats on lotus leaves are the results of the micro/nano-structure-induced lotus effect. Self-cleaning windows and high-efficiency solar cells are only two examples of products that have benefited from surfaces inspired by the lotus effect.

1.14 SENSORS

Pollutant accumulation, membrane clogging, and degradation of function are all things that sensors must keep an eye on (Zhang et al., 2016). Self-cleaning surfaces inspired by the lotus effect need to be cleaned on a regular basis to avoid membrane clogging. Contaminants, blockages, and membrane degradation all require sensors to be monitored. This study was carried out by Zhang et al. (2016). It is possible to reduce membrane clogging by using self-cleaning surfaces that mimic the lotus effect. Membrane performance has been hampered in the past by obstruction. Lotus-inspired self-cleaning surfaces remove dirt and other pollutants from the surface to prevent membrane obstruction. In bioimaging and nanosensors, it has a high degree of sensitivity. Normally, membranes become clogged, resulting in a decreased ability to perform. Self-cleaning surfaces reduce membrane blockage by continuously and passively removing dirt and pollutants from the membrane surface. Self-cleaning coatings are used in bioimaging and optoelectronic systems with highly sensitive infrared nanosensors.

1.15 THE LOTUS EFFECT ON THE BUILDING MATERIALS INDUSTRY

According to Marks (2009), the Lotus effect offers a huge array of benefits now that it has been adopted and finally marketed. This material is used in bioimaging and optoelectronic nanosensors with a high level of sensitivity. Normally, membranes become clogged, resulting in a decreased ability to perform. Surfaces inspired by the lotus prevent membrane clogging by continuously and passively cleansing the membrane surface. Bioimaging and optoelectronic devices with sensitive infrared nanosensors use self-cleaning coatings. These coatings can be used in a variety of different contexts. The construction materials sector can take advantage of the lotus effect in a variety of ways.

Roof tiles that are self-cleaning, solar modules and windows self-cleaning nanofaçade paint and plaster Hygiene with nano-sealing (shower glass wall, tiles, etc.) and Lotus blossoms' pearl-off effect is used by nano-sealers. The NPs in these products have components that firmly adhere to the surface during the sealing process. The surface is still water-resistant, but it can become so saturated that dirt is washed away.

1.16 LOTUSAN®, THE INTELLIGENT FACADE PAINT WITH LOTUS EFFECT

The invention of Lotusan paints is remarkable in its services. One can expect the appearance of the buildings to be more attractive. Lotusan®, the intelligent facade paint, can achieve it. Generally, buildings are highly exposed to vehicles and other types of pollutants. Due to the availability of moisture in the atmosphere and rain, algae and fungi grow. Self-cleaning facades show signs of wear and tear. Repairing these facades in this manner will save money (How the Lotus Effect is Revolutionizing Façades and Roofs by Lead Innovation, http://wengyugi.blogspot.com/2016). For the first time, Willhelm Barthlott understood the importance of the paint. It has a microtexture that is reminiscent of a lotus leaf's surface.

1.17 NASA'S LOTUS COATING REPELS DIRT AND DUST TO KEEP SPACE EQUIPMENT CLEAN AND OPERATIONAL

NASA's Goddard Space Flight Center Materials engineers (https://www.aerospacemanufacturinganddesign.com/article/lotus-dust-shield-to-protect-space-gear-and-prevent-odors) are working to develop a material that decreases contamination on the surfaces of space sensors and spacecraft, as reported by Aerospace Manufacturing and Design. Every piece of equipment that is utilized to collect scientific data or conduct exploratory activities during a space flight must be thoroughly sanitized at all times. Windows are kept spotless using a coating that utilizes Lotus effect technology. Because of the spikes on the lotus leaves, water and dirt have a much smaller surface area to adhere to. Water droplets fall off the Lotus leaf's surface and carry mud, small insects, and other pollutants.

Qu et al. (2010) quoted that Researchers at NASA Goddard (https://www.nasa.gov/centers/goddard/news/topstory/2009/lotus_coating.html) tried to develop a transparent coating. Similar to how water is shed by the lotus plant, it prevents debris from adhering to the surface (Anonymous, 2009). A joint partnership from different organizations synthesized the coatings. They are the Linthicum, Maryland-based Northrop Grumman Electronics Systems. The transparent structure functions similarly to the lotus flower's ability to expel water. Experts believe that once prototypes are created, layers of coatings might be used to prevent bacteria and dust from adhering to surfaces. Gears will be utilized on key spacecraft to clear the dust. Spacecraft and other equipment can be severely damaged by surface dust collection. When anything adheres to a surface, a motion akin to friction is generated.

Caterpillars, water striders, and the lotus are among the few naturally occurring animals and plants dubbed hyper hydrophobic. In order to create a superhydrophobic surface, minute hair-like structures extend from the waxy surface. Despite their diminutive size, they are completely covered in hair. Smaller hairs significantly cover them. Water droplets can't attach to organisms if their surface area is increased (http://www.robaid.com/bionics/biomimicry-could-help-development-of-self-cleaning-materials.htm).

Window cleaning on spacecraft is reduced by a layer of silica, zinc oxide, and other oxides. The new coating was tested in a variety of environments, including severe temperatures, UV light, solar wind, and electrostatic charge. Different coatings have different functions. For a covering to stick to spacesuits, it would need to be flexible. Moving parts of a spaceship or lunar equipment must be protected from wear and tear by special coatings. The biocide-infused coating was employed by NASA.

1.18 LOTUS LEAVES INSPIRE A NEW MATERIAL WITH BIOMEDICAL APPLICATIONS

Wilhelm Barthlott, a German botanist, discovered the superhydrophobic surfaces. Researchers in biomedical engineering have made further advances in the development of a blood-repellent, self-cleaning material. Using a water-repellent coating inspired by biomimetic nano-sized lotus leaves can reduce water damage. This results in nano-lotus leaf bio-mimetic coatings with favorable properties, including self-cleaning mechanisms and high surface tension. The bio-inspired nano-lotus leaf covers have been shown to have a variety of applications. Roads, bridges, and drainage systems can all benefit from the application of this approach. Safiuddin et al. (2018) lauded the several benefits of nano-lotus leaf bio-mimetic coatings. The self-cleaning capabilities of the micro-lotus leaf bio-mimetic covering were one of the key objectives of this work.

1.19 LOTUS EFFECT–MILITARY APPLICATIONS

According to Gu et al. (2017), superhydrophobic surfaces for military equipment such as surface ships, submersibles, and torpedoes are ideal (http://www.robaid.com/bionics/biomimicry-could-help-development-of-self-cleaning-materials.htm).

Scientists at NASA have developed a novel lotus leaf-inspired coating, which repels bugs and boosts aircraft performance. In a wind tunnel, Lotus Technology was employed to develop more than 200 coatings. For NASA and Boeing engineers in Louisiana, US, the finest five surface coatings have been selected. To test the Coatings, 15 flights and a number of takeoffs and landings were selected for the experiment. NASA's coating technology and the lotus leaf phenomena have been discovered by Surface Technology to be the top-performing coating, reducing the bug count by 40%. There are several NASA-led ERA programs that focus on developing new materials to reduce aircraft weight and drag.

1.19.1 MITIGATING DUST ACCUMULATION AND REPELLING LIQUIDS (NASA PATENT)

The Dust Mitigation Gap Assessment Team stated in their 2016 report that dust remains the greatest important obstacle to returning to the lunar surface for long-duration missions. It has been determined which technologies can facilitate travel to both the moon and Mars, but these technologies must mature before they can be used. The airborne dust suspended inside a spacecraft or future celestial homes must be adequately checked. Space trips to the moon were plagued by eye irritation and respiratory difficulties due to the lunar dust transported into the capsules. After returning to "microgravity," the team realized what they'd been missing. Lithium hydroxide, a reactive and poisonous dust, is present on all shuttle flights, posing a health concern to the astronauts. Fine particles from combustion events can pose a significant health hazard. Because of the lack of gravity settling in space, airborne dust and particles tend to be larger than on Earth. Various efforts have been undertaken by scientists to resolve this issue.

It is important to conduct this inquiry because dusty space environments (such as the moon, Mars, comets, and asteroids) contain undesirable chemicals. Lotus-inspired nano-textured hydrophobic dust mitigation covering has been developed by NASA researchers at Goddard Space Flight Center. A NASA patent (GSC-TOPS-19) is available at https://technology.nasa.gov/patent. The unique quality of Lotus Coating is that it is both lightweight and passive, as well as very water-repellent. This coating method was also utilized in further space and aviation applications in addition to terrestrial uses.

1.20 THE TECHNOLOGY

In terms of characteristics, the nano-textured coating includes a wide range of long-lasting and transparent options. A spin coating, brush application, or spray application can be used to apply this coating to a variety of different surfaces, including those used in space flight and aviation applications. According to Cheng et al. (2006), the nano-textured surfaces of the coatings are unique in a manner comparable to the Lotus plant. The coating's brilliance is that it is vacuum-stable and can resist the severe conditions of space travel. As a result, a greater variety of applications can benefit from the Lotus system.

It's possible to use nano-textured coating in other ways as well. Nano-textured coatings can be utilized in the following ways:

- This technology works in air as well as vacuum systems;
- Surfaces that can't be cleaned easily are ideal candidates for this product;

- Radiators, mirrors, and solar arrays are just a few examples of where this material might be put to good use;
- Textiles, automobiles, health care, pharmaceuticals, electronics, and aeronautics all have great potential;
- The building and construction sector, solar arrays, and other potential clean room surfaces;
- Equipment upgrades inspired by the lotus leaf.

It is possible to create antibacterial surfaces using a novel metal manufacturing method (Figure 1.15). For the first time, a group of European researchers has created metal surfaces that are both fluid-repellent and antimicrobial. Novel photonics technology and the leaves of a lotus flower were the inspiration for this new idea. This method is cost-effective for dairy farmers. Food processing and preparation equipment that is self-cleaning is the primary focus of this research.

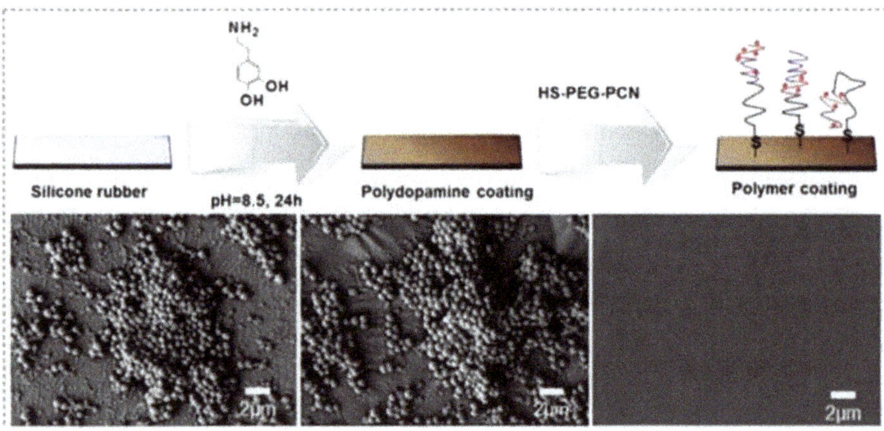

FIGURE 1.15 Bactericidal and bacteriostatic effects of anti-adhesion coatings: preparation process and antibacterial effect of coating.

Source: Reprinted with permission from Leijie Wang et al. (2022). Copyright: © 2022 by the authors. Licensee MDPI, Basel, Switzerland. http://creativecommons.org/licenses/by/4.0/.

The TresClean project (www.tresclean.eu) or high throughput laser texturing has made self-cleaning sheet metal production easier (HTLT). Small spikes and ridges must be created in metal sheets using laser-cutting equipment. In order to keep liquids away from the Lotus leaf's surface, these structures have been included. A smaller amount of liquid comes into contact with the roughened surface, which reduces the amount of air pockets that form.

Initially, cutting-edge technology will be put to use in the food processing industry to create antibacterial surfaces. This method is utilized by plants that produce natural food items such as milk, tomato sauce, and yogurt to improve production while decreasing expenses. As with lily pads, lotus leaves can be beneficial to your health. The Lotus leaf has inspired the creation of alloys that clean themselves. For the creation of novel photonics technologies, a European research group drew inspiration from nature.

1.21 NON-FOULING AND MECHANICAL BACTERICIDAL PROPERTIES INSPIRED BY THE HIERARCHICAL STRUCTURE OF A LOTUS LEAF

Drug-resistant bacteria have become resistant to antibiotics, making antibiotics ineffective in the fight against pathogenic bacterial infections (Jiang et al., 2020). To avoid antimicrobial resistance, nanostructured and mechanical bactericidal surfaces might be used. This adoption would have a substantial impact on the antibacterial properties of the bacterium. Antibiotics, a once-effective weapon in the fight against harmful bacterial infection, are being rendered ineffective by antibiotic resistance (Jiang et al., 2020). Surfaces with nanostructures and mechanical bactericidal capabilities can be used to kill germs without causing antimicrobial resistance. There is a significant impact on the bacterium's antibacterial properties if this is adopted.

The lotus leaf, known for its superhydrophobicity, was discovered for the first time by Jiang et al. (2020). A cell-rupturing process makes it highly resistant to bacteria and bactericidal. In the course of their research, the above scientists created a hierarchically organized surface. These microcylinders had nanoneedles placed on top of them. This surface has been perfluorinated to make it extremely hydrophobic.

Lotus leaves are recognized for their superhydrophobicity, as Cheng et al. (2006) noted. This character's cell-rupturing mechanism has a significant repelling effect on bacteria and bactericidal activity. These researchers have described antimicrobial action in the hierarchically organized surface against *Escherichia coli* (Cheng et al., 2006). More than 99.9% of the surface microorganisms were eliminated (non-fouling). In the end, the tenacious bacteria were eliminated by direct physical contact with the surface. Over a standard superhydrophobic or mechanical bactericidal surface, the newly constructed one offers a huge advantage when it comes to maintaining long-term antibacterial activity.

1.22 PLANTS INSPIRE WATERPROOF 'NANOFLOWER' FOR MEDICAL USES

There are natural superpowers in the leaves of plants. They generate coverings that repel water (called superhydrophobic surfaces). These organisms remove dust particles from their bodies (https://www.sciencedaily.com/releases/2019/07/190702184608.html). It is possible to apply a lotus leaf superhydrophobic coating to a variety of substrates using solvent evaporation. Superhydrophobic coatings are frequently used, according to research. Self-cleaning surfaces are used in a variety of applications, including nanoelectronics and biomedicine. Others employed colors we could not see to attract pollinators to their flowers. They created a water-repellent surface with superhydrophobic properties, which leaves with the ability to self-clean dust particles.

Biomedical engineering experts at Texas A&M University have created a "lotus effect" by inserting atomic flaws in nanomaterial. According to a university press release, this "superhydrophobic nanoflower" can be used in biomedical applications. These include blood-repellent, antifouling, self-cleaning, and biosensing applications within medical technology. Multiple scientific and technological sectors can benefit from this method's multiple applications. Using the solvent evaporation procedure, superhydrophobic coatings can be applied on glass, tissue paper, rubber, and silica. Discovered is the widespread use of superhydrophobic coatings. In nanoelectronics devices and healthcare applications, self-cleaning surfaces are just two examples. Water and pests were kept away from their leaves, while other plants used invisible colors to attract pollination insects. Water-repellent surfaces with superhydrophobic qualities were developed, allowing them to be self-cleaning and dust-proof.

1.23 A BIODEGRADABLE WATER-RESISTANT SUBSTANCE INSPIRED BY THE LOTUS LEAF

This biodegradable lotus leaf-inspired polymer has been developed by researchers from India, Switzerland, and Italy. A new organic and biodegradable substance has been developed. The water-repellent qualities of the lotus leaves are attributed to their composition of cellulose and spontaneously produced wax. Mishra (2019) exhibited superhydrophobic or water-repellent materials. There are numerous applications for these materials, including industrial, oil spill management, healthcare, sensor fabrication, and 3D printing. Petroleum derivatives, fluorinated chemicals, and textured inorganic substances are the most prevalent applications for these materials. Using a method known as soft lithography, extremely water-repellent micro-pillars of cellulose are produced.

1.24 WATER-REPELLENT NANOSTRUCTURES

A hydrophobic polymeric matrix called cutin and hydrophobic integrated and overlaid waxes make up the cuticle in water-repellent plants (*Nelumbo nucifera*), which is composed of the two materials. Lotus (*Nelumbo nucifera*) leaves and flowers are water-resistant because each epidermal cell is covered in nano-sized wax papillae. The papillae's high contact angle causes raindrops to roll off the surface with debris and dust adhering to them. Hydrophobic leaves' "self-cleaning property" is also linked to their super-hydrophobicity, which is the ability of hydrophobic leaves to remain clean even while submerged in dirty water. Superhydrophobic surfaces may now be created for a wide range of items thanks to the lotus effect, a self-cleaning property of extremely hydrophobic surfaces.

A leaf replica coated with water-repellant nanostructures was demonstrated in a study by Karthick & Maheswari (2008). Created superhydrophobic films are responsible for the Lotus effect. In one of the methods, polydimethylsiloxane (PDMS) has generated a rough surface. Using the negative template, a positive PDMS replica sheet is manufactured in order to reproduce the original lotus leaf exactly. The surface structures of the positive and negative PDMS templates are identical. The very hydrophobic surface of the lotus leaf makes it highly water resistant.

Each plant's epidermal cell secretes wax molecules, which solidify into nanoscale pillars when permitted to do so. Due to this, the water-repellent cuticle is composed of two distinct layers: cutin, polyester, and crystalline wax. The peaks of the bumps on a water-repellent leaf cuticle have minimal soil contact, much like a "fakir resting on a carpet of nails" on the plant's surface. On a dirty surface, a rolling water droplet rapidly picks up a particle of dirt.

1.25 LOTUS LEAF INSPIRES WATERPROOFING SCHEME

Barthlott realized in the 1980s that if rough, waxy surfaces could be created, an artificial lotus effect might be used in a variety of ways. In 1980, he obtained a patent on the concept and trademarked the name "Lotus Effect." Self-cleaning properties are achieved by using it to build surfaces with microscopic raised regions. When Barthlott came up with the "honey spoon" in the 1990s, no honey was left behind because of the spoon's homemade microscopically rough silicone surface. Some of the world's largest chemical companies have finally come around to this product's merits. The method worked, and their research muscle quickly discovered new methods to exploit it.

For self-cleaning purposes, Titania's super-hydrophobicity is a good fit: the water tends to spread across the entire surface, resulting in a sheet that can transport debris away as it runs. In addition, the surface is resistant to fogging because condensed water spreads out instead of creating hundreds of small droplets that constitute a fog.

Deodorizing results from Titania mix's photocatalytic effect. By decomposing organics and destroying germs, it disinfects coated products and enhances their self-cleaning capabilities. An explosion is taking place in the titania-coating industry at this point in time. TOTO, for example, manufactures and distributes outdoor ceramic tiles and other photocatalytic self-cleaning products under license to customers across the world. If you compare Titania with the lotus, you will find that it is extremely water-repellent. A wide range of kitchen and bathroom applications can benefit from the antibacterial and deodorizing capabilities of doped Titania. For self-cleaning fabrics, Titania can also remove odors, making it an ideal material. Direct chemical bonding is one method of attaching it to the fabric.

Most water-repellent surfaces that scientists have developed take inspiration from nature's lotus petals. According to Wright (2003), these kinds of coatings can minimize ship hull drag, stain, and water-proof fabrics and improve mining separation operations. Anti-wetting coatings can be easily and cheaply made using this method. Raincoats and umbrellas are used to keep us dry throughout the rainy season thanks to their waterproof and super-hydrophilic materials. It is because of a particular coating on the fabric that raincoats work to keep us dry during stormy weather. Because water molecules are more attracted to one another than to the waterproof surface, rain condenses and rolls off more easily. More rain falls off the roof because the water molecules are more attracted to each other and not to the waterproof surface than they would have been otherwise. The shape of the droplets that settle on a surface determines its water-repellency. The "contact" angle, which is the angle generated by imaginary lines intersecting the droplet and fabric surfaces, is a more precise means of determining the angle of contact.

As a result of the superhydrophobic feature, structures such as power transmission and communication towers or airplane wings do not become iced over. When ice isn't an issue, water repellency can nevertheless extend the life of metal structures by preventing rust and corrosion. Canadian scientists have created a coating method for cotton fibers that makes the normally absorbent material extremely water-repellent. Superhydrophobic clothes, tents, and other soft goods could be made by applying this technology to the fabric. The water droplet's surface tension and the coating's macro, micro, and nano-surface profiles are examples of these. When it comes to surface tension, the elastic properties of liquids allow them to take up as little space as possible. Hydrophilic (water-loving) coatings are one extreme, whereas superhydrophobic coatings are another extreme in terms of water repellency, depending on the surface features. Nano-lotus leaf Biomimicked Coating for self-cleaning applications has been highlighted by Safiuddin et al. (2018) in several construction sectors.

Lotus leaves have a slightly bumpy surface. On the micrometer scale, it resembles a sponge or bird's nest with a porous texture. Water cannot adhere to the solid because of the air trapped in the cracks. This naturally superhydrophobic surface was re-created in the laboratory by Turkish researchers led by Erbil Yildirim of Kocaeli University (Erbil et al., 2003) by first dissolving polypropylene in a solvent. This solution was then applied to a glass slide by way of a precipitating agent. It was discovered that the very porous gel coating developed in a vacuum oven after the solvent mixture had been evaporated possessed water-repellent properties comparable to those of lotus leaves. Similar results were reached when the researchers applied their coating to different substrates, including aluminum foil and stainless steel.

1.26 LOTUS-PAPER THAT REPELS THE LIQUIDS

Biomedical diagnostics and more cost-effective packaging materials could both benefit from a new type of modified paper developed by the Georgia Tech (US) group (https://www.futurity.org/%E2%80%98lotus-effect%E2%80%99-paper-repels-oil-and-water/). Biomedical diagnostics and more affordable, environmentally friendly packaging could both benefit from the modified paper developed by the Georgia Tech (US) group. Using a combination of cellulose fiber etching, chemical coating, and surface "fluff" modification, these researchers have generated an entirely new type of paper that is impervious to a wide range of liquids (Figure 1.16).

FIGURE 1.16 The lotus effect paper is an excellent barrier against oil and water because it is impermeable to both.

Source: Adapted from https://www.futurity.org/%E2%80%98lotus-effect%E2%80%99-paper-repels-oil-and-water/

In a variety of industrial applications, the ability to wet and spread liquid on porous materials is crucial. Lin et al. (2021) demonstrated how a coating on the reverse side of cellulose sheets containing superhydrophobic NPs may control the dispersion of water droplets. The scientists examined the water resistance of PDVB NPs. This coating is applied by spin-coating to a variety of paper types. This strategy has applications in anti-icing, self-cleaning, and water resistance (https://pubs.acs.org/doi/10.1021/acs.langmuir.0c02991). The size of PDVB NPs affects water droplet dispersion on untreated surfaces. Energy barriers are reduced for water droplets that are spread across three different types of paper (heavy-weight, lightweight, and medium-weight paper) (Figure 1.17).

Lin et al. (2021) indicate that the paper's porous structure impacts water droplet dispersal, permeability, and compressibility. This method adjusts the wettability of sheets by applying a coating to their reverse side. The paper's propagation of water droplets, permeability, and compressibility are affected by its porous structure. This method uses the coating on the reverse side of the paper to adjust its wettability.

1.27 HEALTH BENEFITS OF LOTUS SEEDS (*NELUMBO NUCIFERA*)

Because of its nutraceutical properties and historical significance, *Nelumbo nucifera* is becoming increasingly popular. In many parts of the world and

India, *Nelumbo nucifera* is grown for its curative and nutritional value. Roots, leaves, stems, flowers, seedpods, and seeds are all utilized for medicinal and culinary purposes. Since times immemorial, the medicine of the Lotus has been in use. Still, it continues as a contemporary medicine virtually. Seven different components of the Lotus plant can be used for medical purposes, and each one has a unique purpose. In their investigations, Sridhar & Bhat (2007) reported the pharmacological importance of sacred lotus. Antioxidant, hepatoprotective, immunomodulatory, anti-infective, hyperlipidemic, and psychopharmacological properties are all included. Diarrhea, dysentery, disorientation, hemorrhoids, sunstroke, pregnancy promotion, burning sensation treatment, skin improvement, infection management, fever, cholera, hepatopathy, and hyperdipsia are only a few of the conditions for which the lotus flower is used in traditional Chinese medicine. Many scientists have also proved the relevance of *Nelumbo nucifera*'s leaves, rhizomes, seeds, and flowers (Mukherjee et al., 2009; Sheikh, 2014; Moro, 2013; Mehta et al., 2013). In Ayurveda, Lotus seed oil or a paste prepared from Lotus petals is recommended for hydrating and rejuvenating the skin. Using this supplement to lessen the frequency of bowel movements will help relieve diarrhea. Eating big amounts of any Lotus part–petals, blooms, seeds, etc.–might cause gas and constipation.

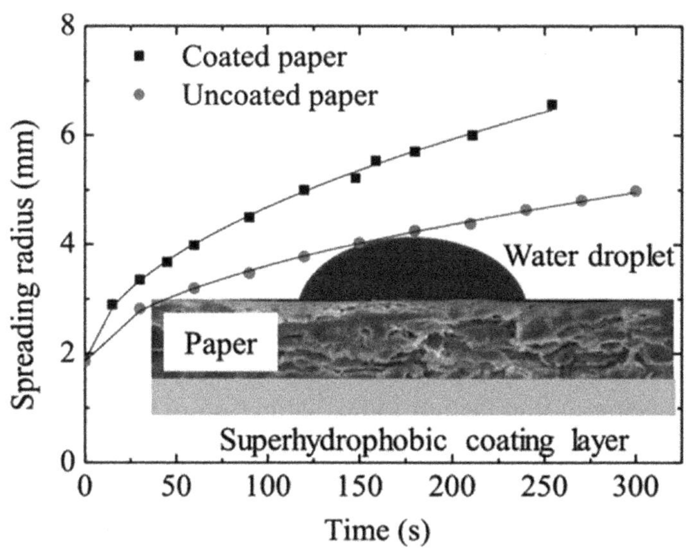

FIGURE 1.17 Relation between the superhydrophobic coating layer and spreading radius.

Source: Reprinted with permission from Lin et al. (2021); Copyright © 2021, American Chemical Society

The anti-obesity and anti-hyperlipidemia benefits of lotus leaf extracts on rodents have been documented. Adipogenesis was inhibited, mice with increased cholesterol levels were brought down, and the phospholipids and triglycerides were lowered. Leaf extracts also have an effect on lipolysis activity. The ethanolic extract of lotus leaves yields two anti-HIV principles (Moro, 2013). Coughs can be alleviated by combining honey with seed powder. Powerful ingredients such as ghee, milk, and gold enhance strength and sexual prowess. Some disorders may benefit from Lotus's astringent properties. Many remedies rely on the honey of the Lotus (Menon, 2021). The petals of the Lotus and the petals of the rose can be dried and applied to the body to eradicate body odor. When it's cold, it'll keep you warm and help you get rid of any remaining sweat. To boost your physical stamina, combine one teaspoon of dried Lotus flowers with one cup of milk. Apply a mixture of Lotus, gooseberry, and sandalwood to the skin to heal any wounds. Hemorrhoids and dehydration in women can be treated with lotus syrup.

Numerous studies have demonstrated its therapeutic value (https://www.emedicinehealth.com/lotus/vitamins-supplements.htm; https://www.healthline.com/nutrition/uses-for-lotus). Also, Lotus has pharmacological qualities such as anti-dermatophytes and antiparkinsonian and hepato-protective hypoglycemic, and immunomodulatory. *Nelumbo nucifera* is a source of numerous bioactive chemicals, including alkaloids, flavonoids, glycosides, triterpenoids, vitamins, and more.

Phytoconstituents, physical and chemical properties of *N. nucifera*, and its therapeutic benefits gained scientific importance. In the future, it can be used as a low-cost protein and energy supplement (Hajela et al., 2018). It is used as a medicine to treat some diseases such as tissue inflammation, skin disease, diarrhea, cholera, liver illness, poisoning, etc. In recent experiments, embryonic lotus seeds show potential as a therapy for lung cancer. It has been demonstrated that embryonic alkaloid inhibits lung cancer cell proliferation. The conjugation of phenolic compounds with food components, such as proteins, lipids, and carbohydrates, has led to substantial research into the nutritional properties of phenolic compounds. According to Limwachiranon et al. (2018), phenolic chemicals found in *Nelumbo nucifera* may be responsible for the plant's well-documented health benefits. All the lotus's phenolic compounds have different types, quantities, and metabolic aspects. This is why it has such a wide range of nutritional and functional properties.

A variety of phenolic chemicals, in various forms, concentrations, and metabolic properties, provide the lotus plant its diverse array of nutritional, organoleptic, medicinal, and functional features. Phenol oxidation, which is still a problem, limits the lotus's application in a broader range of food and cosmetic industries. Never before has the demand for research studies that detail evidence-based hypotheses of what causes these effects been greater. The development of functional foods has the potential to generate substantial economic value (Paudel & Panth, 2015).

1.28 LOTUS COSMECEUTICAL USE

These days, it's nearly impossible to live without some form of cosmetics. In order to achieve radiant skin and healthy-looking hair and nails, practitioners of Ayurveda turn to a variety of herbs. Ayurvedic literature describes over 200 herbs, minerals, and fats for preserving and increasing the health and look of the skin. From the roots, leaves, seeds, flowers, stigmas, and stems, cosmetics are produced. Lotus is commonly used in anti-aging cosmetics because it has been demonstrated to be a highly effective bleaching and anti-wrinkle agent. Lotus has a number of skincare benefits (Namita & Swati, 2013). In addition to vitamins B and C and minerals like copper and iron, the lotus flower contains many other nutrients that can help the skin repair externally. Because of a decrease in humidity, the skin's natural goodness diminishes throughout the winter months. The lotus flower oil can help you get rid of the dead skin cells on your face and restore a healthy glow. According to Namita & Swati (2013), a reduction in humidity might dry skin as well. Keeping the skin moisturized at all times is vital. The lotus flower oil can be used to treat and hydrate the skin. Face and neck wrinkles and dark spots can be reduced with the daily application of Lotus Oil, which has anti-aging properties.

Cosmetics are enriched with petals' essential oil. You can build a face pack out of it, in addition to applying it directly to the face like you would apply any other essential oil. Simply mix the oil with a small amount of milk and honey before slathering it on your skin. Acne-free and healthy skin requires twice-weekly facial cleansings. Applying the Blue lotus flower to the skin is also known to have numerous health advantages! Even though many people who have used blue lotus flowers swear by their efficacy, more research is needed to back up these claims. Each component of the Indian lotus plant is used for distinct cosmetic purposes. Jaychander (2018) has emphasized that the extract of the petals possesses qualities suitable for Ayurvedic skincare, which is frequently used to improve the skin's texture and elasticity. It's packed with antioxidants like vitamin C, which helps brighten skin and boosts immunity, as well as B vitamins, which have anti-aging properties.

The lotus flower carries fatty acids and proteins that will maintain the skin nourished throughout the day. Lotus extracts help the skin balance maintenance and also help the glory of the Face. It has extraordinary power to remove blackheads. Lotus hair oil will help with premature graying, hair fall, and dry scalp (Jaychander, 2018). The notable advantages of employing Lotus are moisturized dry skin, fights inflammation, promotes smooth skin texture, calms and soothes irritated skin, and balances oil production. Lotus will reduce acne, prevent free radical damage (due to its antioxidant content), and boost radiance. The blue lotus flower component reduces hair fall. Blue lotus flower used in shampoos and conditioners helps in hair treatment (Nadkarni, 1982). The extracts of the flower repair unhealthy hair and restore its natural shine. There are a lot of applications for Lotus in cosmetics, according to Namita (2013). Anecdotal evidence suggests that it can help reduce the appearance of wrinkles and acne. Veet hair removal gel, AHAVA mineral botanical body lotion, Clarisonic daily acne cleanser, several Neutrogena products, and Lotus sunscreen are some of the marketed items containing Lotus and its components or extracts.

Rich minerals and nutrients are present in the stem of the plant. The mineral potassium present in the Lotus plant helps to streamline BP. For a healthy dose of nutrients, boil lotus roots for 10 minutes and then eat. In many Asian dishes, Lotus root is a common ingredient. Lotus may help manage digestive problems such as indigestion. Lotus root contains alkaloids that have antispasmodic activity. Tastes harsh and astringent, like the root of the lotus. It's because of the tannins in there. Because cooking takes away some of the harshness, it tends to taste better prepared.

Lotus root (*Nelumbo nucifera*) is a well-liked cuisine in various tropical Asian countries and Australia, according to Li (2021). Lotus root is extremely nutritious, making it an excellent food choice. As a whole food, lotus root is a powerhouse of phytonutrients, vitamins, and minerals. In addition to potassium, phosphorus, copper, manganese, thiamin, pantothenic acid, and zinc, It's a good source of protein and a good source of dietary fiber. Furthermore, polyphenolic chemicals and polysaccharides abound in lotus roots.

KEYWORDS

- **eco-friendly house paint**
- **Health benefits of lotus seeds**
- **lotus effect**
- **lotus leaf-inspired self-cleaning surfaces**
- **lotus-inspired nanomaterials**
- **superhydrophobic surfaces**
- **waterproofing scheme**

REFERENCES

Ajayaghosh, A., (2008). Bioinspired superhydrophobic coatings of carbon nanotubes and linear π systems based on the "bottom-up" self-assembly approach. *Angewandte Chemie, 47*(31), 5750–5754. doi: 10.1002/anie.200802097.

Alhilo, E.A., Kuba, S.A., & Dirweesh, A.F., (2021). Nanotechnology use to preserve the durability of archaeological brick buildings in Al-Najaf city. *IOP Conf. Ser.: Mater. Sci. Eng. 1067*, 012044.

Anonymous, (2009). *Lotus Plant-inspired Dust-Busting Shield to Protect Space Gear.* https://www.nasa.gov/centers/goddard/news/topstory/2009/lotus_coating.html (accessed on 17 August 2023).

Barthlott, W., & Neinhuis, C., (1997). Purity of the sacred lotus, or escape from contamination in biological surfaces. *Planta, 202*(1), 1–8.

Bhushan, B., (2012). Bioinspired structured surfaces. *Langmuir, 28*(3), 1698–1714.

Boreyko, J. B., & Chen, C. H., (2010). Self-propelled dropwise condensate on superhydrophobic surfaces. *Physical Review Letters, 103*(17), 184501.

Buczynski, B., (2013). *Eco-friendly House Paint Inspired by the Self-Cleaning Lotus Flower.* http://ecosalon.com/eco-friendly-house-paint-self-cleaning-lotus-flower (accessed on 17 August 2023).

Cai, Y., Bing, W., Chen, C., & Chen, Z., (2021). Gaseous plastron on natural and biomimetic surfaces for resisting marine biofouling. *Molecules, 26*(9), 2592; https://doi.org/10.3390/molecules26092592

Carbonea, G., & Mangialardi, L., *(*2005*).* Hydrophobic properties of a wavy rough substrate. *Eur. Phys. J. E., 16*, 67–76.

Cheng, Y. T., (2005). Is the lotus leaf superhydrophobic? *Applied Physics Letters, 86*(14), Paper No. 144101.

Cheng, Y. T., Rodak, D. E., Wong, C. A., & Hayden, C. A., (*2006*). Effects of micro-and nano-structures on the self-cleaning behavior of lotus leaves. *Nanotechnology, 17, 1359–1362.*

Collins C.M. & Safiuddin, M., (2022). Lotus-leaf-inspired biomimetic coatings: different types, key properties, and applications in infrastructures. *Infrastructures 7*(4), 46; https://doi.org/10.3390/infrastructures7040046

Darmanin, T., & Guittard, F., (2015). Superhydrophobic and superoleophobic properties in nature. *Materials Today, 18*(5), 273–285.

Erbil, H.Y., Demirel, A. L., Avci, Y., Mert, O. (2003). Transformation of a simple plastic into a superhydrophobic surface. *Science, 299*(5611), 1377–1380. doi: 10.1126/science. 1078365.

Gu, Y., Zhang, W., Mou, J., Zheng, S., Jiang, L., Sun, Z., & Wang, E., (2017). Research progress of biomimetic superhydrophobic surface characteristics, fabrication, and application. *Advances in Mechanical Engineering, 9*(12), 1–13.

Hajela, S., Srivastava, R., & Kaur, J., (2018). A review on health benefits of lotus seeds (*Nelumbo nucifera*). *J. Appl. Sci. Computations, 5*(4), 92.

Heckenthaler, T., Sadhujan, S., Morgenstern, Y., Natarajan, P., Bashouti, M., & Kaufman, Y., (2019). Self-cleaning mechanism: Why nanotexture and hydrophobicity matter. *Langmuir, 35*(48), 15526–15534.

Hans J. Ensikat, Petra Ditsche-Kuru,, Christoph Neinhuis,and Wilhelm Barthlott Beilstein J. Nanotechnol. 2011, 2, 152–161.doi:10.3762/bjnano.2.19.

http://ecosalon.com/eco-friendly-house-paint-self-cleaning-lotus-flower/ (accessed on 17 August 2023).

http://togwt1980.blogspot.com/2014/12/the-lotus-effect-super-hydrophilic.html (accessed on 17 August 2023).

http://wengyugi.blogspot.com/2016/ (accessed on 17 August 2023).

http://www.p3.co.uk/collateral/StoLotusan_Colour.pdf (accessed on 17 August 2023).

http://www.robaid.com/bionics/biomimicry-could-help-development-of-self-cleaning-materials.htm (accessed on 17 August 2023).

http://www.robaid.com/bionics/lotus-flower-biomimicry-for-dirt-and-bacteria-free-surfaces.htm (accessed on 17 August 2023).

http://www.sto.co.uk/bionics/en/bionics-uk/lotusan/lotusan.html (accessed on 17 August 2023).

https://asknature.org/strategy/surface-allows-self-cleaning (accessed on 17 August 2023).

https://biodesign.berkeley.edu/2020/03/08/news-microstructures-in-paint-create-self-cleaning-and-anti-fouling-surfaces (accessed on 17 August 2023).

https://discovery.ucl.ac.uk/id/eprint/10099454/ (accessed on 17 August 2023).

https://engineering.tamu.edu/news/2019/06/researchers-design-superhydrophobic-nanoflower-for-biomedical-applications.html (accessed on 17 August 2023).

https://nanografi.com/blog/lotus-effect-in-nanotechnology/ (accessed on 17 August 2023).

https://salvatorebattaglia.com.au/essential-oils/26-pink-lotus-monographhttps://www.ehow.com/list_7175963_pink-lotus-effects.html (accessed on 17 August 2023).

https://technology.nasa.gov/patent/GSC-TOPS-19 (accessed on 17 August 2023).

https://www.aerospacemanufacturinganddesign.com/article/lotus-dust-shield-to-protect-space-gear-and-prevent-odors/ (accessed on 17 August 2023).

https://www.emedicinehealth.com/lotus/vitamins-supplements.htm; https://www.healthline.com/health/8-uses-for-lotus (accessed on 17 August 2023).

https://www.femina.in/beauty/skin/flowers-and-their-skincare-benefits-108249.html (accessed on 17 August 2023).

https://www.globalspaceexploration.org/wordpress/docs/Dust%20Mitigation%20Gap%20Assessment%20Report.pdf (accessed on 17 August 2023).

https://www.lead-innovation.com/english-blog/how-the-lotus-effect-revolutionizes-facades-and-roofs (accessed on 17 August 2023).

https://www.scientificamerican.com/article/lotus-leaf-inspires-water/ (accessed on 17 August 2023).

https://www.stocorp.com/lotusan (accessed on 17 August 2023).

https://www.technicaltextile.net/articles/acquiring-natures-inspiration-lotus-effect-on-textiles-5174 (accessed on 17 August 2023).

https://www.tumblr.com/bioinspiredteacher/90455612375/lotusan-paint-inspired-by-lotus-leaf (accessed on 30 September 2023).

Jaychander, N., (2018). *Flowers and Their Skincare Benefits*. Femina.

Jiang, R., Hao, L., Song, L., Tian, L., Fan, Y., Zhao, J., Liu, C., & Ren, L., (2020). Lotus-leaf-inspired hierarchical structured surface with non-fouling and mechanical bactericidal performances. *Chemical Engineering Journal, 398*, Article 125609. 10.1016/j.cej.2020.125609.

Johnson, R. E., & Dettre, R. H., (1964). Contact angle hysteresis. I. Study of an idealized rough surface. In: Gould, R. F., (ed.), *Contact Angle, Wettability and Adhesion: Advances in Chemistry* Series (Vol. 43, pp. 112–135). Washington, DC: American Chemical Society.

Karthick, B., & Maheshwari, R., (2008). Lotus-inspired nanotechnology applications. *Resonance, 13,* 1141–1145.

Koch, K., Bhushan, B., & Barthlot, W., (2009). Multifunctional surface structures of plants: An inspiration for biomimetics. *Prog. Mater. Sci., 54*, 137–178.

Latthe, S. S., Terashima, C., Nakata, K., & Fujishima, A. (2014). Superhydrophobic surfaces developed by mimicking the hierarchical surface morphology of lotus leaves. *Molecules, 19*(4), 4256–4283. https://doi.org/10.3390/molecules19044256.

Li, X., Guo, Y., Chen, L., *et al.* (2021). Texture and bio-functional characteristics of a Chinese steamed bread prepared from lotus root powder partially replacing wheat flour. *Sci. Rep., 11,* 16338. https://doi.org/10.1038/s41598-021-95926-3.

Lian, Z., Xu, J., Wang, Z., & Yu, H., (2020). Biomimetic superlyophobic metallic surfaces: Focusing on their fabrication and applications. *J. Bionic Eng., 17*, 1–33 doi: https://doi.org/10.1007/s42235-020-0002.

Limwachiranon, J., Huang, H., Shi, Z., Li, L., & Luo, Z. (2018). Lotus flavonoids and phenolic acids: Health promotion and safe consumption dosages. *Comprehensive Reviews in Food Science and Food Safety, 17*, 458–471.

Lin, C. B., Chang, H. S., Zhang, Y., Yang, F., & Lee, S., (2021). Spreading of water droplets on cellulose-based papers: The effect of back-surface coating, *Langmuir, 37*(1), 376–384.

Lotus-Leaf-Inspired Biomimetic Coatings: Different Types, Key Properties, and Applications in Infrastructures by Christopher M. Collins and Md. Safiuddin, Infrastructures 2022, 7(4), 46; https://doi.org/10.3390/infrastructures7040046.

Marks, P., (2009). *Radiator Roads Too Hot for Ice to Handle*. New Scientist, Technology.

Marmur, A., (2006). Super-hydrophobicity fundamentals: Implications to biofouling prevention. *Biofouling, 22*(2), 107–115.

Mehta, N. R., Patel, E. P., Patani, P. V., & Shah, B., (2013). *Nelumbo nucifera* (Lotus): A review on ethanobotany, phytochemistry and pharmacology. *Indian J. Pharmaceut. Biol. Res., 1*(4), 152–167.

Menon, P., (2021). *Amazing Health Benefits of Lotus You Should Know.* Health and Life Style. https://krishijagran.com/health-lifestyle/10-amazing-health-benefits-of-lotus-you-should-know/ (accessed on 17 August 2023).

Mishra, U., (2019). *Lotus-Inspired Biodegradable Water Repellent Material Developed.* THE Hindu Business line. https://www.thehindubusinessline.com/news/science/lotus-inspired-biodegradable-water-repellent-material-developed/article26026456.ece (accessed on 17 August 2023).

Moro, C. F., Yonekura, M., Kouzuma, Y., Agrawal, G. K., & Rakwal, R., (2013). Lotus – A source of food and medicine: Current status and future perspectives in the context of the seed proteomics. *Int. J. Life Sci.*, *7*(1), 1–5 http://nepjol.info/index.php/IJLS/index (accessed on 17 August 2023).

Mukherjee, P. K., Mukherjee, D., Maji, A. K., Rai, S., & Heinrich, M., (2009). The sacred lotus (*Nelumbo nucifera*)– phytochemical and therapeutic profile. *J. Pharm. Pharmacol.*, *61*(4), 407–422.

Nadkarni, A. K., (1982). *The Indian Materia Medica* (Vol. 1). Bombay. Popular Prakashan.

Namita, V. S., & Swati, A. P., (2013). A review on lotus: Use in herbal cosmetics. *Res. J. Topical Cosmetic Sci.*, *4*(1), 81–83.

Neinhuis, C., & Barthlott, W., (1997). Characterization and distribution of water-repellent, self-cleaning plant surfaces. *Annals of Botany, 79*(6), 667–677.

Onda, T., Shibuich, S., Satoh, N., & Tsujii, K., (1996). Super-water-repellent fractal surfaces. *Langmuir, 12*, 2125–2127.

Paudel, K. R., & Panth, N., (2015). Phytochemical profile and biological activity of *Nelumbo nucifera. Evidence-Based Complementary and Alternative Medicine, 2015*, Article ID 789124. https://doi.org/10.1155/2015/789124.

Qu, M., He, J., & Zhang, J., (2010). *Superhydrophobicity: Learn from the Lotus Leaf.* Intech Open. doi: 10.5772/8789.

Recent Advances in Superhydrophobic and Antibacterial Coatings for Biomedical Materials, Coatings 2022, 12(10), 1469; https://doi.org/10.3390/coatings12101469.

Sachidanandham, A., & Thamima, S., (2019). Sustainable textiles from Lotus. Asian Textile Journal, 28(10), 56–59.

Safiuddin, Md., Hossain, K., & Collins, C., (2018). *Potential Applications of Self-Cleansing Nano Lotus Leaf Biomimicked Coating in Different Construction Sectors*. Conference: CSCE 2018 Annual Conference, At: Fredericton, New Brusmwick. https://www.researchgate.net/publication/324827998_Potential_Applications_of_Self-Cleansing_Nano_Lotus_Leaf_Biomimicked_Coating_in_Different_Construction_Sectors (accessed on 17 August 2023).

Sandle, T., (2018). *Review: All Weather 'Lotushirt' Made by Nanotechnology.* Digital Journal, in Technology, https://www.digitaljournal.com/tech-science/review-all-weather-lotusshirt-made-by-nanotechnology/article/518774 (accessed on 17 August 2023).

Sheikh, S. A., (2014). Ethno-medicinal uses and pharmacological activities of lotus (*Nelumbo nucifera*). *J. Med. Plants Studies, 2*(6), 42–44.

Sridhar, K. R., & Bhat, R., (2007). Lotus–A potential nutraceutical source. *J. Agric. Technol.*, *3*(1), 143–155.

TOGWT, (2014). *The Lotus Effect–Super-Hydrophilic Surfaces.* http://togwt1980.blogspot.com/2014/12/the-lotus-effect-super-hydrophilic.html (accessed on 17 August 2023).

Tomar, S., & Yadav, N., (2019). Lotus fiber: An ecofriendly textile fiber. *Int. Arch. App. Sci. Technol.*, *10*(2), 209–215.

Wang, L., Guo, X., Zhang, H., Liu, Y., Wang, Y., Liu, K., Liang, H. & Ming, W., (2022). Recent Advances in Superhydrophobic and Antibacterial Coatings for Biomedical Materials, *Coatings 12*(10), 1469. https://doi.org/10.3390/coatings12101469

Wong, C. F., (1987). Origin and historical record of *Nelumbo nucifera.* China Institute of Botany (ed.), *China Lotus* (pp. 1–8). Beijing, Academia Sinica, Science Publishers.

Wright, L., (2003). *Lotus Leaf Inspires Waterproofing Scheme*. Scientific American, https://www.scientificamerican.com/article/lotus-leaf-inspires-water/ (accessed on 17 August 2023).

Zhang, M., Feng, S., Wang, L., & Zheng, Y. (2016). Lotus effect in wetting and self-cleaning. *Biotribology, 5*, 31–43. https://nanografi.com/blog/following-the-natures-lead-lotus-effect-self-cleaning/ (accessed on 17 August 2023).

CHAPTER 2

Biotechnology Advancements in the Field of Bionic Plants

ABSTRACT

Plants work as the channel of energy into the biosphere. They give food and materials used by humans and shape our environment. MIT researchers announced that they gained success in their glowing plant experiments. MIT scientists have produced greens that can both detect and electronically transmit information about bombs. Electricity generated by plants can be used to power LED light bulbs. The use of microfluidics in the cleanup of oiled spills. Advances in biotechnology are expected to emerge from–bionic plants. Contaminant identification will be carried out by plants. Wildland firefighting will undergo significant change as a result of significant technological development. A forest fire alert system has been created by Michigan scientists. Gold nanoparticles (NPs) were placed in the leaves by a team of Taiwanese scientists. The trees began to shine a brilliant reddish orange. Greener lighting methods permit the discovery of bioluminescence's outstanding efficiency. Scientists from many universities have developed a novel technique for eliminating contaminants. A plant-based sensor is used to measure soil arsenic concentrations. Researchers at MIT have developed nano-bionic plant optical sensors to prevent arsenic toxicity. Leaf thickness and capacitance can be measured using plant-based sensors. In the future, it could help farmers choose when to activate their irrigation systems. With the rise of IoT and smart sensors, remote sensing, and agricultural applications such as weather analysis, crop monitoring, harvesting robots, and weeding drones have all seen a boost in interest from the industry.

Bio-Inspired Technologies for the Modern World: Emerging Trends and Applications.
R. Ramakrishna Reddy & T. Pullaiah (Authors)

2.1 INTRODUCTION

Astonishingly complex structures can form throughout time from the evolution of plants that survive. The scientific discipline of biomimetics focuses on the replication of natural phenomena in artificial systems. Studies of animal behavior, such as how gecko feet travel on vertical surfaces or termite mounds cool themselves, have dominated most research in the past. In addition to providing inspiration for new materials and surfaces, plants have also recently entered the commercial manufacturing sector (Gruber, 2009). This is just the beginning: there are bound to be many more creative solutions in the works. If we want to preserve this natural library of surfaces that can inspire new materials, we must preserve the world's plant diversity.

2.2 GLOWING TREES–LIGHTING CITY STREETS

Fernando (2012) was of the opinion that humans have existed along with plants. The food that ancient humans ate indicates their abilities and knowledge. Plants were employed in a variety of ways by humans. It went from a hunter-gatherer society to an agrarian society in a few of decades. Humans and plants have shared a special relationship since the beginning of time. We can utilize the nutritional, therapeutic, and aesthetic benefits of plants by cultivating them. Humans and plants have developed a close interaction throughout the history of evolution. People and animals will acquire fiber, medications, and energy from the Plants. Human society allowed plant domestication and agriculture for their settlements comfortably (Schaal, 2019). In today's advanced cities and civilizations, secure and reliable food production and delivery is a major aspect of their infrastructure. As a food source, as well as an important part of our culture, religion, and medicine, plants play an important role in our well-being. People living in the forests gather plant materials for medicinal or spiritual purposes. Plants provide us with food and play an important role in our society's growth.

Naturally, every plant has a growth cycle, but the vegetative and flowering growth stages are influenced by light. Natural sunshine is the only source of light and nutrition that can't be replicated. Permanent growth and rapid production are enabled by artificial lighting throughout the year. Artificial lights have their drawbacks compared to Natural light. Without natural light, there would be no greenery, and no flowers or food gardens would blossom (https://en.wikipedia.org/wiki/Photosynthesis). The process by which plants manufacture food and energy from sunlight is known as photosynthesis. Indirectly or directly, plants offer energy and nourishment to the biosphere. They will provide nourishment for animals who consume them. Agriculture, habitat loss, and water pollution are all closely connected to global warming. When humans figured out how to generate and control fire, it was a huge step forward for human civilization. It is nearly as vital to our lives as fire to know the history of artificial lighting. A hollowed-out rock, shell, or other material that could securely sustain a small fire was presumably used to make the first ancient lamp around 70,000 years ago.

To generate new light sources, scientists are now learning how to manipulate light in the same way nature does. Plants are a part of one of these approaches. Algae, jellyfish, and sea stars are just a few examples of organisms that naturally produce light. Bioluminescence can be found in all living things, including humans. Researchers at MIT announced and achieved success in their luminous plant tests, as Bland (2010) emphasized. The same enzyme that fireflies produce was introduced into them. When luciferase and luciferin are used together, plant leaves light up. Scientists are working to develop plants with the ability to emit as much light as a table lamp so that you may replace the table lamp in your home with a plant.

2.3 STREET LAMPS WILL BE REPLACED BY GLOWING PLANTS

Tree leaves' green fluorescence could eventually alter the golden hue of streetlights (Bland, 2010; Callaway, 2013). Gold nanoparticles (NPs) fashioned like sea urchins, called bio-light-emitting diodes or bio-LEDs, have been injected into plant leaves and are now lighting brightly. Tree leaves' green fluorescence could eventually replace the golden glow of streetlights, according to Bland (2010). Researchers from Academia Sinica and National Cheng Kung University are implanting sea urchin-shaped NPs, also known as bio-LEDs, into the leaves of plants in Taiwan. Nanotechnology advances are having a tremendous influence. Infected plants with NPs generate electricity that powers street lighting (Figure 2.1). This light is utilized to eliminate CO_2 from the atmosphere.

Bio-LED trees that glow at night will be a common sight in the future, according to one of the researchers, Yen-Hsun Su. The bio-LED luminescence will save energy and reduce CO_2 emissions, will lead to photosynthesis in the chloroplasts (Bland, 2010). He believes sea-urchin shaped gold NPs hold the key to turning an absorbent substance into one that emits light.

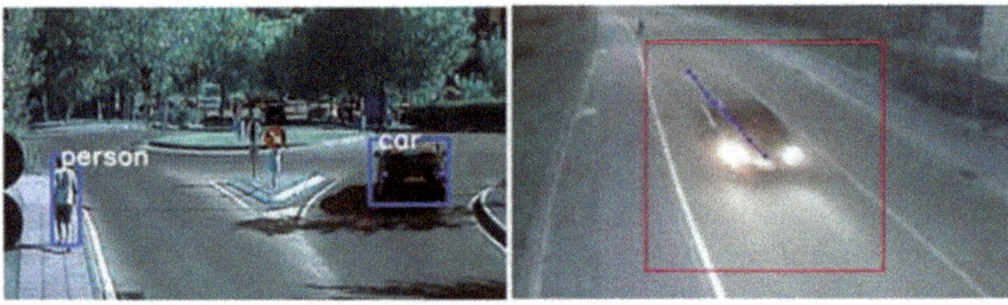

FIGURE 2.1 Biologically based LEDs could be used to make trees illuminate city sidewalks. Advanced Adaptive Street Lighting: NPs generate electricity that powers street lighting: Site n.day time (a) and night time (b) working modes.

Source: Reprinted from Gagliardi et al., 2020. Copyright © 2020 by the authors. Licensee MDPI, Basel, Switzerland. http://creativecommons.org/licenses/by/4.0/.

Pigments, which are light-absorbing chemicals, are found in the leaves of plants. The color of a pigment is determined by the wavelengths of light that it does not absorb or reflect. Chlorophyll (the green pigment in plant leaves), which absorbs all visible light wavelengths except green, is responsible for photosynthesis. It's what gives plants their distinctive green color. Chlorophyll and other photosynthetic pigments absorb electromagnetic waves to fuel photosynthesis. All photosynthetic organisms include chlorophyll-a, which absorbs violet-blue and orange-red wavelengths. In plants, chlorophyll reflects the hues of green and yellow green.

It is well-known that chlorophyll absorbs specific wavelengths of light, which gives plants their green color. Light having a wavelength of less than 400 nanometers will cause the chlorophyll to turn red. A few exceptions exist to this rule, such as when chlorophyll is exposed to violet light. Glow-in-the-dark leaves would make nighttime driving and walking safer, but that isn't always the case. The NPs of gold utilized by the scientists in their investigation contained a source of violet light. It is typical for scientists to explain photosynthesis by illustrating how various colors are noticed.

The Redlight production compared to Violet is accessible and can be scientifically explained straightforwardly. When activated by wavelengths of light undetectable to the human eye, the gold NPs emit a violet glow. When violet light excites neighboring chlorophyll molecules, it emits red light. According to the researchers behind this bio-LED idea, trees covered with gold NPs may be able to replace streetlights that use electricity or gas. According to Krishanu Ray, a University of Maryland scientist, the impact is exclusive to the *Bacopa caroliniana* aquatic plant that was the target of the scientist's experiments. There's no reason why we couldn't expand to streetside plants like those found on sidewalks.

2.4 ENGINEERS CREATE PLANTS THAT GLOW

It's amazing to think that, instead of turning on a lamp, you could keep an illuminated plant on your desk when it becomes dark. MIT engineers (Beato, 2013; Trafton, 2019) have taken a challenge to make plants that glow. A watercress plant's leaves were embedded with specific NPs, and the plants emitted weak light for nearly four hours as a result (Figure 2.2). They have the confidence that one day the trees will illuminate enough bright light to the workspace. It serves as a desk lamp and gives the vision. But this lamp doesn't have to be the plug-in. MIT scientists say that the plant's energy metabolism is ultimately what drives the light. It's a ground-breaking development. Low-intensity lighting or streetlights supplied by trees could be provided by this technology developed by MIT researchers. This technology could replace streetlights one day and also save energy.

2.5 WHAT IS NANOBIONICS?

Plant nanobionics (Khatri & Rathore, 2018) is a new area of research. It has many novel characteristic features gained by a plant through NPs. Plant nanobionics is a new bioengineering discipline that involves inserting NPs into living plants' cells and chloroplasts. An emerging field of plant nanobionics has been inspired by the way plants communicate with and function in their surroundings. Nanobionics engineering may significantly develop biomimetic materials with enhanced efficiency (Bose, 2020). Research into nanobionics (Mitiouchkina et al., 2020; Kwak et al., 2017) could considerably improve biomimetic materials. In light-harvesting and plant biochemical pathways, it is essential.

2.6 A NANO-BIONIC LIGHT-EMITTING PLANT

Biologists anticipate the capacity to target the distribution of nanomaterials to specific plant tissue regions (https://tlo.mit.edu/technologies/nanobionic-light-emitting-plant). Examples include light-harvesting apparatuses, photonic devices, and, in this instance, an emission source for indirect lighting or near-infrared (NIR) communication with external electrical equipment. Nanobionic plants have several agricultural applications, including the monitoring of pesticides and other chemicals, as well as other growth metrics.

- **Advantages:**
  Method for the delivery of nanomaterials at scale, targeting specific plant tissue parts, communication in the NIR spectrum – communication via NIR, crop condition monitoring with minimal use of energy.

2.7 THE IMPORTANCE OF BIOLUMINESCENT FUNGI

When a chemical process releases energy as light, bioluminescence is expected to occur. The chemical molecule luciferin undergoes oxidation, which is catalyzed by the enzyme luciferase, resulting in luminescence. In the field of biotechnology, these bioluminescent mushrooms have a lot of promise. Biotechnological applications include environmental monitoring, extraction of heavy metals, and many more (Kwak et al., 2017; Cooper, 2014). Plants are capable of synthesizing luciferins on their own. They generate luminous trees as an innovative kind of street lighting, replacing conventional streetlights that consume significant amounts of electricity. These trees will provide illumination for the street signs and interior lighting. The trees were illuminated at night and darkened throughout the day. To fulfill their urban lighting obligations, the trees would just need access to air, water, and soil nutrients.

2.8 NANOBIONIC PLANTS

To put it another way, the study of biological systems as a nanoscale model for the creation of nanoscale materials and devices is known as nanobionics. There are several applications for nanobionics across a wide range of industries. Artificial intelligence (AI), nanobiotechnology, nanorobotics, medicine, and the military fields utilize the benefits of Nanobionics. The use of nanobionics to harness solar and photosynthetic energy in plants has great promise. An emerging subject in bioengineering is modifying plant tissues and organelles by introducing NPs into living cells and chloroplasts. Nanobionic engineering of plant functionalities would usher in a new era. It functions as an interface between plant biology and nanotechnology. To illuminate low-intensity indoor areas and transform trees into self-powered streetlights, Cooper (2014) recommends that nano-bionic plants with light-emitting capabilities be used. It is possible for self-repair plants to generate energy and adapt to their surroundings due to the cumulative effect of all of their individual repairs. As a result, plant nano-bionics get a much-needed lift. He said that their findings could contribute to reducing the world's reliance on energy, given that lighting accounts for 20% of the world's energy usage (Kwak et al., 2017).

Plant nanobionics is a new scientific topic that aims to enhance the qualities of plants by embedding them with NPs. In the opinion of engineers, plants should be able to communicate with each other via electricity. Many of the tasks currently performed by electrical equipment can be performed by engineers. In order to promote science and technology, the plant is essential. It's necessary to learn and integrate NPs with plant-based components when using living plants as technological platforms. It is possible to envision self-repairable photonic systems based on plants that could function as communication networks or sensors that sample their environment through transpiration and report the results via radio-frequency signals.

People's well-being can be improved through advances in plant technology (Li et al., 2021). A team of scientists has developed plants that can detect explosives and send the data to a smartphone, in addition to plants that can monitor drought conditions. Due to the lack of energy, plants emit light, which accounts for roughly 20% of the world's energy usage. As a result of population expansion, it is a significant target. Plants can generate their own energy using the most recent technology. The ability of live plants to generate and store energy on their own, as well as to self-repair, makes it an appealing candidate for use in lighting systems that emit visible light and are long-lasting.

2.9 LUCIFERIN IS A CHEMICAL COMPOUND FOUND IN BIOLUMINESCENT ORGANISMS

When oxidized with the help of an enzyme called luciferase, the energy released is light. The MIT team (Beato, 2013) selected luciferase, the enzyme that gives fireflies to glow, similar way the trees ability to give light. As a result of scientific inquiry, the enzyme-substrate system, is also known as luciferase and luciferin (from the Latin *Lucifer*, "Light-bringer"). The MIT researchers used NPs as a delivery system. In this study, the FDA-approved NPs are being employed, which are all manufactured of FDA-approved materials. Components that could be harmful to the plants are also prevented from accumulating in high enough concentrations by the filters. For this study, researchers used 10 nanometer-sized silica NPs to contain luciferase (the enzyme). In order to transport the luciferin and coenzyme A in larger particles, the polymers PLGA and chitosan were utilized in this study. Before injecting the particles into the plants, it was required to suspend them in a solution using the polymers PLGA and chitosan. High pressure is used to cause the particles to penetrate the leaves via the stomata.

Particles that release luciferin and coenzyme will be carefully monitored. The mesophyll and a leaf inner layer's extracellular area are designed to accumulate this design. In the mesophyll, the luciferase-carrying smaller particles are taken up by the cells. This research relies heavily on Luciferase. Luciferin from the PLGA particles progressively enters the plant cells, releasing luciferin. Luciferin glows as a result of a chemical reaction catalyzed by luciferase. It was discovered that plants could glow for 45 minutes at the commencement of the research effort. After tedious efforts of research, they have improved 3.5 hours of glow. An enhanced 10-centimeter seedling of watercress provides light equivalent to one-tenth of the quantity required for reading. In theory, researchers believe they can increase the amount of light released and extend the duration of light.

2.10 PLANT TRANSFORMATION

Plants that have been genetically modified to express the luciferase gene undergo a laborious process that results in a very faint light. Engineers are contemplating this idea. Tobacco and *Arabidopsis thaliana,* two extensively utilized plant genetics models, were employed to begin the research. Michael Strano and others at MIT (Trafton, 2019; Kwak et al., 2017) elaborated on methods to be employed on any plant. These scientists experimented with arugula, kale, watercress, and spinach. NPs could one day be painted or sprayed onto plant leaves by researchers in the future. Light-emitting trees and other huge plants would be transformed into power plants (Collins, 2017). "The research cooperation opens up very seriously the doorway to streetlamps that are nothing but treated trees and to indirect lighting surrounding residences," said Strano, a scientist from the MIT team. Adding NPs containing a luciferase inhibitor, the researchers say, will likewise dim the light. They hoped to develop plants that could turn off their light emissions, much like sunshine does. If this experiment is a success, scientists expect to be able to do away with the use of lighting completely. This method may help minimize carbon dioxide emissions into the atmosphere. Carbon dioxide emissions in some cities can reach tens of thousands of tons per year. It will save a lot of money in the long run.

2.11 PLANTS PRODUCE ELECTRICITY TO GLOW LED LIGHT BULBS

Science has always been intrigued by the relationship between humans and plants. Positive results have been observed in recent studies. Engineers at MIT have gone a step further, modifying the chemical makeup of plants to enable them to perform a wide range of unusual, even bizarre, jobs. One-tenth of the amount of light required to read is merely a start. Strano believes that the technique might be used to power interior lights or perhaps turn trees into self-powered streetlights in the future. The lab's long-term goal is to create a one-time application of the technology that may be sprayed on leaves or saplings and last the life of the plant (Davison, 2018). A "switch" for dimming the glow when the plants are exposed to sunlight is also being investigated by the engineers (https://edition.cnn.com/style/article/glowing/index.html). It is a difficult undertaking to find environmentally acceptable and pollution-free sources of electricity. Scientists (Italy's IIT-Instituto Italiano di Technologic in Pontedera) have demonstrated that living plants may help to generate electricity. Meder et al. (2018) have shown that renewable energy systems should reflect the fact that living plants are actually "green" power sources. Researchers revealed that plants can generate more than 150 Volts from a single leaf. A "hybrid tree" comprising natural and artificial leaves was also shown by the researchers. It launches a new "green" electrical generator that uses wind power to generate electricity. A substitute for the future's electrical supply, it will also be available worldwide.

2.12 PLANTS WITH GENETICALLY ENCODED AUTO-LUMINESCENCE

Bioluminescence genes from bacteria can be targeted to build plastids that emit light. This technique is time-consuming and tedious, and it does not provide sufficient light. As a result of their efforts, they were able to see a self-sustaining light. Bacterial bioluminescence gene clusters in plastids have not been widely utilized because of their inability to create significant light output (Mitiouchkina et al., 2020). The Group of Scientists' results could be used to produce a suite of imaging tools for plants. Four genes that make a fungus glow-in-the-dark have been adopted by worldwide scientists, who have designed tobacco plants that generate green light, inspiring fanciful visions of our future. It is possible to make bioluminescent plants by engineering a plant's DNA with genes from a fungus that glows in the dark. Plants were poisoned by the molecular products of luciferases from bacteria in earlier attempts. In contrast, the fungus luciferase functions in the caffeic acid cycle, which is present in all plants, and does not appear to affect them.

2.13 GLOW SWITCH

Researchers have discovered that the similarity between plants and bacteria allows them to manipulate or exchange segments of their DNA. In both organisms, there is a genetic component. These will allow the organism to glow by enabling the proteins that do so. To flip on this protein production process, plants (particularly chloroplasts) lack the gene (the lux operon) that does so (https://askabiologist.asu.edu/plosable/glow-plants). Luciferase is a protein that can be produced in the presence of the lux operon. Other unique chemicals called luciferins can be produced by them. Chemical reactions occur when luciferins are exposed to luciferase, and energy is released. Light is the shape taken by this liberated energy.

2.14 GREEN FLUORESCENCE PROTEINS (GFPS) AND GENERATION OF LIGHT

To learn more about how bacteria glow, scientists have looked into how it might be applied to other organisms. GFPs are an example used by scientists to determine if gene exchange operations have been successful. GFP genes incorporate other genes into the DNA of an organism. If the organism begins to light following the insertion of these genes, the experiment was a success (Figure 2.2).

It is possible to alter the plant's blueprint through the creation of luciferins. Luciferase, like some microorganisms, should make the plant shine. The lux operon (from bacteria) was inserted into the sex cells of plants using a genetic engineering technique known as transposons. According to their calculations, some of the seeds from that plant would mature into plants that would glow if they had the necessary DNA inserted. The blueprint of a plant can be changed by synthesizing luciferins. As with some bacteria, the plant should sparkle with luciferase. Transposons, a genetic engineering approach, was used to insert the lux operon (from bacteria) into the sex cells of plants. They calculated that if the necessary DNA were introduced into the seeds from that plant, some of the seeds would mature into plants that illuminate. For testing purposes, scientists introduce regions of DNA that code for antibiotic resistance; if the transformed plants are exposed to antibiotics, and they survive, then they know that the new genes were correctly inserted (https://askabiologist.asu.edu/plosable/glow-plants).

(a) (b)

FIGURE 2.2 A plant with a dazzling hue. In the dark, this is what the modified plants should appear like if they were effective.

Source: Ask A Biologist. Arizona State University. https://creativecommons.org/licenses/by-sa/3.0/

2.15 THE FUTURE MAY BE ILLUMINATED BY BIONIC PLANTS

In the modern world, mobile phone masts and streetlights may be found anywhere. To rephrase, picture a world where trees are both antennas and beacons, with the former acting as the later. A landmark study by MIT professor Michael Strano and his team set the framework for the utilization of plants as a source of energy. NPs were injected into plants to improve or provide non-native traits, according to a study published in Nature Materials.

Research on the chloroplasts of baby spinach leaves and thale cresses showed that the researchers could increase the sensitivity of the plants to certain substances, paving the way for their potential use as sensors. The researchers also demonstrated that they could increase the photosynthesis activity of plants three times more than control plants and that they could improve plant repair, paving the way for the potential use of plants as a source of energy.

According to Kwak et al. (2017), these advancements are dependent on the development of two new methods for delivering NPs to plants (Figure 2.3).

When water is pushed into the leaf's pores, NPs are dissolved and vaporized. Lipid exchange envelope penetration, or LEEP, is a second technique for nanoparticle absorption via the chloroplast membrane that uses NPs coated with a strongly negative or highly positive polymer coating. To paraphrase Strano, he claims that following absorption, NPs become trapped inside of the chloroplast, which he refers to as the "plant's engine."

FIGURE 2.3 Photonic biochemical detector made from an Arabidopsis plant with carbon nanotubes embedded in the leaves.

Source: https://phys.org/news/2014-03-bionic-synthetic-nanoparticles-photosynthetic.html.

2.16 TO TURN TREES INTO STREETLIGHTS, TRY BIO-LED: GOLD NANOPARTICLE INJECTION!

We rely on streetlights to help us get home at night and keep the roadways safe for pedestrians and drivers alike. What if, however, we were able to manufacture naturally powered streetlights? The crimson light that trees give off when gold NPs are embedded in their leaves was discovered by a team of Taiwanese scientists. The concept of utilizing trees to replace streetlights is a novel one since it would reduce light pollution in big cities while saving money and reducing CO_2 emissions. Professor Shih-Hui Chang asserts that

LEDs have supplanted conventional light sources in many display panels and streetlights. The emphasis of research in this field is the development of high-efficiency lighting that does not contain toxic compounds such as phosphor powder. Phosphor powder can be used to replicate light from a white light-emitting diode. Toxic and expensive, phosphor powder is a no-brainer. Dr. Yen-Hsun Su (2010) came up with the notion of using the above to find a less harmful alternative to phosphor powder (Quick, 2010). For him, it's a big part of what spurred him to begin his investigation in the first place.

Plant leaves of *Bacopa caroliniana* (Figure 2.4) were infused with gold NPs, causing chlorophyll to glow red. Fluorescence from gold NPs caused chlorophyll in the surrounding area to glow red when exposed to UV light at a high wavelength. To put it another way, Dr. Yen-Hsun Su has made an astounding discovery about the use of plants as illumination, particularly roadside trees (Su et al., 2010), at Academia Sinica's Research Center for Applied Science (Beck, 2010). Gold nanoparticle experiments by Yen-Hsun Su led to the discovery of bioluminescence, a fantastic efficiency for more environmentally friendly lighting technology.

FIGURE 2.4 Plant leaves of *Bacopa caroliniana*.

Source: User:BotBln. Wikimedia Commons. https://creativecommons.org/licenses/by-sa/3.0/.

Academia Sinica, Taiwan's Beck (2010) investigated how gold NPs placed in *Bacopa caroliniana* plants responded to high wavelength UV radiation. As the gold NPs are injected, they emit a blue-violet fluorescence that induces the chlorophyll in the area to glow red. There is a technique to make other biomolecules more efficient through bioluminescence research in Taiwan (Figure 2.5).

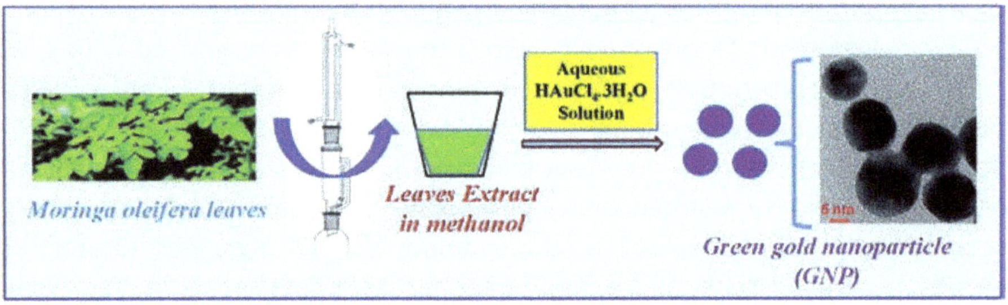

FIGURE 2.5 A pictorial representation of the preparation of *M. oleifera* leaves extract in methanol followed by synthesis of green gold nanoparticles, reduced and stabilized by methanolic extract of *M. oleifera* leaves.

Source: Reprinted with permission from Jayanta et al. 2021.

2.17 PLANTS ARE NOW CAPABLE OF DETECTING EXPLOSIVES AND SENDING AN E-MAIL NOTICE

It is now possible to wirelessly send explosive detection data from a smartphone-like handheld device to MIT engineers' spinach plant sensors (Massachusetts Institute of Technology (MIT), 2016; India Today, 2021). MIT scientists virtually showed their experiment. One of the Scientists from MIT, Professor Micheal Strano, said that they considered it a Novel experiment that serves as a barrier between plants and humans. These *engineers* applied Nanotechnology and managed to send signals sending emails from spinach plants. It is a beautiful and exciting invention. It is one of the admiring applications that come under the plant nano-bionics subject. Scientists tried this experiment to make spinach roots detect nitroaromatics in groundwater. Its use is in explosives such as landmines. Spinach is well known as a nutrient plant, now, it serves as a sensor. Wonder-material carbon nanotubes are used to give signals and detect explosives and wirelessly to a mobile device.

The undersides of spinach leaves were injected with a nanoparticle solution by MIT technologists. These spinach leaves allowed them into the mesophyll layer. Here, photosynthesis takes place. The allowed nanotubes serve as sensors and can detect nitroaromatic compounds. Generally, these compounds are used in explosives like land mines. When plants take water through the roots, these compounds already present in the ground will feed. A fluorescent signal is then emitted from the leaves' carbon nanotubes. When a laser is shined on the leaves, an infrared camera can pick it up.

The scientists hooked up a camera to an inexpensive Raspberry Pi system. An email is sent to the user through this set when the compounds are detected (Figure 2.6). The system has replaced a Cellphone and the right kind of camera. The time taken to conduct this experiment is only 10 minutes. Explosive molecules from the plant's roots travel to the leaves at this period of time. Carbon nanotubes and bee venom have also been used by MIT researchers for bomb-sniffing research.

According to an MIT scientist, plants can be used as analytical chemists. Investigators can receive and read the signal from the plants at a distance of roughly one meter using this method. Future bionic plants may be able to receive radio signals or change color thanks to new engineering approaches (O'Heir, 2017).

2.18 PLANTS WILL SERVE AS TECHNOLOGY

Kwak et al. (2017) argue that this is only the beginning of the process of unlocking plants' full potential. One must have respect for nature in order to really understand the function plants play in the creation of technology (Figure 2.5). Among other things, they are weather-resistant, have their own water supply, and can be repaired by the user themselves. Using a living plant as a chemical sensor is being considered by Strano, one of the scientists involved. Water is a great location for a sensor since plants can sample a wide range of their surroundings through gas exchange. Using nano-bionics, pollutants such as nitrous oxide, which plants are particularly sensitive to, can be injected into the plants. NIR light is emitted by the nanobionic plant, which imaging cameras can detect. Strano claims that plant photosynthetic activity has increased by a factor of three (Kwak et al., 2017). The improvement is due to the NPs' ability to collect light in a wider range of wavelengths. He thinks it's too early to tell if this will lead to better plant growth rates for agriculture.

FIGURE 2.6 In order to detect bombs, MIT scientists and engineers modified spinach plants into sensors that can detect them

Source: Adapted from Raspberry pipod and micro: bit base. https://www.recantha.co.uk/blog/?p=15783.

2.19 SUPERCHARGE

Kwak et al. (2017) discovered a novel use for the method they describe. They argue that "accelerating plant development" could offer extra energy that could be utilized for other plant purposes. If plant maintenance is improved, "bags of chlorophyll" may be used as a fuel source. Using the energy of the sun, chlorophyll transforms carbon dioxide gas into a plant-usable fuel

(Kwak et al., 2017). Given our current situation, this is probably one of the most important chemistry experiments we can conduct. However, despite the convenience of harvesting chlorophyll, it ceases to function and no longer produces sugars [fuel] after approximately one hour outside of the plant. There are numerous opportunities available for this research. The limbs of a tree near a woodland region in the United States interfere with a cell phone tower. These branches are capable of absorbing and emitting gigahertz radiation. Similarly, the scientist envisions tree-lined boulevards in place of unattractive traffic lights. "If we could place NPs [in trees] that catalyzed chemiluminescence, we could create trees that glow in the dark and are powered by the plant."

2.20 CROP AND FOOD SYSTEMS CAN BENEFIT FROM NANOTECHNOLOGY

One of the most promising methods for ensuring a sustainable food supply is to use nanotechnology in food production. There are numerous examples of agriculture benefiting from nanotechnology (Sanzari et al., 2019), such as crop development schemes and nutrient management (Sanzari et al., 2019). Interestingly, nanotechnology is used throughout the entire food industry's operational systems, which is rather fascinating. However, at this time, the development and growth of nanotechnologies are mostly focused on their good impacts, with minimal emphasis on nano-toxicology (Ziarati et al., 2018). The usage of NPs does not necessitate the establishment of any regulatory norms. It is preventing the general population from fully embracing and accepting nanotechnology. For this reason, it is possible to close the information gap and provide comfort to customers by developing regulations that cover particle size, measurement, and range, as well as particular physicochemical and processing property criteria.

2.21 USING THE POWER OF TREES, WE CAN CREATE A REMOTE FIRE ALARM SYSTEM

Many sections of the world are plagued by forest fires, whether they are caused by natural or man-made causes. Although summers are hot and dry, they can nevertheless pose a threat to health. In addition to destroying thousands of forest hectares, wildland fires have a devastating effect on property and public resources. Forest fires are a problem in many parts of the world, whether caused by natural or man-made factors. The most fragile areas of the landscape can support significant vegetation growth thanks to high pluviometry and moderate temperatures. Wildland fires have a severe impact on property and public resources, in addition to destroying thousands of acres of forest. In addition, firefighters and civilians are put in harm's way, with a devastating toll on human lives each year, as a result of manmade global warming, the frequency and intensity of these calamities will grow. Progress in research has led to a variety of new wildfire-fighting techniques during the past few decades. All forest fire monitoring systems require better real-time data sharing and early warning systems. In the future, technology advancements will have a significant impact on wildfire fighting. To illustrate the impact, take a look at Figure 2.7 (Sanzari et al., 2019). Already, forest fire detection systems have a significant influence on information and tree communication technologies. Wildfires are easy to notice once they've erupted into raging inferno. Keeping flames under control is a matter of catching them as they enter a forest. This is something that new tree sensors are capable of, and the only source of energy they require is the wind passing through the trees (https://www.sciencenewsforstudents.org/article/trees-power-this-alarm-system-for-remote-forest-fires).

A forest fire detection system has been developed by Michigan researchers (Ornes, 2020; Molina-Pico et al., 2016). This battery-free contraption can be hung from trees to harvest the energy it needs from the movement of the leaves. As a result of a fire, authorities can be alerted as well. For a decade, an unattended sensor of this type would be able to keep an eye on the forest. "It's a terrific idea," said Newcastle, Australia engineer Chris Knight (Ornes, 2020). Engineer Chris Knight served in Australia's national research agency, CSIRO. He has depth in wildfire sensors and developed a detector.

A small amount of power is all that is needed to detect smoke or increasing temperatures with this type of detector (Wang et al., 2018; Molina-Pico et al., 2016). According to one of the engineers at Michigan State University in East Lansing, high energy participation in manufacturing forest-fire monitoring is less significant (Hefeeda & Bagheri, 2009). Cao has led the team in developing this device. For the purposes of taking measurements, this device recharges itself every 10 minutes or so using minimal power. Even if the fire is a mile distant, the equipment can detect the fire signals. It's possible for that hub to act as a relay for an alarm signal. Cao claims that it is capable of gathering this data from a large number of sensors, possibly as many as hundreds (https://msutoday.msu.edu/news/2020/self-powered-alarm-fights-forest-fires-monitors-environment). There is a pressing need for rapid monitoring of isolated woods, according to Sahin & Ince (2009a). There are several fires in the United States each year, resulting in the destruction of millions of acres. Nearly a hundred fires were still burning in Oregon, California, and Washington as of mid-September of the year 2020. Thousands of individuals were forced to leave with their goods. Detection of disease early on can save lives, as well. Satellites and fire towers are currently used to identify forest fires. It is Cao's opinion that the cost of tree-mounted sensors is minimal.

FIGURE 2.7 Wildfires.

Source: Reprinted with permission from Pixabay. https://www.pexels.com/photo/forest-on-fire-51951/ https://creativecommons.org/publicdomain/zero/1.0/

2.22 PLANTS INSPIRING DESIGN AND TECHNOLOGY

Plants need to shield themselves from wind, rain, harsh temperatures, and fluctuating light levels in order to survive. Unlike animals, plants cannot simply relocate to a new location. In the early stages of land plant life, a protective outer layer was essential. Plants' protective outer layers undergo a range of changes over time to help them thrive in harsh conditions. There have previously been a number of creative solutions to complex issues generated by evolution, such as Velcro. As a source of ideas for novel materials, scientists are researching plant surfaces. Under a microscope, the surface of a plant leaf, stem, or petal may initially appear flat, but a surprisingly complex landscape can be observed. The surface cells may be pointy, ridged, or jigsaw-like, among other shapes, and hairs or protrusions may form fragile structures. Plants have had millions of years to develop responses to the daily challenges they face. There are a number of other issues that can arise, including the leaves drying out or getting too wet, the plants absorbing enough light for photosynthesis, and pollination and seed dispersal. Random modifications to a plant's genetic coding that give it an edge over other plants will help it survive. This evolutionary process occurs gradually over very long periods of time. As a source of ideas for novel materials, scientists are researching plant surfaces.

2.23 PLANT SURFACES AND ANALYZE THE MICROSTRUCTURE OF WOOD (*POPULUS TREMULA*)

Wood has been used in construction for millennia since it is inexpensive and offers a wide range of advantages. It is easy to work and readily available everywhere. However, since the advent of skyscrapers and more significant buildings, wood utility has been reduced. Nowadays, construction work involves steel and concrete. Wood requires the structural properties required to support such large buildings. Similar techniques were followed to investigate plant surfaces and analyze the microstructure of wood by scientists.

2.24 HOW DO PLANTS GENERATE ELECTRICITY TO POWER LED LIGHT BULBS?

Future human society will place a premium on the development of pollution-free and ecologically friendly energy sources. Plants can be utilized to generate electricity, according to researchers from the IIT-Instituto Italiano di Tecnologia in Pisa, Italy. Power sources identified by Fabian and colleagues at IIT (Meder et al., 2018) could be one of the future's electrical supplies that are completely integrated into their natural environments and available worldwide. Researchers have discovered that a single leaf can produce more than 150 volts. In addition, researchers showed a unique "green" electrical generator based on a "hybrid tree" with both natural and synthetic leaves.

IIT at Pontedera's Center for micro-bio-robotics (CMBR) (Pisa, Italy) is where Barbara Mazzolai led the research team. With the support of bio-inspired techniques, robots and technology better suited to today's solutions for unstructured environments will be developed. Their mission is to do cutting-edge research and develop new methods, robots, and materials that are inspired by nature. Using funding from the European Union and a team led by Barbara Mazzolai, Plantoid was able to create the world's first plant robot in 2012. When plants were studied, the researchers discovered that the leaves generated electricity when they were exposed to certain materials or the wind.

Some leaf designs can convert mechanical energy supplied to the leaf surface into electrical energy because of the unique components provided by most plant leaves. The leaf's surface can gather and transmit electrical charges to the plant's core. The plant's tissue acts as a "cable" for the created energy, transporting it to other parts of the plant.

For the first time, researchers detail how plants might harness this effect to generate electricity from wind power. To mimic the native *Nerium oleander* leaves, researchers inserted fake leaves into a *Nerium oleander* tree. Electricity is generated by a "hybrid tree," which is a tree with leaves that can be moved by the wind. The more leaves that are touched, the more electricity is generated. It is possible to easily scale it up by employing the complete leaf surface of a tree or even a forest. In 2019, Barbara Mazzolai led a European-funded project called Growbot, which aims to create bioinspired robots that mimic plant growth. This study is the first critical step. Plants could become a major source of future electricity, as demonstrated by the fact that the new plant-derived energy source will help power new robotic devices.

2.25 PLANTS WILL TEACH ABOUT OIL SPILL CLEAN-UP MICROFLUIDICS

Oil spill clean-up and microfluidics can be learned from plants, according to researchers from Bonn and Aachen universities and the Heimbach-GmbH (Science News, 2018; Zeiger et al., 2016). The oil poses a significant threat to marine life. Using improved technologies, contaminations like this could be eliminated. Skimming the oil from the water is done using textiles with certain surface properties that passively transport it to a floating container. Researchers used plant-based surfaces as a paradigm for their work.

2.26 FUNCTIONAL TEXTILES AS "SUCTION TUBES"

Furthermore, this makes superhydrophobic textiles relevant for environmental technologies. Ultimately, they promise a novel strategy for addressing the urgent environmental issue of growing oil spills in water bodies. Floating oil coatings in the ocean have caused a variety of problems. When they come into contact with plants and animals, they pose a serious threat to their well-being. The fast spread of oil coatings across large areas puts entire ecologies at risk (Science News, 2018; Zeiger et al., 2016).

2.27 OIL SPILLS ARE TREATED BY NATURE

Researchers highlighted environmentally beneficial Tcleanup oil spills as well as the procedure of removing invasive weeds. It is common knowledge that aquatic vegetation repels water and attracts oil. Researchers highlighted (Zeiger et al., 2016) environmentally friendly To cleanup oil spills as well as the technique of removing invasive weeds. Existing approaches for addressing oil spills have considerable disadvantages. Chemical dispersants and combustion can disperse contaminants, yet environmentally friendly items such as sawdust and wheat straw retain both water and oil, making cleanup difficult and inefficient.

The aquatic fern *Salvinia molesta* and the water lettuce *Pistia stratiotes* were the focus of research by German scientists from the Karlsruhe Institute of Technology and the Nees Institute for Biodiversity of Plants. Trichomes, which are minute hair-like wax structures, cover the leaves of these plants. Using this leaf form, water will be repelled, and air will be captured. As a result, it floats on the water's surface. In order to keep water out and air in, this leaf shape was created. Because of this, the plant is able to float on the water's surface. It turns out, though, that it can also hold a lot of oil. The dried leaves of both plants (Zeiger et al., 2016; Siciliano, 2018) absorbed more oil than Durex Pure and Ol-Ex, two commercial oil sorbents used to clean up oil spills (Zeiger et al., 2016; Siciliano, 2018). In order to keep water out and air in, this leaf design was chosen. This causes the plant to float on top of the water. However, it has also been found to have a high capacity for oil absorption as well. Scientists at the Karlsruhe Institute of Technology have produced an experimental product called nano-fur, which they claim to be superior to others. Less leaves of *Salvinia molesta* can absorb an oil puddle in less than 30 seconds. It turns out to be able to soak up a significant quantity of fat as well. It's also better than nano-fur, an experimental product developed by scientists at the Karlsruhe Institute of Technology. *Salvinia molesta* leaves can soak up an oil puddle in under 30 seconds. Lotus (*Nelumbo nucifera*) leaves absorb more oil than *Salvinia molesta* or *Pistia stratiotes* leaves because of their waxy coating and lack of hairs. Durex Pure and nano-fur both have more hairy textures and hence absorb more oil than l-Ex, which is a hairless synthetic oil sorbent. Oil absorption is greatly enhanced by the presence of the trichomes, according to these findings. A variety of hair trichome forms and sizes were tested on *Salvinia* leaves, including short and bent, united at the tips in a group of 2, and tall, connected in a group of 4 at the tips. If you're looking for the best oil-absorbing trichome, look no further than *Salvinia molesta.* The joined ends of the trichome, according to the researchers, are important in capturing oil in its egg-beater-like structure.

Researchers may be able to use these findings to improve nano-hair or develop synthetic materials for oil cleanup. An ideal hairy surface is one that is interconnected and forms pockets that hold oil rather than one that is randomly distributed. An even simpler solution to oil spill cleaning has been suggested by the study. Leaves such as *Salvinia molesta* or *Pistia stratiotes* are commonly used as mop-up materials. Some issues could be resolved by harvesting large numbers of these plants. Both species are strangling subtropical streams around the world as invasive weeds. It is advised that heavy metals, nitrate, and phosphate pollution be removed by the plants mentioned. A serious threat to aquatic life is oil spills. Using innovative technology created by experts (Bonn and Aachen Universities and Heimbach GmbH), it is now possible to eradicate these toxins. A container composed of textiles with certain surface qualities passively skim oil from the surface. The scientists modeled their research on plant surfaces (Barthlott et al., 2020).

2.28 ROBOT CURLS AND CLIMBS (LIKE A VINE) FROM PLANT-INSPIRATION

Plant tendrils served as the inspiration for a soft robot constructed by scientists from Italy's IIT Institute of Technology (Irving, 2019). As water moves through plants, the same physical principles that govern the movement of water may be used to make it bend and climb. Wearable devices such as soft braces that can change their shape could be inspired by this tendril-like soft robot in the future. Researcher Barbara Mazzolai has published in Nature Communications and leads the team. Creating robots based on plants may not seem like the most interesting idea at first glance, but the end result may be high-tech decorations or self-sustaining home plants. Italian researchers have now created a soft robot that moves like plant tendrils, clinging to and encircling objects (Irving, 2019). An IIT-Istituto Italiano di Tecnologia team led by Barbara Mazzolai has worked on the Plantoid project, a robotic tree filled with electronic sensors, earlier (Figure 2.8).

FIGURE 2.8 The IIT (Istituto Italiano di Tecnologia) has built a soft robot that can climb and coil like a plant tendril. The tendril-like soft robot is able to curl around Passiflora caerulea plant stalk. It is able to curl and climb, using the same physical principles determining water transport in plants. Credit: IIT-Istituto Italiano di Tecnologia

Source: Reprinted from Must et al. 2019. https://creativecommons.org/licenses/by-sa/3.0/

The new robot is just a flexible PET tube, and 1.3-volt batteries power it. Like its biological cousin, it uses a hydraulic mechanism to move. Tiny particles in the plant's intracellular fluid move and flex the plant's tendrils. Then, the robot fills with liquid containing ions. The power that activates these ions functions as flexible electrodes. Electrodes attached to the base of the robot are to blame for its failure. The charged energy source of these ions functions as a bendable electrode. After turning off the Power Switch, the fluid redistributes. This action induces the expansion of the tendril. The researchers developed the robot to be small enough to move quickly, and they landed on a 1 mm-wide foil vine.

2.29 SOIL ARSENIC MONITORING USING A PLANT-BASED SENSOR

Plant-based sensors are physiologically-dependent mechanisms capable of detecting the presence of specific chemicals, diseases, and radiation stimuli on land, whether they are already there or purposefully manufactured to make brilliant plants. Sensors based on plants have two components. The plant's physiology-based one is included in the first. In response to a certain substance or environmental condition, the physiological responses are either genetically changed or kept undisturbed ("natural"). It is possible to estimate the physical reactions of the plant from afar. Data can be interpreted by a technician who uses artificial technology to track the multicolored plants' reactions. It is possible to keep track of the physiological mechanisms, responses, and physical reactions of plants using currently available technology. Different methods already exist to keep tabs on plant development, leaf coverage, and chemical composition, as well as the reflectance of visible and infrared light. Data from reporting bright plants can only be gathered with the help of measurement devices. It is a fresh and potential long-term source of intelligence that is durable, stealthy, discrete, and easy to distribute with these biosensors.

Real-time monitoring of toxic heavy metals. Detection of nanoscale devices embedded in plant leaves is possible. Arsenic may be detected and monitored in real time using a plant micro-bionic optical sensor created by MIT scientists (Lew et al., 2020). Arsenic in the environment can be measured using methods that are centuries old. Because it is a natural component of the earth's crust, arsenic can be found in all three major ecosystems: air, water, and land. It is extremely poisonous in its inorganic form. Environmental monitoring and agricultural applications are expected to benefit greatly from the use of this novel sensor, which has been developed by experts.

Our food supply must not be affected by arsenic in the soil. Preventative actions to keep arsenic levels as low as possible can now be taken in real time, increasing our food safety safeguards (Lew et al., 2020). Arsenic levels in the soil can be detected and monitored using a new form of plant nanobionic optical sensor. It is extremely poisonous in its inorganic form. Environmental monitoring and agricultural applications are expected to benefit greatly from the use of this novel sensor, which has been developed by experts. Identification of Soil Arsenic Contamination in Rice Paddy Field Based on Hyperspectral Reflectance Approach has been enlightened by Saha et al., (2022).

Arsenic levels in the ground can be detected and monitored in real time using a new form of plant nanobionic optical sensor. Tedrick Thomas Salim Lew, a doctoral student at MIT, remarked that their plant-based nanosensor is unique not just for being the first of its kind but also for its major advantages. Arsenic levels in the subsurface environment can be measured using this technology, which requires less time, equipment, and personnel than standard procedures (Lew et al., 2020).

In the case of the new optical nanosensors, changes in fluorescence intensity are used to quantify arsenic. Arsenic's internal dynamics can be monitored in a non-destructive manner using these devices. The optical nanosensors enable plant tissues to become self-powered arsenic detectors in their natural environment when arsenic is introduced into plant tissue.

2.30 MARKING A SIGNIFICANT UPGRADE OF THE PLANT-POWERED AIR PURIFIER

A more serious concern is that we all spend the more significant portion of our lives thinking that we're safe since we're indoors. Is it more intoxicating inside your house than it is outside? We need to find the person who came up with the concept. Rice and spinach plants, as well as the fern *Pteris cretica,* were used to test the optical nano-sensor. It is not harmed by large concentrations of arsenic that it can absorb and tolerate. Arsenic may be detected using optical nanosensors in all tests. Arsenic levels as low as 0.2 parts per billion can be detected using this device.

On the surface, here's a shortened list of the common pollutants in modern homes: ammonia from cleaning products, carbon dioxide from your heating system and benzene are all things that you don't want your child inhaling, but they do almost every day. The nature smart natural air purifier is a no-brainer, considering how many breaths of purportedly fresh air you breathe and the fact that your children are exposed and at a greater risk of exhibiting a reaction than you are (Milan & San Francisco-PRWEB, 2018). The first thing that springs to mind is purchasing low-cost air filters that capture contaminants like cigarette smoke and fireplace gases. You should be able to tell by now that these are inapplicable to the previously stated contaminants. Since its launch on Kickstarter, Natede, the most incredible smart natural air purifier, has wowed backers, raising more than 10 times its initial investment goal of $60,000 in the first 24 hours.

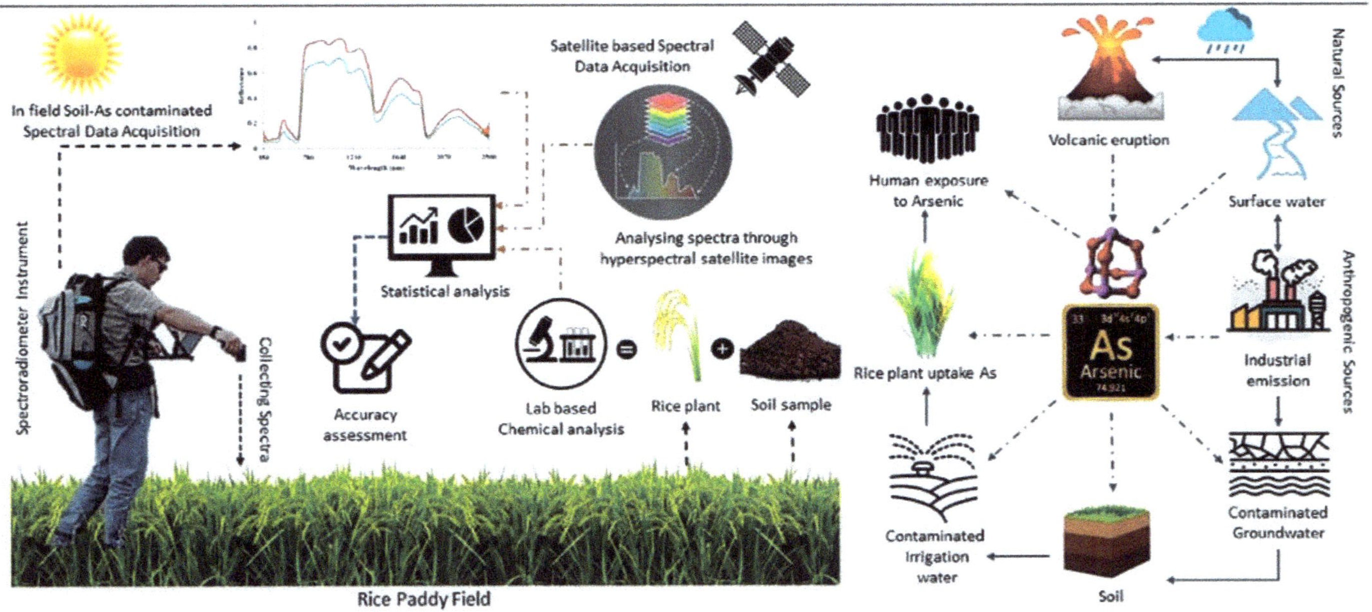

FIGURE 2.9 Identification of Soil Arsenic Contamination in Rice Paddy Field Based on Hyperspectral Reflectance Approach has been enlightened by Saha et al., (2022).

Source: Reprinted with permission from Saha et al., 2022.

2.31 NATEDE: SMART NATURAL AIR PURIFIERS

The Smart Natural Air Purifier has sensors built into its design that can detect changes in your living environment that your skin is unaware of. Air temperature, pressure, humidity, gas composition, and refined particle concentration are all examples of environmental factors. The new photo-catalytic filters in NATEDE (NAture, TEchnology, and DEsign) serve as a natural air purifier, eliminating germs, viruses, fine particles, and volatile organic compounds (VOCs) (Figure 2.10). As a form of phytoremediation, (https://www.kickstarter.com/projects/3rdparty/vitesy/natede) utilizes the inherent capacity of ordinary plants to remove airborne pollutants. With its self-watering system, the plant can be cared for up to one month. In just nine hours, it can purify the air in a 36-square-meter room. Sensors and the internet of things (IoT) are integrated into the plant holder.

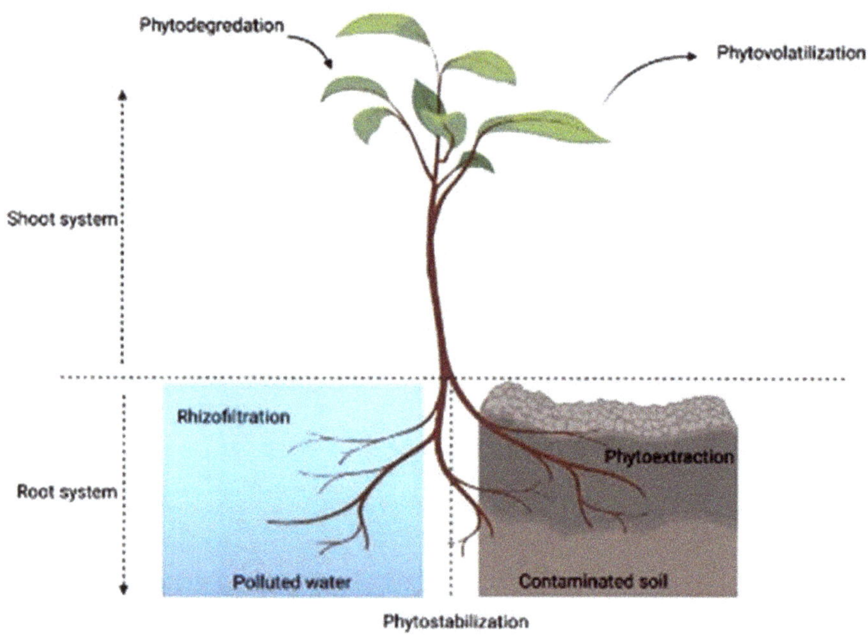

FIGURE 2.10 The air purifier was created by Clairy and is based on phytoremediation, which makes use of the plant's natural capacity to remove pollutants from the air.

Source: Reprinted from El-Tanbouly et al., 2021. https://creativecommons.org/licenses/by/4.0

An indestructible photocatalytic filter and a self-watering pot hold the chosen plant in the NATEDE, which was created by Clairy. A key component of the system's technology is phytoremediation, which makes use of a plant's natural capacity to remove pollutants from the atmosphere. It is more efficient because the NATEDE system allows air to move more freely through the plant's roots. Temperature and humidity levels in the room may also be measured by NATEDE's sensors, which allow the user to alter the system via a companion app as well as monitor their air quality. NATEDE uses a unique technology that enhances the natural characteristics of plants to purify the air in the home (during the eighties, NASA learned that through their roots, certain species of plants were able to reduce toxins). Phytodepuration was used by the Venetians to create a double-vase filtration system.

Essentially, it is a heart for the plant that uses cutting-edge technology to improve airflow. Air pollutants are consumed by soil microorganisms, which connect to the roots and remove them. Air quality, temperature, and humidity are constantly monitored by sensors in the system. If you'd like to keep tabs on the process, you can use a live feed, the app, or your Wi-Fi connection. NATEDE has a self-watering system with an approximate 1-liter capacity, is manufactured from sustainable materials, and consumes little electricity. To top it all off, NATEDE has been endorsed by the International Laboratory of Plant Neurobiology (LINV) and PNAT (a spinoff from the University of Florence).

It's a flowerpot with a unique design that incorporates popular houseplants with cutting-edge air purification technologies. It is not necessary to update the photocatalytic filter in NATEDE in order to get rid of germs and viruses because plants can naturally eliminate contaminants. Photocatalysis and NATEDE design technologies work together to significantly boost the purification power of the plants. NATEDE has a five-fold impact in reducing indoor pollution compared to traditional methods. A special app for the device's five sensors allows users to keep tabs on the air quality. The NATEDE filtration system is unique. Phytoremediation, the ability of plants to clean the air, is enhanced. The double-pot filtering system is designed to improve airflow and has a technical heart that significantly boosts airflow through the roots of the plant. The bacteria that live near the root's base are very interesting. They purify the air by feeding the plant with pollutants in the air that would otherwise harm it.

Organic matter can be decomposed at the molecular level using the natural process known as photocatalysis, which occurs when a mineral, titanium dioxide (TiO_2), is struck by light and activated by a chemical reaction. NATEDE is a non-toxic, environmentally friendly, and biodegradable brand of plastic. At the end of their lives, the pieces can be disassembled and recycled. In terms of performance, NATEDE's five sensors are unrivaled

(CO_2). NATEDE's LED-based user interface makes it simple to operate. It pulses at different rates depending on the air quality. The NATEDE smartphone app lets you keep track of the air quality in your home or office in real time. The measured contaminants can also be broken down in great detail. Smart home gadgets like Amazon Echo and Google Home can be used with this. In addition to saving water, NATEDE's self-watering technology ensures the survival of plants while people are away for long periods of time. The water stored in the tank will be adequate to hydrate the plants for approximately one month. Because we spend the majority of our time indoors, the air we breathe, whether at work or at home, is quite important. Natural air purifier NATEDE can help persons with allergies and asthma breathe clean air and enhance their health.

2.32 PLANT-BASED MEASUREMENTS FOR IRRIGATION PLANNING

Plant-based sensors are used to determine leaf capacitance and thickness (Figure 2.11). In the case of farmers, it might help them avoid water waste and parched plants by letting them know when to start their irrigation systems (Afzal et al., 2017). In arid regions, it is difficult to measure "water stress" on a continual basis, as is normally done. Technology that better identifies when plants need water, however, could improve water efficiency. Soil moisture content will be utilized in this process. Models of evapotranspiration can also be employed. This approach is used to compute the total evaporation and transpiration of the ground and plants. We can forecast when crops need to be irrigated. It has the potential to save water and money. It also stimulates the growth of crops that are healthier and more productive and reduces crop failures. There are two primary approaches to measuring plants, particularly in greenhouse conditions. These include measurements of stem diameter and plant weight. A terrestrial laser scanner was used to establish crop surface models for a pixel-wise calculation of plant height and manual measurements of plant height (Figure 2.11).

2.32.1 MEASURING THE DIAMETER OF THE STEM

The diameter of the stem can be used to determine the health of a plant, as well as to monitor irrigation and other environmental conditions, such as soil quality and light intensity. These environmental variables can be carefully measured using a stem diameter sensor by farmers. Predicting environmental changes and adaptations that may be required to keep plants healthy is made possible by this technology. Farmers can precisely measure these environmental variables using a stem diameter sensor. Sensor technology advancements aid farmers in their efforts to produce more. Intelligent sensor technology linked to an industrial transmitter can now be utilized. Some of the benefits of this new technology include a stronger connection to Industry and more accurate, useful information about crops that affect their livelihoods as well as a company's food supply. When assessing the health of a crop, stem diameter is typically the least invasive method.

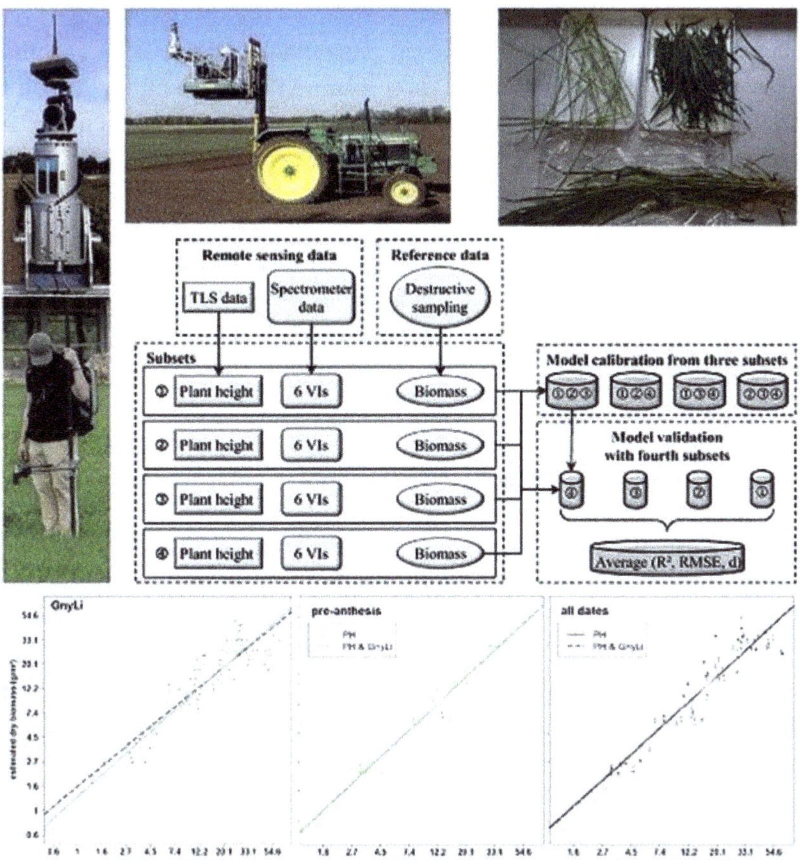

FIGURE 2.11 A terrestrial laser scanner was used to establish crop surface models for a pixel-wise calculation of plant height and manual measurements of plant height.

Source: Reprinted with permission from Tilly et al. 2015. Copyright © 2015 by the authors; licensee MDPI, Basel, Switzerland. http://creativecommons.org/licenses/by/4.0/.

2.32.2 THE MEASUREMENTS OF PLANTS

Plant weight can be used as an alternative to assessing stem diameter. Measurements of entire plants might be made with greater precision using this method than with stem diameter measurements. To avoid interrupting the plant's growth, if possible, weight measurements are taken after the plant has been harvested, but they can also be taken in real time as the plant grows. Weighing a plant while it is still growing poses the risk of interfering with its growth, but it provides farmers with a greater understanding of the plant's health.

A leaf's electrical capacitance tells us how much charge it can hold. The electrical capacitance decreased during the night and increased during the day. Photosynthesis and electrical capacitance are linked. The diurnal changes in electrical capacitance are lessened when wilting moisture is reduced. It was revealed that water stress affects electrical capacitance when soil volumetric water content is less than 11% (Afzal et al., 2017). Scientists have created sensors that can simultaneously measure the thickness of a leaf and its electrical capacitance. Daily electrical capacitance fluctuations are diminished by a decline in withering moisture. The threshold for water stress to influence electrical capacitance was determined to be less than 11% of the soil's volumetric water content (Afzal et al., 2017). The sensor provides a more precise knowledge of what is occurring within the plant since it monitors the thickness and electrical capacitance of a leaf, the most sensitive portion of the plant, for monitoring differences in water content. Climate change and water scarcity urge agricultural systems to cut their irrigation rates.

In the future decades, it is vital to increase agricultural productivity. Poor soil quality, drought, flooding, rising temperatures, and the emergence of new plant diseases pose a threat to the global agricultural supply as the global population and demand for food increase. If the health of plants is checked, agrochemicals, fertilizers, and water may be utilized more efficiently. It has been shown that genetic engineering, imaging, spectroscopy, and electrical sensors can all be used to monitor plant health *in vivo*.

Global food supplies are threatened by pathogenic microbe-caused crop diseases. To ensure agricultural sustainability and food security on a worldwide scale, it is crucial to have diagnostic techniques that can quickly identify plant diseases. The gold standard for identifying plant diseases is nucleic acid and antibody-based molecular assays. However, the analysis procedures are complex and time-consuming. Due to the remarkable physical or chemical features of NPs, new, high-performance diagnostic procedures for a wide variety of plant diseases and other essential disease biomarkers are now feasible. Diseases induced by pathogenic microbes pose a hazard to global food supplies. Diagnostic technologies that can quickly identify plant diseases are vital to ensuring global food security and agricultural sustainability. For the first time, nanomaterials can be used to detect plant infections and other critical disease biomarkers because of their distinct physical or chemical features. Engineered nanomaterials can significantly improve the sensitivity and selectivity of laboratory molecular tests and sequencing methods. As a result of the advent of portable imaging devices like smartphones and nanostructure-supported noninvasive detection tools, such as those described by Li et al. (2020), it is now possible to quickly diagnose plant diseases and keep track of their progress over time, even in areas with limited access to resources.

2.33 REMOTE SENSING AND AGRICULTURE APPLICATIONS–DEVELOPMENTS IN IoT AND SMART SENSORS

Food supply is in high demand due to a variety of factors, including an expanding population, climate change, and the wants of individual consumers. The availability of food, water, and land is causing concern for many people. Keeping in mind that farmers have traditionally used technology advances to adapt agricultural operations to changing times is important. In this day and age, especially with the development of Smart Agriculture, there is no exception to this rule. In our existing farming, advanced technologies like the IoT, sensors, location systems, robotics, and AI in order to implement smart farming. Crop quality and quantity can be improved while human labor is utilized to its full potential. Smart Agriculture is not an exception to this rule in the current era. Our farm uses Smart agriculture, which includes IoT, sensors, positioning systems, robots, and AI. By Gao et al. (2020), remote sensing of forest degradation was explained. Crop quality and quantity can be improved while human labor is utilized to its full potential.

Physiological data can be measured with sensors and self-driving automobiles can be implemented with them. According to Ullo & Sinha (2021), many of today's cutting-edge computer vision research and development applications use sensors. The IoT is a fascinating new application that uses wireless networks and dispersed sensors to collect real-time data and provide specific results of interest through proper processing. For IoT devices to be truly intelligent, sensors and AI are essential. An important function of AI in enabling sensors to operate as intelligent sensors can be found across many different fields. This includes weather forecasting and satellite imagery.

Modern sensors will be used in robotics, automation, remote sensing, underwater photography, and a variety of other applications. In recent years, AI and other high-tech sensors have become increasingly significant in remote sensing and smart agriculture. The sensors equipped with AI fulfill the functions of intelligent sensors. By making available sensor-based equipment and devices, the IoT has produced beneficial agricultural tools. Ullo & Sinha (2021) examined the IoT and smart sensors for remote sensing and agriculture applications (IoT). Examples include the use of drones and other unmanned aerial vehicles (UAVs) for harvesting and for spraying pesticides and fungicides. Many different types of people, from scientists to farmers to remote sensing experts to policymakers, could benefit from advances in the IoT.

2.34 SMART AGRICULTURE SENSORS

Smart farming is a management concept that employs big data, IoT, and AI to benefit the agriculture industry. A bright future for the agriculture business is possible because of this new technology, which makes farming more efficient by allowing it to monitor, track, automate, and analyze processes. Technology helps the Farms in a variety of ways. Data from sensors, equipment, and analytics are all brought together in one place. Farmers may analyze their yields using GPS-tagged data and web-based tools (Ravindra, 2020). The software helps drones, robotics, and other systems to work. Precision agriculture is a method of optimizing the management of crops with various technologies. Using these drones, robotics, and other systems coordinate.

Agricultural intensification cannot be achieved without the help of technology. Sensors in agricultural farming, according to researchers, have enhanced growth and yield by a factor of 10 or more and cut production costs even further (Bose, 2020). Due to advancements in IoT technology and sensor applications in agriculture, sensors may now collect data and convert it as necessary to electrical signals or other outputs. Achieving agricultural intensification without technology is impossible. According to Bose (2020), agriculture sensors have both advantages and downsides. Increasing agricultural productivity is only possible with the aid of technical solutions. Remote sensing and agricultural applications were investigated by Ullo &

Sinha's (2021) research on the IoT and smart sensors. IoT advancements could be advantageous for scientists, farmers, remote sensing specialists, and politicians, among others.

2.35 AGRICULTURAL SENSORS

Precision agriculture makes use of a variety of important sensing technologies to assist farmers in monitoring and optimizing crops and responding to shifting environmental variables. Li et al. (2020) discussed recent progress and challenges in agricultural nanodiagnostics for plant diseases:

- An altitude and latitude can be determined within feet using data from GPS satellites. Only with three satellites can a position be triangulated. Precision agriculture requires accurate positioning. GPS integrated circuits, such as the NJR, NJG1157PCD-TE1, are examples of position sensors.
- Optical sensors are capable of analyzing soil characteristics. Sensors that measure light reflectance at NIR mid-infrared, and polarized wavelengths can be mounted on drones or satellites. Soil reflectance and plant color can be measured and analyzed using optical sensors. Optical sensors can be used to assess soil moisture, organic matter, and clay concentration. Vishay makes photodetectors and photodiodes, which are the building blocks of optical sensors.
- Precision farming relies on the use of electrochemical sensors to monitor soil pH and nutrient levels. Soil ions can be detected using sensor electrodes. Soil chemical data is currently collected, processed, and mapped using sensors mounted on specially built "sleds."
- Ground compaction can be measured using mechanical sensors. For this purpose, load cells or strain gauges are used to measure resistance in the soil. Large tractors employ a variant of this technology to anticipate the amount of pulling power needed by ground-engaging machinery. During irrigation, teniometers like the Honeywell FSG15N1A may measure the root water absorption force. Electrified moisture in the soil, the dielectric constant of soil, can be measured with the aid of sensors.
- Airflow–Air permeability in the soil is determined via sensors. It is possible to take measurements at certain points or while moving. Air must be pushed at a specific depth into the ground at a specific pressure to achieve the desired output. Various aspects of a soil's

composition and structure, as well as its kind and moisture content, each leave their own imprint.

Agricultural weather stations are small, self-contained weather stations dotted over farmland to collect data. Each of these stations is equipped with sensors specifically designed for the crops and climate of the area in which it is placed. Variables like precipitation, leaf wetness, chlorophyll content, and wind speed and direction are also considered in addition to air and soil temperatures. At specified intervals, this data is wirelessly transferred to a central data logger. Weather stations can be beneficial to farms of all sizes because of their portability and affordability.

2.36 SMART AGRICULTURE

The end goal of research into intelligent agriculture is to establish a farm management decision-support system. As population increases, climate change affects our ability to grow crops as well as our ability to harvest them, smart farming considers these issues to be of paramount importance (Zhang et al., 2020). The utilization of sensor data incorporated into technology by farms is reaping the rewards. They are both web-based and GPS-tagged. Using software commands, drones, robotics, and other devices may automate precision farming.

2.36.1 FIELD AND WEATHER MONITORING

Sensors that collect productivity and environmental data are often distributed across farms in wireless networks (Ullo & Sinha, 2021). Through GPS location tagging, this data can enable variable rate technology: the selection of agricultural inputs based on site-specific yields and conditions for maximum profitability.

2.36.2 GEOLOCATION

GPS and GIS are both used in precision agriculture for many purposes including farm planning, field mapping, soil sampling, crop scouting, and yield mapping. It enables yield mapping and variable rate technologies by allowing agricultural data to be tagged to a specific area.

2.36.3 SMART IRRIGATION

Crop water uptake, root health, and necessary irrigation adjustments are all determined by soil sensors. Soil sensors offer data for use. Irrigation systems can be automated using cloud analytics. Using soil sensors, researchers can collect and evaluate data.

2.36.4 LIVESTOCK MANAGEMENT

Farmers can use biometric and production data from wearable and internal sensors for livestock to improve animal feeding, breeding, and comfort management. Automation can be achieved by integrating robotic equipment with this data set.

2.36.5 DRONES

Drones equipped with sensors and other payloads can be used to study crop health, field evaluations, yield mapping, planting, spraying, and irrigation.

2.36.6 ROBOTICS

Using high-resolution cameras, the robots construct a 3D model of the surface, allowing them to freely navigate. Because the factors they are trained to avoid or maneuver through robotics in precision agriculture (RPA) are so complex, movement is not completely autonomous (Zhang et al., 2019). The above-mentioned methods of agriculture, if automated, might reduce labor and material costs while also increasing output. Robots may perform a variety of jobs, such as milking and feeding cattle.

2.36.7 SMART GREENHOUSE

Indoor farming is a viable option for growing crops in a cost-effective manner. Sensors, vertical farming, and automated control systems, among other innovations, can produce satisfactory yields. LED lighting and sensor-based systems are commonly used in smart greenhouses in an effort to increase yields.

2.36.8 FISHERIES MONITORING

Consumers expect seafood goods to be traceable and transparent. Direct fisheries monitoring is being implemented to improve the efficiency of the supply chain. In order to gather information on animal health and business activities, sensors are attached to vessels such as boats, underwater nets, or robots.

2.36.9 PRODUCT MONITORING

While food is being transported from the farm to the shop, sensors and asset monitoring systems on pallets and vehicles keep an eye on its freshness and quality. Using real-time data, products can be dynamically routed to the best location.

2.37 REMOTE SENSING TO FOREST ECOLOGY AND AGRICULTURE

It is the goal of Remote Sensing to collect data and information about objects and substances without having to make physical touch with them first. In order to collect data from the sensor, electromagnetic radiation or sound waves emitted by the target item are used. Aerial photographs are a common example of remotely sensed (camera and film or now digital) items. Large areas can be observed with higher spatial and temporal precision using remote sensing. Remote sensing offers a wide variety of applications in civil engineering. Through education on watersheds, simulations of fluxes, and hydrological modeling, one can obtain knowledge about hydrology. Among the many useful uses of disaster management are flood and drought warning and monitoring, damage assessment in the event of natural disasters, environmental monitoring, and urban planning. Numerous remote sensing uses include monitoring deforestation, the effects of climate change on glaciers, and the Arctic and Antarctic regions of Antarctica.

Using remote sensing photography, it is feasible to detect, locate, and define the surface attributes of the Earth. In addition, although each image represents the environment, it is normal to accumulate photographs over time (Lechner et al., 2020). It is used for a variety of purposes, including mapping invasive species or monitoring land-cover changes like habitat fragmentation in forest ecology. Due to the necessity to determine the location and distribution of species, such research typically requires spatially explicit data. Using remotely sensed photos, direct ecological remote sensing allows researchers to observe vegetation types and animal populations in a way that's both straightforward and direct (Aplin, 2005).

Remote sensing and geographic information systems (GIS) are becoming increasingly important in the study of plants (Sudhakar Reddy et al., 2015). The combination of spatial and non-spatial data produces forest, phytodiversity, and conservation data. Modern remote sensing can be used to discover areas of high biodiversity, predict species distribution, and model species reactions to environmental and anthropogenic changes.

It focuses on distant sensing techniques that are not based on the Earth. This list excludes terrestrial laser scanning or field spectroscopy. Remote sensing is a useful tool for analyzing environmental issues from a practical standpoint (Lechner et al., 2020). It's an essential tool for moving things forward. Remote sensing is an important topic in forest ecology and management. Forest ecosystem dispersion and three-dimensional structure are highlighted in this study. Nowadays, remote sensing provides critical data. Sensors and platforms (such as multispectral and synthetic aperture radar) are used to map the diversity of forest characteristics (such as UAVs and satellites). Forest management data is likely to benefit from recent advances in remote sensing technology, methods, and platforms.

Many trees and a wide variety of wildlife and plants are characteristics of a forest. Plant and animal species diversity must be preserved for the sake of the planet's climate stability. Through the collection of data, remote sensing provides information about the woods. The forest data can be gathered without the use of instruments (Abad-Segura et al., 2020). Ecological sensitivities, sustainable rural development, ecological conservation zones, ecological control buffer zones, air pollution mitigation, cross-border biophysical forest phenomena, satellite fire mapping, topological acoustic sensing, and unsupervised treetop detectors are all new research fields. Researchers Abad-Segura et al. (2020) have come up with 10 new research directions that have appeared in the last few years. These discoveries, which aid in decision-making, add to our understanding of the relationship between technology and forests.

When forests are preserved, the roles of production, ecology, and enjoyment are all maintained in equilibrium. Forest structure and its biophysical features must be precisely characterized in order to assist well-informed decision-making and long-term conservation efforts. For international reporting and national forest monitoring, governments can employ the REDD+ process (Calders et al., 2020) to improve and produce forest geospatial products.

Remote sensing and GIS are becoming increasingly important in the field of plant sciences. Laser scanning is a combination of digital aerial photogrammetry and high spatial resolution (HSR)/very HSR satellite optical imaging. Forest inventories can be produced with the required spatial resolution and accuracy across large areas and a wide range of forest types, thanks to this new method. Knowing what information you need and where to find it is key to making the most of these tools.

Wójtowicz et al. (2016) have listed various applications of remote sensing technologies in agriculture. Its applications are as follows:

- Agronomic forecasting for crops;
- Crop damage and crop progress assessment;
- Identification of plants;
- Estimation of crop acreage;
- Analysis and estimation of crop yields;
- Insect and disease detection and prevention;
- Estimation of the soil moisture;
- Mapping the soil;
- Droughts are always being monitored;
- For the purpose of mapping water resources.

Among the five examples presented by Atzberger (2013) were:

- Estimation of biomass production and yields of biomass products;
- Monitoring of plant vigor and drought stress;
- Crop phenological development is evaluated;
- Cropland mapping and acreage estimation; and
- Disturbance and land use/cover (LULC) change mapping.

Using a GIS-based mapping program, it is possible to determine where crops are grown across the country, alter numerous variables, monitor the health of individual crops, anticipate yields from a given area, and maximize crop productivity (Gebeyehu, 2019). It is possible to identify places in need and the underlying causes of hunger by using information on land use and primary food crops combined with mobile device data. The government of a country with a large agricultural sector needs fast and reliable information on the types of crops planted, the regions where they are grown, and the expected output. For crop issues, remote sensing technology is needed.

Remote sensing is used extensively in civil engineering. Hydrological investigations, hydrologic states and fluxes simulation, and hydrologic modeling are their areas of expertise. Remote sensing data is used in disaster management, including flood and drought warning and monitoring, damage assessment in natural calamities, environmental monitoring, urban planning, and so on. This includes remote sensing monitoring for deforestation in Amazon basin areas, glacier, and arctic and Antarctic consequences of climate change as well as seafloor exploration. Earth observation data can be monitored for misuse. A wide range of environmental variables include water quality, land and sea surface temperature, as well as vegetation biomass and phenology. Global changes in biogeochemical cycles, land use, and climate may all be tracked using satellite data from the International Space Station (ISS).

Satellite data is presently used in a plethora of mission-critical applications. Atmospheric observations are used for a variety of purposes, including weather forecasting, pollution monitoring, and tracking the effects of global warming. Many other applications can be made use of in this way: tracking the movement of ships and fishing nets in the sea; mapping water currents in shallow places and the underlying terrain; studying the ocean ecology; measuring sea surface temperature, salinity, and carbon biomass (Fu et al., 2019). It is possible to use space-based studies to help a number of different industries: mining, flood, and drought monitoring, forest fire detection, agricultural monitoring, and urban planning (Gao et al., 2020; Lentile et al., 2006).

Finally, satellite remote sensing datasets that apply a variety of targeted visualizations to identify human settings increase social science studies on global disasters (the COVID-19 pandemic). In addition, satellite remote sensing has provided a powerful instrument for obtaining global data, such as:

- Topography of the planet;
- Temperatures, water vapor, CO_2, and other trace gases in the atmosphere;
- Mineralogy and the composition of the atmosphere;
- Sea ice, glacier, and cryosphere characteristics;
- Its ability to transmit and receive electromagnetic waves.

Earth-based remote sensing aids in the growth of science and technology. Data from deep space remote sensing missions such as Voyager and Cassini-Huygens is dependent on it (Kohlhase & Penzo, 1977; Matson et al., 2002; Howell, 2017).

Remote sensing datasets that use many targeted visualizations to detect human settings and then connect this information with numerous socioeconomic data sets, like as the COVID-19 pandemic, aid in global tragedies. Human contexts are being identified using satellite remote sensing data on the COVID-19 pandemic, which is then linked to various socioeconomic data sets and other global concerns. Satellite remote sensing data on the pandemic COVID-19 is being utilized to find human settings and then link these observations to a number of socioeconomic data sets, among other global concerns: (i) the topography of our planet; and (ii) the temperature profiles in the atmosphere; Cryosphere properties, such as snow, sea-ice thickness, and glaciers, as well as particle and electromagnetic properties of the thermosphere are all covered in this section. It is possible to study the thermosphere's electromagnetic properties using satellite data.

KEYWORDS

- **bionic plants**
- **glowing plants**
- **nanobionic plant**
- **natural air purifier**
- **plant-based sensor**
- **plants inspiring design and technology**
- **smart greenhouse**

REFERENCES

Abad-Segura, E., González-Zamar, M. D., & Belmonte-Urena, L. J., (2020). Effects of circular economy policies on the environment and sustainable growth: Worldwide research. *Sustainability, 12*(14), 5792.

Afzal, A., Duiker, S. W., Watson, J., & Luthe, D. S., (2017). An electrical capacitance measurement with leaf thickness as a metric for plant water status. *Trans. Amer. Soc. Agric. Biol. Engineers, 60*(4), 1063.

Aplin, P., (2005). Remote sensing: Ecology. *Progress in Physical Geography, 29*(1), 104–113.

Atzberger, C., (2013). Remote sensing in agriculture: A review of the context, current operational monitoring systems, and the most pressing information requirements. *Remote Sens., 5*(2), 949–981 https://doi.org/10.3390/rs5020949.

Babaeian, E., Sidike, P., Newcomb, M. S., Maimaitijiang, M., et al. (2019). High-resolution soil moisture mapping using a new optical remote sensing technology. *Front. Big Data,* https://doi.org/10.3389/fdata.2019.00037.

Barthlott, W., Moosmann, M., Noll, I., Akdere, M., Wagner, J., Roling, N., Koepchen-Thomä, L., et al., (2020). Surface transport of oil over biological and bionic hydrophobic materials: A unique method of separation of oil from water. *Philosophical Transactions* A., *378*, 2167. doi: https://doi.org/10.1098/rsta.2019.0447.

Beato, G., (2013). *Replacing Street Lights with Glowing Trees*. Reason, Biotechnology. https://reason.com/2013/09/24/replacing-street-lights-with-glowing-tre/ (accessed on 17 August 2023).

Beck, J., (2010). *Streetlights Could be Replaced by Trees Impregnated with Luminescent Nanoparticles.* Popular Science. https://www.popsci.com/technology/article/2010-11/trees-could-one-day-serve-streetlights-thanks-gold-nanoparticles/ (accessed on 17 August 2023).

Bland, E., (2010). *Glowing Trees to Replace Street Lamps*. Source: Discovery Channel.

Bose, P., (2020). *What Is a Nano Bionic Light-Emitting Plant?* Azonano, https://www.azonano.com/article.aspx?ArticleID=5512 (accessed on 17 August 2023).

Calders, K., Jonckheere, I., Nightingale, J., & Vastaranta, M. (2020). Remote sensing technology applications in forestry and REDD+. *Forests, 11, 188*. https://doi.org/10.3390/f11020188 (accessed on 17 August 2023).

Callaway, E., (2013). Glowing plants spark debate. *Nature, 498*(7452), 15, 16. doi: 10.1038/498015a.

Collin, T., (2017). *Transplant Production: Light Considerations*. Michigan State University Extension, https://www.canr.msu.edu/news/transplant_production_light_considerations (accessed on 17 August 2023).

Cooper, D., (2014). News in Science, ABC Science. *Bionic Plants Could Light the Future*. https://www.eurekalert.org/news-releases/776479 (accessed on 17 August 2023).

Davison, N. (2018). *Could Glowing Plants Replace Light Bulbs*? CNN, Design, CNN, Design. https://edition.cnn.com/style/article/glowing-plants-mit/index.html (accessed on 17 August 2023).

El-Tanbouly R, Hassan Z, El-Messeiry S. The Role of Indoor Plants in air Purification and Human Health in the Context of COVID-19 Pandemic: A Proposal for a Novel Line of Inquiry. Front Mol Biosci. 2021 Jun 30;8:709395. doi: 10.3389/fmolb.2021.709395. PMID: 34277711; PMCID: PMC8279815.

Fernando, W. G. D., (2012). Editorial. *Plants, 1*, 1–5. doi: 10.3390/plants1010001.

Fu, J., Chao, C., & Yanli, C., (2019). According to data from remote sensing instruments, the East China Sea's Zhoushan sea area shows spatial and temporal fluctuations in oceanographic characteristics. *Regional Studies in Marine Science, 28*, 100626.

Gagliardi, G., Lupia, M., Cario, G., Tedesco,F., Gaccio, F.C., Scudo, F.L., & Casavola, A. (2020). Advanced Adaptive Street Lighting Systems forSmart Cities. *Smart Cities, 3*(4), 1495-1512; https://doi.org/10.3390/smart-cities3040071

Gao, Y., Skutsch, M., Paneque-Gálvez, J., & Ghilardi, A., (2020). Remote sensing of forest degradation: A review. *Environ. Res. Lett., 15*, 103001.

Gebeyehu, M., (2019). Agriculture and natural resource management can benefit from the use of remote sensing and geographic information systems (GIS). *Int. J. Environ. Sci Nat Res., 19(2),* 556009. doi: 10.19080/IJESNR.2019.19.556009.

Gruber, P., (2009). *Conference: 6th Plant Biomechanics Conference.* At: Cayenne, French Guyana. www.botanic.cam.ac.uk/design-technology (accessed on 17 August 2023).

Hefeeda, M., & Bagheri, M., (2009). Forest fire modeling and early detection using wireless sensor networks. *Ad-Hoc and Sensor Wireless Networks*, *7*(3, 4), 169–224.

Howell, E., (2017). *Cassini-Huygens: Exploring Saturn's System*. https://www.space.com/17754-cassini-huygens.html (accessed on 17 August 2023).

https://nptel.ac.in/content/storage2/courses/105108077/module1/lecture1.pdf (accessed on 17 August 2023).

https://sitn.hms.harvard.edu/flash/2020/glowinthedarkplants/ (accessed on 17 August 2023).

https://www.eurekalert.org/news-releases/776479 (accessed on 17 August 2023).

https://www.kickstarter.com/projects/vitesy/natede-the-most-amazing-smart-natural-air-purifier (accessed on 17 August 2023).

https://www.sciencenewsforstudents.org/article/trees-power-this-alarm-system-for-remote-forest-fires (accessed on 17 August 2023).

https://www.theweek.in/news/sci-tech/2019/01/28/Scientists-design-soft-robot-that-creeps-like-plant-tendrils.html (accessed on 17 August 2023).

India Today Web Desk, (2021). *Spinach Plants Can Now Detect Explosives, Send Email Alert to Researchers*. New Delhi. https://www.indiatoday.in/science/story/spinach-plants-can-now-detect-explosives-send-email-alert-to-researchers-1765314-2021-02-0 (accessed on 17 August 2023).

Irving, M., (2019). *Plant-Inspired Robot Curls and Climbs Like a Vine*. Istituto Italiano di Tecnologia – IIT, Science daily, Robotics. https://newatlas.com/robot-plant-tendril/58242/ (accessed on 17 August 2023).

Khatri, K., & Rathore, M. S., (2018). Plant nanobionics and its applications for developing plants with improved photosynthetic capacity. In: Canedo, J. C. G., & Lizarraga, G. L., (eds.), *Photosynthesis*. IntechOpen, DOI: 10.5772/intechopen.76815.

Kohlhase, C. E., & Penzo, P. A., (1977). Voyager mission description. *Space Science Reviews, 21*, 77–101.

Krulwich, R., (2010). *Bioluminescent Christmas Trees?* Krulwich Wonders. https://www.npr.org/sections/krulwich/2010/12/08/131669011/bioluminescent-christmas-trees (accessed on 17 August 2023).

Kwak, S. Y., Giraldo, J. P., Wong, M. H., Koman, V. B., Salim, L. T. T., Ell, J., Weidman, M. C., & Strano, M. C., (2017). A nanobionic light-emitting plant. *Nano Letters., 17*(12), 7951–7961 https://doi.org/10.1021/acs.nanolett.7b04369.

Lechner, A. M., Foody, G. M., & Boyd, D. S., (2020). Applications in remote sensing to forest ecology and management. *One Earth, 2*, 405, https://doi.org/10.1016/j.oneear.2020.05.00.

Lentile, L. B., Holden, Z. A., Smith, A. M. S., et al., (2006). Techniques for assessing active fire characteristics and post-fire consequences using remote sensing techniques. *Intern. J. Wildland Fire, 15*(3), 319–345. https://doi.org/10.1071/WF05097.

Lew, T. T. S., Park, M., Cui, J., & Strano, M. S. (2020). Plant nanobionic sensors for arsenic detection. *Advanced Materials, 33*(1). https://hdl.handle.net/1721.1/140521 (accessed on 17 August 2023).

Li, B., Chen, R., Zhu, C., & Kong, F., (2021). Glowing plants can light up the night sky? A review. *Biotechnol. Bioeng., 118*(10), 3706–3715. doi: 10.1002/bit.27884.2021.

Li, Z., Yu, T., Paul, R., Fan, J., Yang, Y., & Wei, Q., (2020). Agricultural nanodiagnostics for plant diseases: Recent advances and challenges. *Nano Scale Adv.,* **2**, 3083–3094.

Massachusetts Institute of Technology, (2016). *Nanobionic Spinach Plants Can Detect Explosives: After Sensing Dangerous Chemicals, The Carbon-Nanotube-Enhanced Plants Send an Alert.* Science Daily. https://www.sciencedaily.com/releases/2016/11/161101104731.htm (accessed on 17 August 2023).

Matson, D. L., Spilker, L. J., & Lebreton, J. P., (2002). The Cassini/Huygens mission to the saturnian system. *Space Science Reviews, 104*, 1–58.

Meder, F., Must, I., Sadeghi, A., Mondini, A., Filippeschi, C., Beccai, L., Mattoli, V., et al., (2018). Energy conversion at the cuticle of living plants. *Advanced Functional Materials*, 1806689 DOI: 10.1002/adfm.201806689.

Milan and San Franciscio (PRWEB), (2018). *Introducing NATEDE, the IOT-Ready Natural Purifier That Combines Technology and Design with the Power of Plants to Remove Pollutants and Toxins from Indoor Air.* https://www.prweb.com/releases/natede_the_smart_natural_air_purifier_by_clairy_raises_over_300k_in_one_week_and_trends_towards_1_8_million/prweb15441097.htm (accessed on 30 September 2023).

Mitiouchkina, T., Mishin, A., Somermeyer, L. G., et al., (2020). Plants with genetically encoded auto luminescence. *Nature Biotechnology, 38*, 944–946.

Molina-Pico, A., Cuesta-Frau, D., Araujo, A., Alejandre, J., & Rojas, A., (2016). Forest monitoring and wildland early fire detection by a hierarchical wireless sensor network. *Sensors for Environmental Monitoring, 2016*, Article ID 8325845. https://doi.org/10.1155/2016/8325845.

Must, I., Sinibaldi, E. & Mazzolai, B. (2019). A variable-stiffness tendril-like soft robot based on reversible osmotic actuation. *Nature Communications 10,* Article number: 344.

O'Heir, J., (2017). *Bomb Sniffing Spinach.* Mechanical Engineering Magazine. https://www.asme.org/topics-resources/content/bombsniffing-spinach (accessed on 17 August 2023).

Ornes, S., (2020). *Trees Power This Alarm System for Remote Forest Fires*. Science news for students. https://www.sciencenewsforstudents.org/article/trees-power-this-alarm-system-for-remote-forest-fires (accessed on 17 August 2023).

Quick, D., (2010). Gold nanoparticles turn trees into streetlights. *Science*, https://newatlas.com/glowing-trees-using-gold-nanoparticles/16917 (accessed on 17 August 2023).

Ravindra, S., (2020). *IoT Applications in Agriculture*. https://www.iotforall.com/iot-applications-in-agriculture (accessed on 17 August 2023).

Roper, J. M., Garcia, J. F., & Tsutsu, H. (2021). Emerging technologies for monitoring plant health *in vivo*. *ACS Omega*, *6*(8), 5101–5107.

Saha, A., Gupta, B.S., Patidar, S. & Martinez-Villegas, N. (2022). Identification of Soil Arsenic Contamination in Rice Paddy Field Based on Hyperspectral Reflectance Approach *Soil Syst. 6*(1), 30; https://doi.org/10.3390/soilsystems60100301

Sahin, Y. G., & Ince, T., (2009). Early forest fire detection using radio-acoustic sounding system. *Sensors, 9*(3), 1485–1498.

Sanzari, I., Leone, A., & Ambrosone, A., (2019). Nanotechnology in plant science: To make a long story short. *Front. Bioeng. Biotechnol.* https://doi.org/10.3389/fbioe.2019.00120.

Schaal, B., (2019). Plants and people: Our shared history and future. *Plants, People, Planet, 1*, 14–19. https://doi.org/10.1002/ppp3.12.

Science News, (2018). *What Plants Can Teach Us About Oil Spill Clean-Up, Micro Fluidics*? Source: University of Southern California, https://www.sciencedaily.com/releases/2018/03/180320141315.htm (accessed on 17 August 2023).

Siciliano, S., (2018). *Nature Can Heal Itself After an Oil Spill, it Just Needs a Little Help. University of Saskatchewan.* https://news.usask.ca/articles/research/2018/nature-can-heal-itself-after-an-oil-spill,-it-just-needs-a-little-help.php (accessed on 17 August 2023).

Su, Y. H., Tu, S. L., Tseng, S. H., Chang, Y. C., Chang, S. H., & Zhang, W. M., (2010). Influence of surface plasmon resonance on the emission intermittency of photoluminescence from gold nano-sea-urchins. *Nanoscale, 2,* 2639–2646. https://pubs.rsc.org/en/content/articlelanding/2010/NR/c0nr00330a (accessed on 17 August 2023).

Sudhakar, R. C., Jha, C. S., Diwakar, P. G., & Dadhwal, V. K., (2015). Nationwide classification of forest types of India using remote sensing and GIS. *Environmental Monitoring and Assessment, 187*, Article number: 777.

Trafton, A. (2019). *Engineers Create Plants That Glow.* MIT News Office, ACS News Service Weekly Press Pac, https://news.mit.edu/2017/engineers-create-nanobionic-plants-that-glow-1213 (accessed on 17 August 2023).

Ullo, S. L., & Sinha, G. R. (2021). Advances in IoT and smart sensors for remote sensing and agriculture applications. *Remote Sens., 13*, 2585. https://doi.org/10.3390/rs13132585.

Wang, Y., Han, L., Zhao, Y., Xue, Y., & Wang, L. (2021). Design of smoke alarm disperser based on single-chip microcomputer. *J. Physics: Conference Series, 2113, 4th International Conference on Mechatronics and Computer Technology Engineering (MCTE 2021) 15-17, Xi'an, China.*

White, J. C., Coops, N. C., Wulder, M. A., Vastaranta, M., Hilker, T., & Tompalski, P., (2016). Remote sensing technologies for enhancing forest inventories: A review. *Canad. J. Remote Sensing,* 42*(5)*, 619–641. doi: 10.1080/07038992.2016.1207484.

Wójtowicz, M., Wójtowicz, A., & Piekarczyk, J., (2016). Application of remote sensing methods in agriculture. *Communications in Biometry and Crop Science,* 11*(1), 31–50.*

Zeiger, C., Da Silva, I. C. R., Mail, M., Kavalenka, M. N., Barthlott, W., & Hölscher, H., (2016). Microstructures of superhydrophobic plant leaves – inspiration for efficient oil spill cleanup materials. *Bioinspir. Biomim.*, *11*(5), 056003. doi: 10.1088/1748-3190/11/5/056003.

Zhang, J., Zhao, Y., Wu, J., & Chen, B. (2020). LVPDA: A lightweight and verifiable privacy-preserving data aggregation scheme for edge-enabled IoT. *IEEE Internet Things J.*, *7*(5), 4016–4027.

Zhang, Q., & Karkee, A. T. (2019). The use of agricultural robots in orchard management. In: John, B., (ed.), *Robotics and Automation for a More Sustainable Agriculture.* 1907.13114.

Ziarati, P., Shirkhan, F., Mostafidi, M., & Zahedi, M. T., (2018). A comprehensive review: Toxicity of nanotechnology in the food industry. *J. Med. Discov., 3*, 12. https://doi.org/10.24262/jmd.3.2.1801.

CHAPTER 3

Amazing Technologies Inspired by Insects and Beetles

ABSTRACT

Many ecological processes, such as nutrient cycling, seed dispersal, bioturbation, pollination, and pest control, are made possible by insects in terrestrial ecosystems. Insects play an important part in the sustainability of our agricultural systems and food security in the future if we understand their role in ecosystems. The original inventors of the technology are thought to be insects. First, insects came up with the concept for the technology. Different anatomical qualities and methods of coincidence related to our own inspire new designs or refinements to existing ones. Insects and their cousins are a source of inspiration and materials for humans. Using the locust avoidance reaction, engineers are developing a low-power collision detector that mimics the locust and might be used by robots, drones, and potentially self-driving automobiles to avoid collisions. A new flying robotic system designed to combat aerial insect activity has been developed by MIT researchers. A new free-flying robot that replicates the rapid ascension of insects has been unveiled by scientists. Insect-like drones are now being developed by scientists. A team of experts has come up with some astounding findings after examining the Bombardier beetle in depth. A team of researchers was ready to acquire new technology based on the beetle's perspective. These innovations have inspired innovative nebulizers, needle-free injections, fire extinguishers, and strong fuel injection systems. Anti-theft technology based on the bombardier beetle's defense mechanism was developed by scientists because of its importance. It was Bombardier's motivation for developing a chemical defensive mechanism to prevent vandalism—a self-defending surface made of many sandwich-like plastic layers. Foam sprays in the

Bio-Inspired Technologies for the Modern World: Emerging Trends and Applications.
R. Ramakrishna Reddy & T. Pullaiah (Authors)

attacker's face if the surface is destroyed. Anti-vandalism measures and the protection of expensive commodities are among the uses of the technology. Several countries throughout the world may suffer significant financial losses as a result of forest fires. Hairy, flying insects have been used as a novel sensor by scientists, especially those of Indian descent. With a few simple and affordable components, a jewel beetle can build an infrared sensor. They are capable of quickly and precisely detecting a wide range of potentially harmful compounds in the air. These infrared sensors cover large regions of the forest, alerting authorities to potential fires before they spread. Modern hearing aids are being transformed by the advent of an insect-inspired microphone that can find noises and reduce background noise. Stanford University researchers used creepy crawlies to build a photovoltaic material inspired by insects.

3.1 INTRODUCTION

For more than a century, physicists and biologists have been fascinated by the flight of insects. The flapping insect wing motions and the forces and flows around their wings were previously unquantifiable by researchers. The evolution of flying in insects is largely responsible for many scientific advancements. It's easy for flying insects to find new habitats and food sources when the need arises. Using their technologies has been going on for more than 50 million years with no negative impact on the environment. Pollinators of crops, pest controllers, food for other creatures, and decomposers of decomposing organic materials, insects serve several purposes. Insects' various anatomical traits have been shown by Rinaldi (2007) to serve as inspiration for new ideas and enhancements to existing ones. Insects and their cousins continue to serve as a source of inspiration and raw materials for humans even today. There are numerous advantages to having insects around. They also serve as a source of inspiration for new designs or upgrades because of their unique anatomy. It's hard to believe how similar these methods are to our own. Insects and their cousins continue to serve as a source of inspiration and raw materials for humans, even in the digital age.

In fact, bugs are showing us a lot these days, especially in our technologically advanced society. Insect growth can have a negative impact on our designs if we don't pay attention. Bats may be detected and avoided by moths, mantises, and other arthropods using sonar (Rinaldi, 2007). In the early days of GPS, people observed dung-rolling scarab beetles using celestial objects to navigate. New biomimicry technology makes robot insects that allow humans to "work" live insects as if they were humanoid creatures. It seems that they are picking up on the deep emotions that insects have long since abandoned. A novel way of optical tracking that is 20 times faster than previous software has been developed by dragonflies.

Biomimicry has inspired a wide range of technical advancements, including:

- Mollusk shells, bones, and sponge spicules are all inspirations for novel synthetic materials with improved mechanical qualities.
- Self-cleaning surfaces are based on lotus leaves' microstructure.
- A shallow-water echo receiver based on dolphin sonar.
- Photosynthesis serves as an inspiration for new photovoltaic approaches.

Humans pick these up from insects alone are:

- A sea sickness-fighting spider-themed yacht is on the market.
- Miniature robots can avoid flying into walls by mimicking the optical flow navigation sensors of insects.
- Inspired by Palmetto tortoises, spider-man's hands can stick to glass, stone, and wood.
- Autonomous vehicles and drones could benefit from locust (Figure 3.1) vision, which could assist them in avoiding collisions.
- Robots, drones, and self-driving cars could all benefit from learning from swarms of locusts.

FIGURE 3.1 Locust.

Source: Reprinted with permission from Kiryha_rigel from needpix.com. https://creativecommons.org/publicdomain/zero/1.0/

Future drones, robots, and self-driving cars could benefit from a new sensor inspired by swarming locust collision avoidance neurons. Despite traveling at speeds of 2–3 miles per hour, the millions of locusts in the air can instantly alter their route to avoid colliding with one another. A locust neuron has been electronically re-created by American scientists, allowing the insects to respond swiftly and effectively without expending much energy.

Locusts avoid crashing into one another when migrating in swarms of millions (Figure 3.2). Neurobiologist Claire Rind at New Castle University imagined that a highly evolved visual neural system – was quick to react and have reliable circuits (Yue & Rind, 2012). In the tropics and subtropics, Locusts exhibit legendary mass migrations. It was observed in 1929 swarm covered about a mile wide, over a hundred feet deep, and passed for nine hours at a speed of about six miles per hour. When it comes to engineering issues, "there is no law of physics that can stop them." As said by Michio Kaku.

FIGURE 3.2 Migratory locust swarm.

Source: CSIRO. https://creativecommons.org/licenses/by/3.0/

Modern automobiles have both passive and active sensor systems that can detect the possibility of a collision. Locusts, on the other hand, avoid collisions purely by the use of specialized visual cues in the face of approaching objects. Neurophysiological tests were conducted to examine if locusts' 'collision-detect neurons respond to approaching car collisions in dashboard-camera films. In order to reproduce the brain response recorded during the object approach, scientists developed a simple algorithm (Friedrich, 2014).

3.2 TINY VIBRATIONS

Bristol team's research, led by Daniel Robert (https://physics.org/news/2006/03/locust-eardrum-tiny-microphones), researched the usefulness of this locusts' anticollision. In order to describe how the eardrum membrane of a locust reacted to sound waves emitted by a loudspeaker, they employed a laser Doppler vibrometer. The membrane is formed like a kidney, allowing for greater mobility. Two groups of mechanoreceptor cells—neurons that respond to mechanical stress – are found on its inner surface at two places. The laser was focused on the membrane's surface by the researchers. The sound waves caused minuscule out-of-plane vibrations as small as 10–12 micrometers.

The nanostructure of the membrane was examined using a focused ion-beam mill. A low-frequency sound wave would travel completely over the membrane, according to the data. In order for high-frequency waves to travel the full length of the membrane, they must first be attached to low-frequency neurons. This finding demonstrates that locusts have the ability to discriminate between high and low-pitch sounds. It's also clear from the ion beam that the membrane had fluid-filled chambers inside that could suppress sound based on its frequency.

3.3 MIT INSPIRED BY INSECTS FOR NEXT-GENERATION DRONES

We're constantly swatting away tiny flying insects like gnats despite our best efforts. As they fly through the air, flying insects have a remarkable ability to avoid being swatted by hands and other impediments. Insect activity in the air has been studied by MIT scientist Kevin Yufeng Chen, who has developed a new flying robotic device (Ackerman, 2021). During his time at the Electronics Research Laboratory, Chen invented the same. Chen has created insect-sized drones that are extremely agile and flexible (Ackerman, 2021). A unique soft actuator powers the flying robots. As a result, they are more equipped to resist the rigors of real-world flying. According to Chen, compact and agile flying drones could assist humans in jobs like pollinating crops or inspecting machinery in limited locations.

Most flying drones are unable to operate in restricted regions because of their inexperience. Additionally, they lack the sturdiness needed to endure impacts with other objects. Smaller drones necessitate entirely different approaches to construction than larger ones. If you're looking to fly a little drone, you'll need a different type of motor. Due to their lack of experience, most flying drones are unable to operate in restricted areas. Their lack of toughness also makes them ineffective in protecting against impacts from other items. The production of smaller drones necessitates a completely different strategy than that of larger ones. A different kind of motor is required if you want to fly a little drone. Chen and his colleagues employed soft actuators instead of the sophisticated and delicate parts used in other flying machines to develop a more flexible tiny drone. They use narrow rubber cylinders coated with carbon nanotubes to treat. The electrostatic force of the carbon nanotubes is generated by applying a voltage to them. It causes the rubber cylinder to contract and expand. That voltage causes the drone's wings to beat at a high rate when it is used frequently. According to Chen, the actuators are able to flap 500 times per second, allowing the small drone to bump into objects, rebound, and continue flying like an insect, like a mosquito. Because of its little weight, the drone is comparable in size to a large bumblebee. The current design resembles a miniature cassette tape with wings, but a dragonfly-inspired version is being developed by the same team (Ackerman, 2021).

3.4 HOW DO DESERT LOCUSTS IN BILLIONS FLY IN SWARMS ACROSS THE SKY? WHEN THEY TARGET CROPS, HOW WILL THEY AVOID COLLIDING WITH ONE OTHER AT THE SAME TIME?

Locusts in the solitary stage are quiet and reserved insects. They travel faster and are more attracted to other locusts when they enter their gregarious phase. When the swarms are close together, the sky can turn black, and crops can be wiped off. It's amazing to see how these insects can coexist peacefully in swarms as large as 80 million individuals.

US NSF-funded scientists at Penn State are developing a low-power collision detection system that mimics the locust avoidance response (Jayachandran, 2020). The researchers use their 'insect eyesight' to help these pests avoid collisions for long periods of time. Locust swarms soar around the sky, attacking crops without colliding with one another (Jayachandran, 2020).

Autonomous vehicles, including robots, drones, and potentially self-driving cars, could benefit from this study's findings. Penn State's Saptarshi Das states: "Some animals with extraordinary talents to identify already better than humans" (Jayachandran, 2020). Because insects fly and don't argue, many autonomous systems are based on insect vision. However, locusts' eyesight is unique.

Locusts will only employ a single neuron for this purpose. To avoid collisions, it is termed the Lobula Giant Movement detector. Locusts are "simply unfathomable," Das added. "So, we started researching how it works." "These people's abilities are truly awe-inspiring," Jayachandran (2020), a team member, claims that the specific neuron receives two impulses. The eye of the evading locust is struck by a vision of an approaching locust. Secondly, there is a difference in the invading locust's angular velocity compared to the avoiding locust. An invading locust's picture and excitation signal expand in size and strength as it gets closer. There are two branches to the neuron. The locust analyzes the changes in these two inputs as a result of these two signals and realizes that something will collide (Jayachandran, 2020). Monolayer molybdenum sulfide was used by researchers to create a photodetector for nanoscale collision detection. The photodetector was built on top of a programmable floating-gate memory architecture. Only a modest quantity of energy is required for it to duplicate the locust's neural response. It allows the locusts to stay in one place and not shift direction (Jayachandran, 2020). Using this research, collision-avoidance systems of the future can be developed that are intelligent, low-cost, task-specific, energy-efficient, and small enough to fit in the palm of a hand.

3.5 APPLICATION TO AUTONOMOUS VEHICLES

These systems were found to have significant limitations, such as being too large and requiring too much energy. A breakthrough in this relationship could come from their design, which they claim is denser and more energy efficient. Small flash memory cells were placed on top of a tiny photoreceptor to imitate the LGMD's function, which was constructed by the research team (Jayachandran et al., 2020; Shigang & Rind, 2006). When the amount of light entering the eye increases, the amount of internal inhibitory signals decreases. Due to their massive size and high energy consumption, these systems were determined to have considerable drawbacks. Could their design, which they claim is denser and less energy-intensive, be the catalyst for a breakthrough? To mimic the LGMD's function, the study team layered small flash memory cells on top of a tiny photoreceptor (Jayachandran et al., 2020; Shigang & Rind, 2006). An increase in the amount of light entering the eye leads to a decrease in the number of inhibitory signals within the eye. The system was tested in a simulated setting by the team. It worked effectively, and the car was able to anticipate potential crashes. The car was unable to avoid a collision because of its inadequate depth and rotational sensing. Currently, the researchers are attempting to expand their stimulus environment by adjusting the system's rotation, speed, and light intensity to respond to other things. They intend to develop a collision avoidance system that can be used by self-driving cars and robotics.

3.6 AI FOR SELF-DRIVING CARS POWERED BY THE INTERNET OF THINGS (IoT)

There are now "smart" versions of everything from phones and watches to fridges and thermostats. The IoT is a catch-all term for a wide range of devices that are connected to the Internet. In its most basic form, "things" connected to "the Internet" is what the phrase refers to. Internet-connected gadgets are frowned upon by the IoT. According to Lonari et al. (2020), drivers of self-driving cars could benefit from this link when updating their algorithms. These vehicles require a lot of data gathering and processing. In this scenario, the driverless automobile uses IoT to communicate with other vehicles on the road. The route, traffic, and how to get around any impediments are all included in this report. This data is exchanged wirelessly between cars that are connected to the IoT, and it is then tested and put to use in order to improve automation. Many people have long been intrigued by the prospect of self-driving cars. The field of computer science known as AI deals with the development of intelligent machines that can do tasks that would normally necessitate the use of human intelligence. Machine intelligence (AI) has the potential to revolutionize the way we work and live in the workplace.

IoT is an industry 4.0 theory that incorporates international support for the collection and analysis of data/information from storage, actuation, sensing, improved services, and communication technologies. The IoT revolution is a part of the fourth industrial revolution. AI/IoT technologies and high-speed, flexible, low-latency communication will enable the development of spectacular Autonomous automobiles (AV). It exemplifies the synergistic relationship between physical and digital knowledge in the context of Industry 4.0 (Khayyam et al., 2020).

3.7 WHAT IS A SELF-DRIVING OR DRIVERLESS CAR?

Human drivers are no longer necessary for self-driving automobiles to move. These automobiles are also referred to as “free” or “driverless” automobiles. Autonomous vehicles are becoming more widespread as self-driving car technology becomes more widely available (Lonari et al., 2020; Hancock et al., 2019). Sensors and software are used to control, drive, and guide cars in this technology. The roadways are currently devoid of any completely autonomous vehicles. Partial-autonomous vehicles, on the other hand, are becoming more common in more advanced countries, requiring varied degrees of self-automation from the driver.

3.8 HOW DO SELF-DRIVING CARS WORK?

Self-driving cars are becoming a reality thanks to recent technology advancements. Because of the presence of three technologies, self-driving automobiles are possible today.

3.8.1 IoT SENSORS

Due to the broad array of sensors currently accessible, autonomous vehicles (Hancock et al., 2019) will be possible. An automobile’s self-driving capabilities are made possible by a wide range of different sensors that work together.

3.8.2 IoT CONNECTIVITY

Traffic data, weather, maps, other vehicles, and road surface conditions are all analyzed by self-driving automobiles using cloud computing technology. Individuals can better understand their surroundings and make better decisions when they use this technology. Even if edge computing technology is able to do modest computations locally, self-driving cars will still require an internet connection to function properly.

3.8.3 SOFTWARE ALGORITHMS

An automobile that collects so much data wants to analyze it for its own benefit. Control algorithms and software play a key role in this. Self-driving cars must be able to make the right decisions at all times, making this the most difficult aspect of the vehicle. If a "fault," like Uber's self-driving car accident, occurs, it can be deadly.

3.9 THE NOVEL INSECT-INSPIRED FLYING ROBOT

Dragonflies are agile and efficient flyers that use two pairs of wings for demonstrating exquisite aerial maneuvers. Compared to two-winged insects such as bees or flies, dragonflies leverage forewing and hindwing interactions for achieving higher efficiency (Chen et al., 2022) and efficiency and net lift (Figure 3.3). An insect-like flying robots have been also illustrated in literature (Press Trust of India, 2018; The Economic Times-News, 2018). There are numerous ways in which insects might be beneficial. Additionally, their anatomical features and behaviors, which are very comparable to our own, inspire the development of new technologies or enhancements. Humans have always relied on insects and their kin for inspiration, even in the digital era.

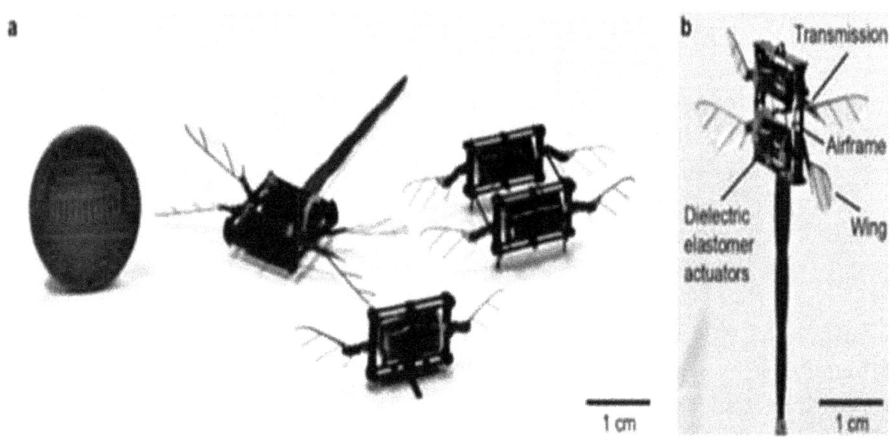

FIGURE 3.3 Illustration of insect-scale flapping-wing robots.

Source: Reprinted with permission from Chen et al. (2022). Copyright © 2022 by the authors. Licensee MDPI, Basel, Switzerland. https://creativecommons.org/licenses/by/4.0.

Unmanned Aerial Vehicle (UAV) propulsion technology is significantly related to the flight performance of UAVs, which has become one of the most important development directions of aviation (Figure 3.4). Insect-inspired flying robots are being developed by Dutch engineers for a number of drone applications. By flapping its wings like a fly, the flying robot could be propelled and controlled. Internet reports claim it can hover and fly in any direction while being incredibly nimble. When the robot's wings flapped, it generated lift force and regulated its flight in the same way that flying insects do.

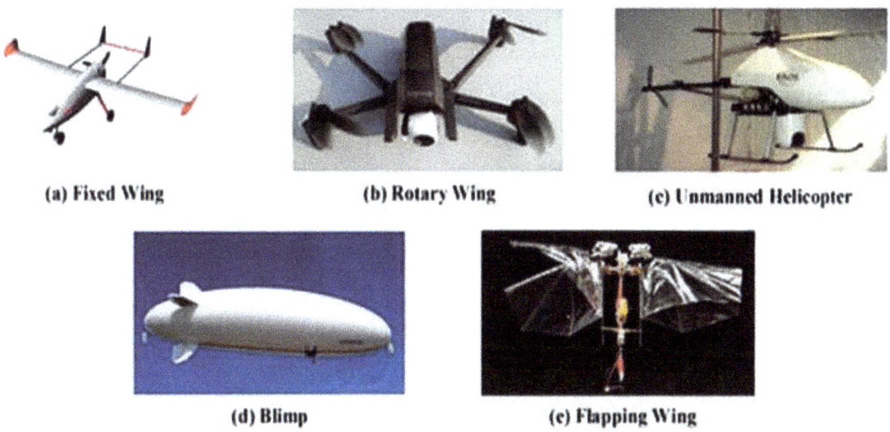

FIGURE 3.4 Classification of UAVs (Unmanned Aerial Vehicle (UAV)).

Source: Reprinted with permission from Zhang et al., 2022. https://creativecommons.org/licenses/by-sa/4.0/

3.10 AS NIMBLE AS FLIES, ENGINEERS CREATE NEW ROBOTS

Until recently, scientists were baffled by insect flight. Scientists found it difficult to observe flight mechanics because of the little insects and the high frequency of their wing beats. Flying is more difficult in insects than it is in birds. Rather than directly moving the wings, the flight muscles alter the shape of the thorax, which in turn moves the wings. Insects are known for their ability to fly. Experiments with animals or tethered robots cannot always prove or disprove assumptions about their flight control. Programmable and flexible, autonomous free-flying robots were created by Karáse et al. (2018) using bio-inspired flapping wing motion settings. When it comes to fleeing from predators, the fruit fly has perfected the art of yawing in the direction of its escape path. A fruit fly according to Karáse et al. (2018), these yaw rotations are caused by a passive, translation-induced aerodynamic interaction between the yaw torque and the roll torque. Pitch torques are generated as the technique develops. The robot opens up new avenues for studying animal flight, and it is capable of carrying out real-world operations.

One of the robot's many maneuvers includes 360° fender benders and barrel rolls. It can reach a top speed of 25 kilometers per hour. For the study's lead author, Karáse and his team, it was a clear assertion. The robot, despite its diminutive stature, packs a punch, considering its weight and wingspan. It has a five-minute hover time and a one-kilometer range on a fully charged battery. There was a definite fruit fly influence on the robot's motions. Another demonstration made by the robot showed how fruit flies adjust their spin angles to maximize their escape velocity. Miniature drones that fly like insects are being developed by scientists (Figure 3.5). Alex Caccia, CEO of Animal Dynamics (Dempsey, 2020), noticed birds take to the skies during Storm Ciara in February. Although the wind was blowing hard, Alex Caccia was amazed to see birds so gracefully soaring through the air. "While airliners are restricted by the weather, the birds couldn't care less," he said on the report.

FIGURE 3.5 Engineers want to replicate the dragonfly's flight, but so far, they've been unsuccessful.

Source: Michael Palmer on Wikimedia. https://creativecommons.org/licenses/by-sa/4.0/.

Animal dynamics manufactures a wide range of drones based on the information gleaned through animal analysis. Two drones have already been delivered to the biomechanics research team, demonstrating their commitment to studying birds and other insects.

The flight of a dragonfly inspired the design of one drone. Since the four-winged drones could function steadily in heavy gusts, the military has already funded the research, which is better than existing small surveillance drones. Skeeter is the codename for a top-secret project that has found a way to control a drone with flapping wings instead of propellers. The study goes on to explain why wings are preferable to propellers. The dragonfly's wings allow it to fly in the face of high gusts, a feat that is nearly impossible for human engineers to do. To put it another way, "making objects with flying wings" is extremely difficult, according to Mr. alex Caccia, according to the story. As a bonus, he explained that helicopters fly by adjusting the angle of their rotor blades to propel them forward. As wind gusts increase in strength, it will be more difficult to keep smaller items in place.

Flying a dragonfly is like flying a kite (Figure 3.6). "Insanely beautiful, there's nothing that could go wrong with that. It has very advanced flight control." Animal Dynamics has read the 300 million years of development the insect has taken. Four years prior, the business had created software that could fly the hand-launched drone at winds of over 20 knots like an insect (23 mph). About 22 to 27 knots is already listed as a "strong breeze," according to the report.

FIGURE 3.6 A whimsical 200 mm wingspan biomimetic dragonfly drone in flight.

Source: Reprinted with permission from Javaan Chahl et al 2021. Copyright © 2021 by the authors. Licensee MDPI, Basel, Switzerland. http://creativecommons.org/licenses/by/4.0/.

The tiny drone will carry a camera and communication links into the skies and be low enough for military workers. Currently, the Skeeter is about eight inches long, but the company is planning to produce smaller drone versions. Removing a lot of navigational and aerodynamic consciousness into the insect drone will be a big test for the company to achieve. The company's 70-strong team of developers relies on electronics provided by the smartphone industry to apply all their knowledge to Skeeter's frame.

3.11 BOMBARDIER BEETLE INSPIRES TECHNOLOGY

Bombardier beetles (Figure 3.7), which exist on every continent except Antarctica, have a pretty easy life. Virtually no other animals prey on them, because of one particularly effective defense mechanism: When disturbed or attacked, the beetles produce an internal chemical explosion in their abdomen and then expel a jet of boiling, irritating liquid toward their attackers. *Carabidae brachinini*, or bombardier beetles, are primarily found in tropical regions such as Asia, Africa, Australia, and the United States (Sugiura & Sat, 2018). Small colonies can be found in southern England, but they can also be found in Europe. They are frequently found around the water's surface and hide out during the day by hiding out in rocks. It identifies it as a powerful creature capable of fending off predators like birds and frogs with a jet of hot, toxic fluid. The venom's molecular reaction was depicted in detail by McIntosh & Lawrence (2018). The beetle's abdomen is filled with hydroquinone and hydrogen peroxide gases. Brachinini or bombardier beetles are prevalent in tropical places such as the United States and Asia (Sugiura & Sat, 2018). It's not just southern England that you'll find colonies, though. During the day, they can be seen hidden between rocks near the water's surface. Using a jet of hot, toxic fluid to fend off predators like birds and frogs, it is identified as a formidable monster. McIntosh & Lawrence (2018) illustrated the venom's molecular reaction in great detail. Hydroquinone and hydrogen peroxide fumes are flooding the beetle's abdomen (Figure 3.7). Toxic benzoquinone can be produced in a connected 'combustion chamber' when necessary for defense. For this beetle to be so effective at protecting itself, it must have both inlet (or entry) and exit valves on the fire chamber. Upon meeting, the chemicals begin to boil, and the inlet valve opens to allow them in. When there is enough gas, the valve closes again. As a result, a jet of heated fluid is emitted, traveling up to a distance of 20 cm. When gases react with each other, they generate heat and increase pressure in the final chamber, which is the purpose of this process. "Flash evaporation" is a term used to describe the sudden release of hazardous steam from a ruptured exit valve when the pressure drops to a critical level. Immediately after the gas has been supplied, the exit valve is closed, the inlet valve is opened, and the chamber is refilled, ready for the deadly ejection that follows.

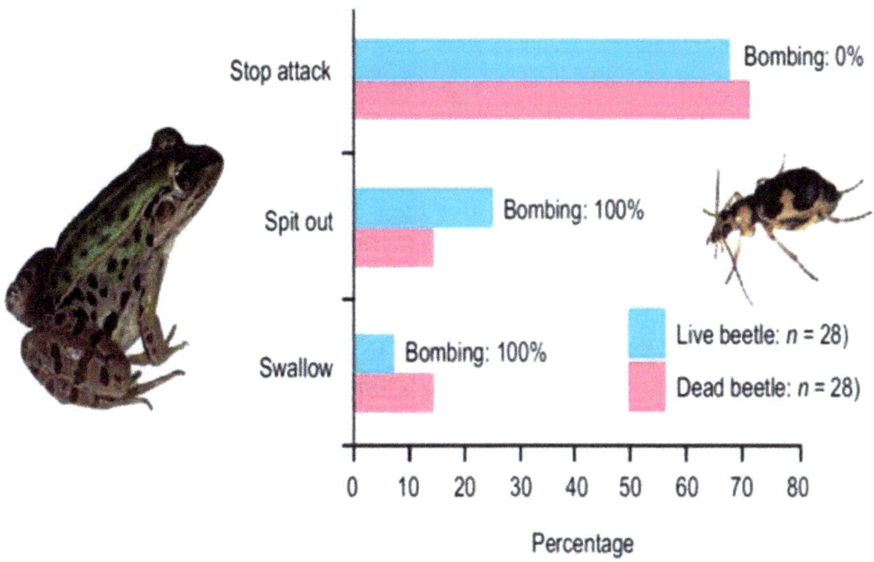

FIGURE 3.7 Behavioural responses of the black-spotted pond frog *Pelophylax nigromaculatus* to live and dead adult individuals of the bombardier beetle *Pheropsophus jessoensis*.

Source: Reprinted with permission from Sugiura, 2018. https://creativecommons.org/licenses/by/4.0/

Two chambers in the bombardier beetle glands are required for the formation of an irritant spray. The chambers are described as follows: One of the beetles' cuticles is made up of proteins and waxed chitin that is capable of withstanding the toxic chemicals, high temperatures, and the other (ii) higher pressures that define the beetle's explosion. Researcher Eric Arndt adds, "It's a really robust and adaptable material." It won't hurt, but the chemicals it sprays will leave your hand looking like a henna tattoo if you get sprayed by it. When scientists studied the Bombardier beetle, they came up with some amazing findings (McIntosh & Lawrence, 2018). To lead the inquiry, the University of Leeds professor Dr. Andy McIntosh, a young-earth creationist, used his expertise in thermodynamics and combustion theory.

In the Bombardier beetle, there is a unique security system. When it feels threatened, it can blast a toxic chemical spray (heated to 212°F) towards its prey. This beetle's chemical-shooting abilities are the result of natural selection. This creature creates a volatile brew (hydrogen peroxide and hydroquinone). Chemicals are present in the combustion chamber of the Bombardier beetle. There is a mechanism that ejects the chemical mixture that was mixed using two catalysts (catalase and peroxidase). All of this happens just as the combination is about to escape the back end, just in time for a powerful reaction.

In comparison to other species of beetles, the Bombardier beetle stands out. Its self-defense mechanism makes it stand apart from the rest of the Beetle species. The bombardier beetle gets its name from its ability to hurl "chemical bombs" when threatened. There are no limits to the number of times Bombardier beetles can use their chemical "bombs." Incredibly, these half-inch beetles were able to emit this lethal spray while avoiding any stinging injuries. In the meanwhile, MIT, the University of Arizona, and Brookhaven National Laboratory have come up with a workaround. At 100°C, a bombardier beetle has the ability to shoot out a deadly liquid when it senses a threat. Using a chamber less than one millimeter in length, it generates and dispenses this liquid from within its own body.

In the following systems: (i) the beetle displays irreducible complexity; (ii) a predator's approach is detected via the sensory mechanism; (iii) both valves in the valve system must operate in unison; (iv) the synthesis of hydrogen peroxide and hydroquinone using chemical means; (v) ejecting an explosive mixture under control using catalytic chemistry; and (vi) it is possible to eject in any direction because to the exhaust turret's mobility and flexibility.

In principle, the same holds true for chemical systems. These and other systems can only function well when all of their components operate together in unison as part of a well-coordinated system. The entire chemical system can only function properly if each component chemical is on a predetermined path.

To improve medicine delivery systems, researchers at Arndt et al. (2015) examined and designed an efficient fuel injection system. Researchers at the University of Leeds have built a 2 cm chamber that can propel fluids up to a few meters in length. As little as 2-micron droplets can be sprayed out of this chamber. The MistTM' is a fitting name for this technology.

3.12 ACID-SPEWING BEETLES INSPIRE (ANTI-THEFT TECHNOLOGY) NEW DEFENSE TACTIC FOR ATMS

Throughout the last two decades, ATMs have become an essential part of many people's daily lives. Since ATMs were introduced by banks some 25 years ago, scammers have also been trying to get into the system. In the last several years, ATM frauds have been on the rise all around the world. Card-holders who don't know how to use their cards properly risk losing them in ATM machines or shattering the glass door to get out after a late-night transaction because banks haven't adequately trained their clients. It's no surprise, then, that ATM fraud and security have become hot subjects among ATM owners and operators. Banks and other organizations that use ATMs should focus on minimizing losses, reducing risk, and preserving customer trust in the ATM channel. When using an ATM, citizens need to know what precautions they can take to lessen their chances of being a victim of fraud. When they open the ATM, they must disguise their PIN and be on the lookout for evidence of manipulation or suspicious activity.

ATM fraud is on the rise, and with it, the number of countermeasures required to keep it under control. By using bombardier beetle spray technology, Zurich researchers constructed a film that discharges hot foam when damaged (Figure 3.8). Manganese dioxide or hydrogen peroxide honeycomb structures are piled on top of each other to create the film. Hot foam can reach temperatures of up to 180°F when the film is pierced. The clip could serve as a deterrence to theft and vandalism at ATMs. From this process, Prof. Wendelin Jan Stark created a film that, if destroyed, ejects heated foam to ward off potential attackers. Several plastic honeycomb constructions were produced by the researchers and filled with hydrogen peroxide or manganese dioxide before being stacked on top of one another. Two substances are mixed together and heated to 80°C (180°F) when the structure is damaged.

Researchers at ETH Zurich have taken inspiration from the bombardier beetle's defense mechanism and developed a film that discharges a heated foam when destroyed (Jain, 2017). ATMs can be protected from theft and vandalism by using this technology. ATM robberies may soon face an interesting new deterrent. Those attempting to commit fraud on the basis of so-called "easy money" may be in for a rude awakening. It was inspired by the bombardier beetle's ability to store different chemicals in distinct sacs in its abdomen, which inspired a new kind of security. The beetle uses both compounds at the same time when it senses danger. As soon as they come into contact with these two compounds, predators are swiftly deterred.

As a result of the Bombardier Beetle's use of revolutionary spray technology, ATM anti-theft technology has been developed. Sweden's Biomimetics 3000 and Leeds University's Andy McIntosh are working together to create this new technology, which they've named 'Mist.' The spray mechanism found by McIntosh could be used in fire extinguishers, gasoline injectors, and aerosol medicine delivery systems. Another beetle spray mechanism is being studied with the help of a high-speed X-ray camera that works really well.

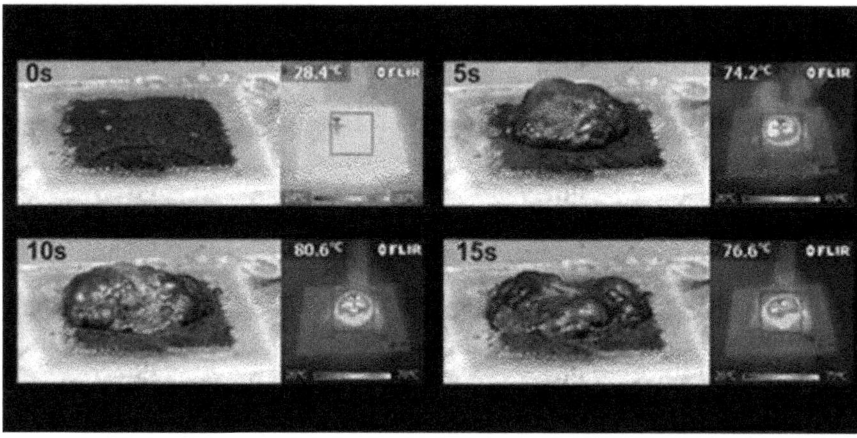

FIGURE 3.8 Foam is released by the self-defense film in a sequence of photographs taken during laboratory tests. The infrared scans reveal a spike in temperature of 80C.

Source: Reprinted with permission from Halter et al., 2014. Copyright 2014. Royal Society of Chemistry

3.12.1 BOMBARDIER BEETLE EXPLOSIONS INSPIRE NEW FIRE EXTINGUISHER

The Bombardier Beetle is equipped with a unique defense system. Once it detects danger, the beetle releases an incendiary spray of gases and liquids. Nearly 100°C is the temperature at which the mixture of benzoquinone and water spontaneously decomposes. To put it another way, the combustion stability of hydroquinones and hydrogen peroxide is to blame. The beetle's abdomen is where this occurs. No damage to the beetle's nerve endings was found throughout the research, suggesting that the defense mechanism may be highly uncomfortable for humans. Researchers from Design and Nature IV highlighted that a pressure discharge valve is a crucial step in duplicating high-velocity sprays from the bombardier beetle's experience (McIntosh & Beheshti, 2008). To better understand how beetles release chemicals like the diaphragm and lung do, researchers at MIT used fast-moving X-ray images of the insects' sac-shaped organs. Beetle spout occurs over 500 times per second compared to the human respiratory rate of a few breaths per minute.

The beetle known as a "machine gun" shoots forth a torrent of spores. From its rear, the "machine gun" beetle launches a stream of scalding hot water, replete with "smoke." Chemicals in the beetle's stomach simulate a machine gun. As for the substances that enter the abdomen, they are broken down by enzymes and then expelled. The foul-smelling liquid is ejected from the rear of the device with each explosion. When the beetle's backside is in motion, it emits a 'pulse' sound.

The peculiar behavior of the bombardier beetle was discovered by scientists from Cornell University and MIT. There are several micro-pulses in each of the bombardier beetle's blasts. Using the machine gun analogy, Assistant Professor Moore explained the outburst pulse. Figure 3.10(B) shows a stock photo of a machine gun.

3.12.2 NEEDLE-FREE INJECTIONS

To overcome obstacles related to needle-based injections, needle-free injection technologies (NFIT) have gained popularity, it was decided to develop needle-free injection technology in an effort to lessen the occurrence of accidental needle sticks and the associated problems. Accordingly, Kale & Momin (2014) conducted an extensive study into the many types of needle-free injection technologies, as well as the advantages and disadvantages of using them over standard needles. As detailed by Kale & Momin (2014), needle-free injection technology employs compressed gas as a source of force in order to deliver drugs at high rates of velocity.

Using a scaled-up experimental equipment that uses heating and flash evaporation techniques to push various liquids up to a height of 4 meters, the spray's biochemical and mechanical properties and the physiology of the insect are replicated (13 ft). A water-based spray system was employed in place of the "damaging propellants inherent in standard aerosol sprays" in the final product.

Bombardier beetle-based technology was developed by Dr. McIntosh and his team of scientists (McIntosh & Beheshti, 2008) after five years of research. "Advancements in different fields where the qualities of the mist are crucial" can be made thanks to this new technology's ability to manage droplet size, temperature, and velocity. We may expect to see innovations like these in future nebulizers, fire extinguishers, and fuel injection technologies. "We were the first to look at the beetle from the perspective of physics and engineering, and we had no idea how much we would learn from it," according to Andy McIntosh, the team's leader. Hays (2018) came up with a new technology based on the design of the Bombardier beetle.

3.13 INFRARED SENSORS INSPIRED BY BEETLES

In many nations, forest fires cause significant economic losses each year. In addition, the lives of humans and countless other species are frequently lost as a result of these disasters. Wildfires are becoming increasingly out of control as a result of global warming. Thousands of hectares of land are burned in these incidents. A fire that is just starting to spread is caught early enough before it spreads out of control. There has never been a greater demand for early warning systems that can identify tiny flames in open spaces. Due to their larvae's ability to grow only in fire-dead trees' wood, *Melanophila* "fire beetles" of the genus *Melanophila* can be considered airborne fire detection devices. According to Schmitz et al. (2011) and Bousack et al. (2012), *Melanophila* beetles can identify big flames from more than 100 kilometers away utilizing optical and infrared cues. For long-range detection of a smoke plume from a fire utilizing a basal infrared emission zone, a concept is developed employing a biomimetic technique that mimics the surveying tactics of "fire beetles" (Figure 3.9).

FIGURE 3.9 About 1 cm long, *Melanophila acuminata* is the "fire beetle."

Image courtesy: AG Prof. Schmitz. https://creativecommons.org/licenses/by-sa/2.5/deed.en

Because of its ability to absorb infrared radiation, the jewel beetle is able to produce an audible signal that may be traced back to an active forest fire (Kahl et al., 2014). In addition, when their favored trees are burned, they catch the compounds that the fire throws into the air. This helps them to survive. To improve forest fire detection systems, scientists are studying them.

In order to find a fire, the *Melanophila acuminata* needs a variety of extremely sensitive stimuli. They can "smell" the charred remains of a fire, even in microscopic concentrations, thanks to sensors on their antennae that are extremely sensitive, as chemical ecologist Dr. Stephan Schutz revealed. Researchers from a variety of fields, including biology, physiology, ecology, and physics, have been working diligently for years to uncover the beetle's infrared-detecting secrets, according to Erik Schneider et al. (2015). Since working on *M. acuminata*'s infrared detectors, the German team has discovered that two more species of fire beetle, this time in Australia, also use infrared detectors to find fire. South American beetles use infrared detectors that are distinct from those found in the northern hemisphere. It is time for scientists to put their infrared radiation-sensing expertise to use. They intend to work with engineers to create a new generation of infrared detectors with their assistance. Currently, they're aiming to improve the sensitivity and reduce the size of a prototype that they already have.

An enormous amount of military potential can be recognized by the US Defense Advanced Research Projects Agency (DARPA). According to Alkhatib (2014), *Melanophila acuminata* contains biological receptors that can detect infrared sources even at a distance. Uncooling infrared indicators were made possible thanks to a little heat-seeking bug. Helmut Schmitz, a scientist at the University of Bonn, Germany, has taken the idea from a small insect. Schmitz and his colleagues at the University of California, Berkeley, have built a prototype that can detect the beetle's sensors as infrared sensors. This technique, according to Schmitz, can be found in a wide range of applications that make use of infrared sensors. In the event of a forest fire, these new infrared sensors might be used to monitor the area and issue an early warning.

An infrared sensor can be built by a jewel beetle using simple and affordable components. Infrared radiation from a forest fire can be detected by the fire beetle's pit organs. The pit organs are heated by the radiation, which prompts the beetle to flee toward the blazing source. These beetles, however, are equipped with a variety of different sensors that can help them detect fires. The beetles' antennas detect tiny amounts of particular compounds released into the air when their favored trees burn down (Alkhatib, 2014; Schmitz & Bleckmann, 1998).

3.14 USING SCARAB BEETLES, SCIENTISTS CAN LEARN MORE ABOUT HOW LIGHT IS POLARIZED

The golden scarab beetle is the only insect that has polarization reflectance. This insight could serve as inspiration for biomimetic optical technologies (Bandari, 2019). The cuticle of golden scarab beetles (Figure 3.10) *Chrysina resplendens* has unique optical properties that have inspired biomimetic applications (Bandari, 2019), ranging from optical filters to phase retardation in lasers. It acts as a circular Bragg reflector, bouncing both left- and right-handed circularly polarized light off its body with high reflectance. To better understand these polarization properties, Mendoza-Galván et al. (2018) have modeled the structure of the cuticle of *Chrysina resplendens.* The beetle's cuticle is a multilayer unit with a total thickness of about 20 micrometers. Beneath a waxy top is the exocuticle, which comprises of birefringent layer sandwiched between two helicoidal layers. The sandwiched birefringent house mirrors reflected the same pattern as the round Bragg reflector. They're responsible for a lot of *C. resplendens'* optical properties.

FIGURE 3.10 Golden scarab beetle.

Source: Reprinted with permission from University of Exeter. https://creativecommons.org/licenses/by-sa/3.0/.

The cuticle's left-handed components are reflected in the light from the outer helicoidal layer. However, it works with right-handed hardware as well. This circularly polarized light is right-handed; therefore it will pass through the birefringent layer. Its handedness is altered by applying a phase shift that varies with both angle and wavelength. Two books address this topic: Pye (1971) and Caveney (1971, 2010). The outer helicoidal layer of the cuticle is responsible for reflecting light off of the cuticle's left-handed components. However, it works with both left- and right-handed hardware. This light, with its circular polarization tilted to the right, can pass through a birefringent medium. The phase shift that occurs as a function of both angle and wavelength makes its handedness somewhat reversible. After the left-handed component has been formed, the lower helicoidal layer of the brain is identified as such by Caveney (1971) and Pye (2010). The function of the beetle's polarization feature is not yet clear–state that the bottom helicoidal layer is visible as a result of the left-handed component. The beetle's polarization property's function is still mostly unknown. The reflected beam receives an additional phase shift when it returns to the source. The birefringent substance and the amount of uric acid in it are likely responsible for the variation in reflectance observed between beetles. The studies of the complex architecture of *C. resplendens* may produce other biomimetic technologies. "There is minimal information about which role these structures play in nature," said the author of the investigation. "A few tries to study intraspecies communication done, but they are not definitive."

The development of self-driving automobiles could benefit from biologically inspired compound eyes. Understanding how insects perceive motion and trajectory is being aided by the approach (Figure 3.11).

Robots, self-driving automobiles, and unmanned aerial vehicles (UAVs) could benefit from an insect eye constructed from their optical systems, which could aid in the development of 3D positioning algorithms (Zheng et al., 2019). The development of a compound eye that is comparable in appearance and function to an insect's (Rob, 2016). Each of the 169 microlenses in the insect's eye has a diameter of around 1 mm. Accuracy decreased with increasing distance from an object, but it was still possible to get a 3D location of it quickly. According to the researchers, the method might be used in small robots that require fast item detection from a lightweight device. In addition, it will aid in the study of insects' vision systems. Researchers are developing new mass-production methods for compound eyes in an effort to cut the per-unit cost.

3.15 HOW INSECTS ARE ADVANCING HEARING AID TECHNOLOGY?

Our six-legged buddies can teach us a thing or two about how to improve technology and our lives. Both mammals and humans have the ability to "hear," but how they do so differs from species to species. A human can hear a variety of frequencies, but a cat or dog can hear sounds that are not apparent to us. Insects can hear despite the absence of ears they appear to have. Scientists may be able to learn more about how to improve hearing aids by studying how these animals perceive sound differently from humans. The three middle ear bones move in response to sound waves traveling through the eardrums and into the stereocilia in the outer ears of humans (hammer, anvil, and stirrup). The brain processes the sound waves in the cochlea before they can be heard by the ear. It sounds like a lengthy and laborious procedure. Because our ears and brains can process sounds so quickly, we don't even realize it. However, if any of these systems are broken, hearing is hampered, and in many cases, people are left with permanent hearing loss as a result.

FIGURE 3.11 Micrograph of the compound eyes of a horse fly.

Source: Reprinted with permission from Leonard et al., 2020. https://creativecommons.org/licenses/by/4.0/

The directional microphone has had a substantial impact on hearing aid technology, according to Kuntzman & Neal (2014). Nobody knows that the developers of this game were inspired by a little yellow insect's ability to bring in sound. The *Ormia ochracea*, a little insect, has a lot to teach us about modern hearing aid technology. Because of its exceptional natural aptitude for sound localization, the hearing aid industry has investigated the *Ormia ochracea* (Figure 3.12), a yellow, nocturnal fly. In *Ormia ochracea*, there is no hearing loss. In terms of hearing, it's amongst the best in the animal kingdom. When compared to other senses, the *Ormia ochracea's* hearing is superior to that of an owl and on par with the extremely adept vision of a hawk.

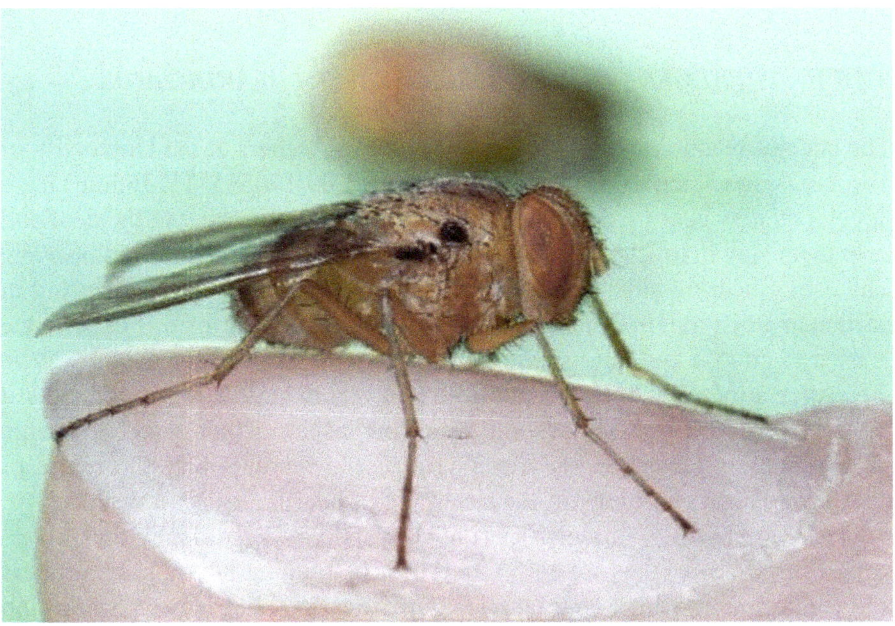

FIGURE 3.12 *Ormia ochracea*. A gravid female *Ormia ochracea* resting on a fingernail.

Image courtesy: Jpaur at Wikimedia commons. https://creativecommons.org/licenses/by-sa/3.0/deed.en

Insect parasites of crickets include *Ormia ochracea*. It is famous for its outstanding directional hearing abilities. Insects like the *Ormia* insect have a high degree of directionality. Speed and crucial response, although it's not obvious how the flies negotiate more tough settings, such as many signals and noise. These flies are great at one thing: collecting cricket noises. There is a cost to this since they have become hyper-focused on this single bit of data. There is a feeble connection between its two ears in the tiniest of spaces. The coupling swings back and forth as the sound travels in one direction. It helps the fly get a sense of where the sound is coming from and how far away it is. It is possible for a fully developed adult fly to detect even the tiniest changes in sound and respond accordingly. Deaf individuals have a hard time hearing above the din of a noisy cocktail party or restaurant, and the fly is no exception. Insects have a hard time dealing with echoes and background noise. However, flies are better at detecting sounds coming from behind than from above or below, such as echoes in a large room. Additionally, the technology could lead to low-power hearing aids that are better able to identify conversations from background noise, as well as possible military uses.

3.15.1 TREATMENT OF HEARING LOSS WITH HEARING AIDS

The state of technology now is impressive. Researchers at the University of Toronto (www.sciencedaily.com/releases/2017/05/170515111136.htm) have developed new hearing devices that may help restore hearing to the deaf and help them stay in touch with friends and loved ones who are hard of hearing. The use of hearing aids can, in certain cases, outperform normal hearing in particular noisy environments. Scientists' understanding of our experience is used to choose and alter sound waves. As they rise and see sound waves, they offer a lot. However, it's possible that others have more to offer. As it turns out, the Researchers got some help from insects, which is an odd source of information. However, the discoveries could revolutionize the future of hearing aids and audiology in the treatment of hearing loss and wellness.

There are two UK universities (the University of Bristol and the University of Strathclyde) that are committed to investigating how insects process and create sound, which could lead to better hearing aids. New speakers and directional microphones may benefit from the findings of this study. There are around 1 million species of insects known to science, and each one has developed a unique set of hearing organs. Larger animals, on the other hand, can use differences in timing and amplitude to determine the direction from which sound waves are being received to determine the source of the sound. It is likely that the distance between the two ears of an insect is too narrow for this to be effective. Engineers are now taking an interest in the inventive tactics used by smaller species.

There are no external ears on the *Ormia ochracea*, but insects can hear a lot. The critters' eardrums are near the tiny ear-like organs. The receptor cells in some insect species' hearing organs are so sensitive that they can detect even the smallest changes in the surroundings. Researchers may have been motivated to design new hearing aids and better sound products after examining how these insects select their preferred noises.

Despite their diminutive size, insects have an amazing capacity to perceive sound waves. A wide range of frequencies that pass through our ears can be heard by them. To put it another way, Insects, unlike humans, have the ability to hear frequencies as high as 15 to 20 kilohertz. They are able to distinguish and filter noises from afar and isolate vibrations based on where they are located. Even while insects' hearing mechanisms appear to be quite simple, their variations allow them to collect and amplify sounds more effectively than humans. Insects' eardrum membranes vary in thickness, allowing them to filter and amplify sound to the precise level they desire. Because of these findings, researchers are striving to reproduce these structures in order to harness these diverse processes to improve hearing technology. In order to work with nanotechnology, it is necessary to connect the fields of biology, medicine, and engineering. It offers a whole new world of possibilities for improving hearing aids for the deaf and hard-of-hearing. Seeing how much these tiny organisms can accomplish is mind-boggling!

3.16 INSECT-INSPIRED MOTION SENSING

The motion-sensing capabilities of consumer-sized applications can be improved with an insect-inspired mechanical gyroscope. It was inspired by flying insects to display a miniature gyroscope (Figure 3.13), which is employed in navigation technology, by researchers at Oak Ridge National Laboratory and the National Renewable Energy Laboratory. It is impossible for signal jamming or other cyber threats to affect the rotational guidance provided by gyroscopes, which is why they are so well suited to aircraft and submarines. It has been difficult to incorporate the devices into minor defense and consumer electronics because of fundamental constraints.

Gyroscopes can suffer from performance issues if their electrical components are too small. A mechanical system with no on-chip electrical components was designed by the researchers in order to preserve microscale performance. The halteres of flying insects are based on a coin-sized design from the Center for Nanophase Materials Sciences. According to Dempsey (2020), a researcher's purpose was to maximize cost and performance in the smallest design possible in order to increase the market for this technology (Dempsey, 2020).

3.17 THE STRUCTURE OF A NEW SOLAR PANEL CELL WAS INSPIRED BY INSECT EYES

The fragile photovoltaic material inspired by bug eyes was developed by Stanford University scientists using creepy crawlies. Small solar cells were arranged like micro-lenses in an insect's compound eye by Stanford researchers. As a photovoltaic material, perovskite can withstand heat, moisture, and mechanical stress. A novel solar cell design inspired by bug eyes could usher in a new era of cutting-edge photovoltaics (Swartz, 2017). The demand for solar energy is rising as more companies and individuals alike seek to reduce their carbon footprints in order to save money. Solar-powered building blocks have recently made headlines. Then there are those who are attempting to develop solar energy systems that are as efficient and long-lasting as possible, even drawing inspiration from the delicate wings of butterflies.

Butterflies are one of several insects that serve as a source of inspiration for researchers. For the first time ever, researchers at Stanford University in California have developed a new type of solar cell based on insect eyes (Figure 3.14). They've found a way to create a new generation of sophisticated photovoltaics by putting together a bunch of tiny solar cells. If perovskite is exposed to high temperatures, dampness, or mechanical stress, the Stanford researchers developed an insect-inspired approach to prevent the material's deterioration despite the fact that most solar devices, including rooftop panels, employ a flat shape, perovskite solar cells do not. It was possible to construct a compound solar cell using perovskite microcells in the shape of an insect's eye as a model. To examine if they could withstand the heat of rooftops, encapsulated perovskite cells were tested for six weeks at 185°F (85°C) and 85% relative humidity. Each is encased in a 0.02-inch-wide hexagon-shaped scaffold. Despite the severe conditions, the cells were still able to generate energy at a relatively high rate of efficiency.

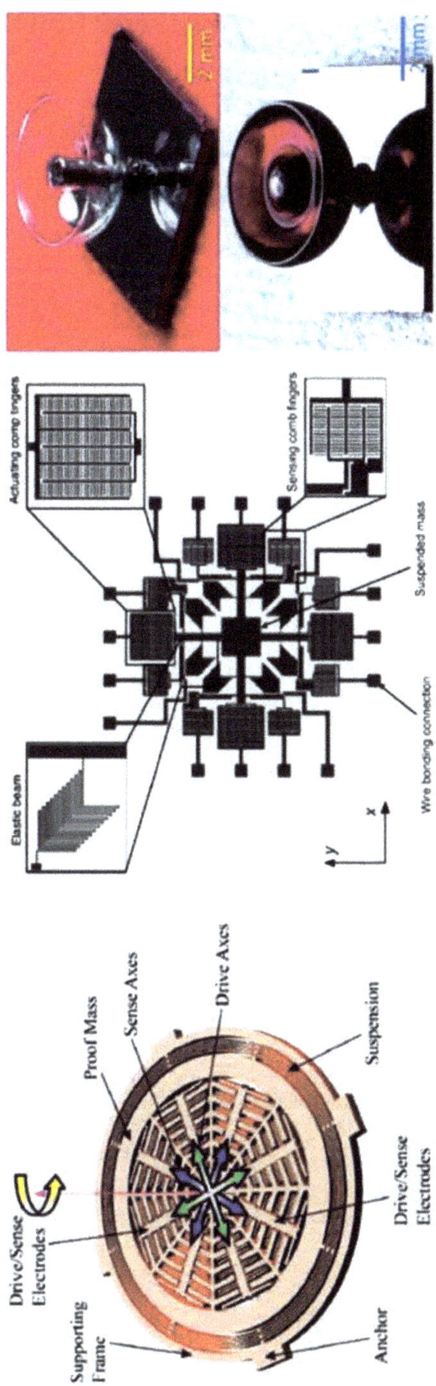

FIGURE 3.13 Various angle gyroscopes. (a) UC Irvine. (b) University of Minnesota. (c) University of Michigan.

Source: Reprinted with permission from Dunzhu Xia et al., 2014. Copyright © 2014 by the authors. Licensee MDPI, Basel, Switzerland. http://creativecommons.org/licenses/by/4.0/

3.18 INSECT-INSPIRED ARCHITECTURE IS FOR BUILDING SUSTAINABLE CITIES

Engineers have a keen interest in the properties of various types of materials, structures, and surfaces. The fundamental understanding of how these systems work is also well-suited to technical fields, such as architecture. Multifunctionality, energy conservation, and environmental sustainability are some of the most pressing issues in today's design. A few of the difficulties that insects have had to overcome over the course of their development. Entomologists have learned a great deal about how and where insects live, as well as how they adapt to their environments. A copy of this data can be utilized to get a job in the architectural industry. In the following domains of technology, insect-inspired concepts can be used: Adhesion and bonding technologies, novel materials, optics, and photonics all fall under this category.

FIGURE 3.14 The eyes of a robber fly, *Holcocephala fusca*. Researchers at Stanford were inspired by the compound eye of a fly to develop a perovskite microcell-encased hexagonal framework for use in a compound solar cell.

Image courtesy: Thomas Shahan. https://creativecommons.org/licenses/by/2.0/deed.en

3.19 INSECT SILK FOR BIOTECH APPROACHES AND MEDICAL APPLICATIONS

Spider silk is one of nature's most adaptable materials because of its unique combination of strength and flexibility. With their efforts, they have made it the most difficult natural fiber material on Earth, even surpassing synthetic materials such as polyester and nylon. Spider silk, on the other hand, is a biodegradable polymer that is almost entirely composed of proteins. As a result, there are far too many advantages to spider silks. Spider silk can be used in a wide variety of fields, from medicine to textiles. Silk is a biopolymer that is good for the environment. In terms of load-bearing applications and biocompatibility, it's been demonstrated in people for a long time. All kinds of different insects produce silk in their own unique ways, from the familiar silkworm (*Bombyx mori)* to wholly unrelated sorts of silk from other insects. Insect-silk-inspired, novel materials have been developed thanks to biotechnological progress in the manufacturing of silk proteins. Such biological applications as drug delivery systems or tissue engineering scaffolds are included.

3.20 DUE TO ITS LOW WEIGHT, SILKY FEEL, AND GLEAMING LOOK, IT HAS A VARIETY OF USES

Interestingly, natural insect silk is in high demand in addition to its monetary worth. In the 4th millennium BCE, the Chinese invented it, which explains its widespread use in the textile industry. Silk's exceptional toughness is the result of its fibers' ability to combine high extensibility with high strength. Additionally, silk fibers are non-toxic and safe for use in the body. There are various silks that have no cytotoxicity no immunological reaction and disintegrate into non-toxic by-products if introduced into the body. Overall, this makes silk a good candidate for use in biological applications in the future.

For a long time, silk studies concentrated primarily on the silks of silkworms. The quality of silk kinds offered by diverse animals is increasingly attracting attention (mainly insects and spiders). A recent evaluation of recombinant silk products and technical systems for processing silk proteins into fibers, films, and hydrogels was conducted by Chung et al. (2012). The survival and spread of many insects are dependent on the production of silk. The majority of arthropods produce silk for pupation cocoons; however, silk is also used for a variety of other purposes. As far as silk research goes, silkworm silks have been the primary focus for decades. A growing number of people are interested in the numerous varieties of high-quality silk offered by diverse animals (mainly insects and spiders). Chung et al. (2012) examined the most recent recombinant silk products and technical systems for processing silk proteins into fibers, films, and hydrogels. Silk is an important survival and dissemination tool for many insects. For pupae to develop into adult forms inside their silk-lined cocoons, the majority of arthropods produce silk, although the material can also be used for other purposes (Figure 3.15).

FIGURE 3.15a Man gathering silk thread for reeling.

Image courtesy: wirestock_creators. https://www.freepik.com/premium-photo/man-gathering-silk-thread-reeling_37452540.htm

FIGURE 3.15b Colours of India — silk yarn waiting to be made into sarees.

Image courtesy: Ekabhishek on Flickr. https://creativecommons.org/licenses/by/2.0/deed.en

Biotechnological production of insect silk proteins can be scaled up with the use of primarily bacterial organisms as expression hosts, resulting in consistent quality and sufficient quantity. Additionally, recombinant protein manufacturing allows for the creation of proteins with specific changes. Cell adhesion can be modulated using a fibronectin-derived motif, for example, and sensors can be produced by adding catalysts to silk films. By altering the ratio of elastic/strong building blocks, silk-based materials can be made to have specific mechanical properties. Fibers can be formed by altering the protein's processing conditions. However, it is possible to create morphologies that deviate from the natural pattern, such as films, coatings, or hydrogels, which will play an important part in future biomedical applications in tissue regeneration in the near future.

3.21 SENSORS BASED ON INSECTS CAN DETECT DANGEROUS COMPOUNDS IN THE AIR

As Heike & Scheibel (2017) put it: "It is inspired by hairy, flying insects—which can promptly and reliably identify a wide range of harmful airborne substances." Airborne pollutants can be detected using this insect-inspired sensor. A new sensor has been delivered to scientists throughout the world, including those in India. The purpose of the microscopic hairs found on the body surfaces of most insects is unknown. Arindam Phani and Thomas Thundat from the University of Alberta in Canada developed tests using a "forest" of small hairs on a thin vibrating crystal chip in an attempt to make sense of what these hairs may be capable of.

When employing resonators as sensors, most people wish to eliminate waste or resistance. Many people don't like it since it obscures what you're attempting to figure out. Vakhtang Putkaradze, a PhD student at the University of Alberta, remarked, "We've made that undesired item worthwhile." It has been difficult in normal situations to detect chemicals (Figure 3.16) without the use of chemical receptors. As Thundat puts it, "The frictional loss of mechanical resonators in motion holds a plethora of data, which is particularly pronounced at the nanoscale" (PTI, 2016).

Tiny variations in air pressure around the resonance chamber may be detected by using nanostructures, he says. Putkaradze said the material's sensitivity makes it ideal for recognizing a wide range of compounds. According to Phani (PTI, 2016; http://www.businessstandard.com/article/pti/stories... an insect-inspired sensor can detect toxic airborne chemicals), relocating nano-hairs could be used for sensing by living organisms. Because friction is so elementary, it varies greatly with even minute modifications to its environment. Because the friction is so simple, it varies so significantly with even the tiniest changes in the surroundings. Finally, a device that plugs into a wall might be created that is the same size or slightly larger than a Rubik's cube. The technology developed by the team is primarily used to detect airborne chemical vapors. It is hoped that this instrument can be used to measure compounds in the parts-per-million range, similar to a spectrometer. In addition to its small size and inexpensive price, scientists feel that its versatility sets it apart from more complex and expensive equipment.

There is no need to attach the molecules to anything in order to make a mechanical response because the sensor does not directly detect any specific chemical to interpret a wide range (PTI, 2016).

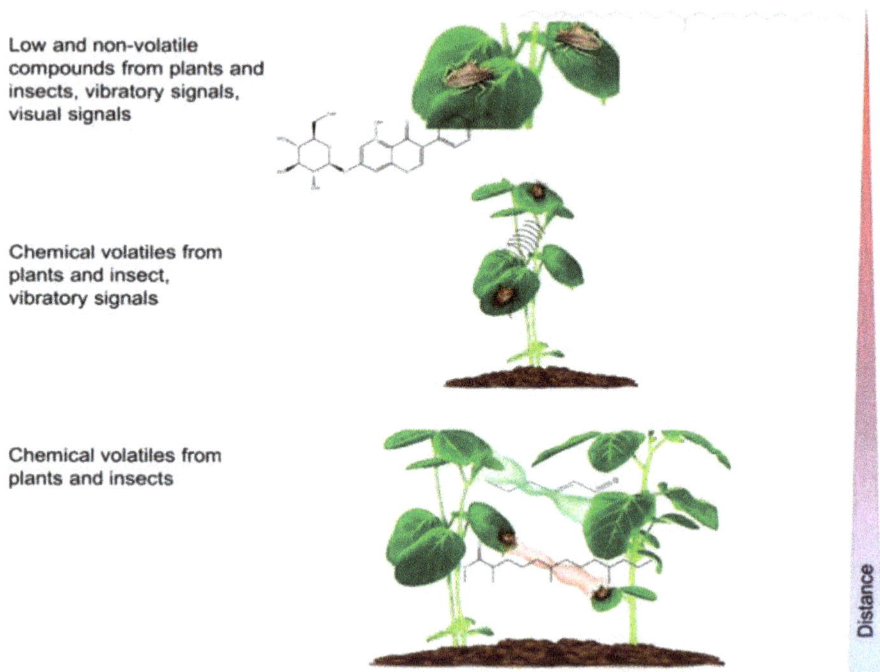

FIGURE 3.16 Stink bugs dispersed in the field congregate on the same plant, attracted by airborne chemical signals, and communicate there during calling, courtship, and rivalry phases by signals of different modalities.

Source: Reprinted with permission from Čokl et al., 2021. Copyright © 2021 by the authors. Licensee MDPI, Basel, Switzerland. https://creativecommons.org/licenses/by/4.0.

KEYWORDS

- **anti-theft technology**
- **bombardier beetle**
- **infrared sensors**
- **insect silk**
- **insects**
- **locust solar panels**

REFERENCES

Abbas, N. J., Wright, A., & Pao, L. (2020). An update to the national renewable energy laboratory baseline wind turbine controller. *J. Phys.: Conf. Ser.*, 1452012002.

Ackerman, D., (2021). *Researchers Introduce a New Generation of Tiny, Agile Drones*. MIT News Office. https://news.mit.edu/2021/researchers-introduce-new-generation-tiny-agile-drones-0302 (accessed on 17 August 2023).

Alkhatib, A. A. A., (2014). A review on forest fire detection techniques. *Int. J. Distrib. Sens. Netw.*, 597368.

Andrej Čokl et al., 2021. Stink Bug Communication and Signal Detection in a Plant Environment. Insects 2021, 12(12), 1058; https://doi.org/10.3390/insects12121058.

Arndt, E. M., Leeland, K., & Ortiz, C., (2015). Mechanistic origins of bombardier beetle (*Brachinini*) explosion-induced defensive spray pulsation. *Science, 348*(6234) 563–567. DOI: 10.1126/science.1261166.

Bandari, A., (2019). *Scarab Beetles Can Help Scientists Understand Light Polarization Mechanisms. AIP* Scilight https://aip.scitation.org/doi/10.1063/10.0000144 (accessed on 17 August 2023).

Biomimetic Drones Inspired by Dragonflies Will Require a Systems Based Approach and Insights from Biology Javaan Chahl et al., *Drones* **2021,** *5*(2), 24; https://doi.org/10.3390/drones5020024.

Bousack, H., Soltner, H., Klocke, D., & Schmitz, H., (2014). Infrared sensors inspired by pyrophilous insects. World scientific series in nanoscience and nanotechnology. *Handbook of Biomimetics and Bioinspiration* (pp. 533–564).

Bowen Zhang et al., *Energies* **2022,** *15*(2), 455; https://doi.org/10.3390/en15020455.

Caveney, S., (1971). Cuticle reflectivity and optical activity in scarab beetles: The role of uric acid. *Proc. Roy. Soc. Lond. B*, *178*, 205. https://royalsocietypublishing.org/doi/10.1098/rspb.1971.0062 (accessed on 17 August 2023).

Chahl, J., Chitsaz, N., McIvor, B., Ogunwa, T., Kok, J.-M., McIntyre, T., & Abdullah, E. (2021). Biomimetic drones inspired by dragonflies will require a systems based approach and insights from biology. *Drones* 5(2), 24; https://doi.org/10.3390/drones5020024

Chandler, D. L., (2015). *How Some Beetles Produce a Scalding Defensive Spray*. MIT News Office. https://news.mit.edu/2015/how-bombardier-beetles-produce-defensive-spray-0430 (accessed on 17 August 2023).

Chen, Y., Arase C, Ren, Z., & Chirarattananon, P. (2022) Design, characterization, and liftoff of an insect-scale soft robotic dragonfly powered by dielectric elastomer actuators. *Micromachines 13*(7), 1136; https://doi.org/10.3390/mi13071136

Chung, H., Kim, T. Y., & Lee, S. Y., (2012). Recent advances in the production of recombinant spider silk proteins. *Curr Opin. Biotechnol., 23*, 957–996.

Chung, H., Kim, T. Y., & Lee, S. Y., (2012). Recent advances in the production of recombinant spider silk proteins. *Curr. Opin. Biotechnol., 23*(6), 957–964. doi: 10.1016/j.copbio.2012.03.013.

Dempsey, M., (2020). *The Stealthy Little Drones That Fly Like Insects*. Technology of Business Reporter. https://www.bbc.com/news/business-51840027 (accessed on 17 August 2023).

Dunzhu Xia et al., 2014, The Development of Micromachined Gyroscope Structure and Circuitry Technology, *Sensors* **2014,** *14*(1), 1394-1473; https://doi.org/10.3390/s140101394

Friedrich, P., Vella, M., Gulyás, A. I., Freund, T. F., & Káli, S., (2014). A flexible, interactive software tool for fitting the parameters of neuronal models. *Frontiers in Neuroinformatics, 8*, 63. https://doi.org/10.3389/fninf.2014.00063.

Hancock, P. A., Nourbakhsh, I., & Stewart, J. (2019). On the future of transportation in an era of automated and autonomous vehicles. *Proc. Natl. Acad. Sci., 116*(16), 7684–7691.

Hays, B., (2014). *Acid-Spitting Beetles Inspired ATM Defense Technology*. Science News.

Herold, H. M., & Scheibel, T., (2017). Applicability of biotechnologically produced insect silks. *Z Naturforsch C. J. Biosci., 72*(9, 10), 365–385. doi: 10.1515/znc-2017-0050.

Holland, C., Numata, K., Rnjak-Kovacina, J., & Seib, F. P., (2019). The biomedical use of silk: Past, present, future. *Advanced Health Care Materials, 8*(1), 1800465.

https://commons.wikimedia.org/wiki/File:A_migrant_hawker_dragonfly.jpg.

https://phys.org/news/2006-03-locust-eardrum-tiny-microphones.html (accessed on 17 August 2023).

https://phys.org/news/2017-06-secret-jewel-scarab-beetles-pure.html/

https://www.business-standard.com/article/pti-stories/insect-inspired-sensor-can-detect-harmful-airborne-chemicals-116091900200_1.html (accessed on 17 August 2023).

https://www.bvdw.org/fileadmin/user_upload/20180509_bvdw_accenture_studie_datadrivenbusinessmodels.pdf (accessed on 17 August 2023).

https://www.business-standard.com/article/pti-stories/insect-inspired-sensor-can-detect-harmful-airborne-chemicals-1160919002001.html).

https://www.frontiersin.org/articles/10.3389/fphys.2015.00391/full (accessed on 17 August 2023).

https://www.modernhealthcare.com/article/20150516/MAGAZINE/305169951/bugs-may-hold-secret-to-better-hearing-aids (accessed on 17 August 2023).

https://www.needpix.com/photo/download/216267/locusts-green-insect-free-pictures-free-photos-free-images-royalty-free-free-illustrations.

Jain, S., (2017). ATM frauds – detection & prevention. *Intern. J. Advances in Electronics and Computer Science*, *4*(10), http://www.iraj.in/journal/journal_file/journal_pdf/12-410-151445396982-89.pdf (accessed on 17 August 2023).

Jayachandran, D., Oberoi, A., Sebastian, A., Choudhury, T. H., Shankar, B., Redwing, J. M., & Das, S., (2020). A low-power biomimetic collision detector based on an in-memory molybdenum disulfide photodetector. *Nature Electronics*, *3*, 646–655.

Kahl, T., Bousack, H., Schneider, E. S., & Schmitz, H., (2014). Infrared receptors of pyrophilous jewel beetles as model for new infrared sensors. *Sensor Review, 34*(1), 123–134. https://doi.org/10.1108/SR-10-2012-716.

Kale, T. R., & Momin, M., (2014). Needle-free injection technology–an overview. *Inov. Pharm., 5*(1), Article 148. http://pubs.lib.umn.edu/innovations/vol5/iss1/10 (accessed on 17 August 2023).

Karásek, M., Muijres, F. T., Wagter, C. D., Remes, B. D. W., & De Croon, G. C. H. E., (2018). A tailless aerial robotic flapper reveals that flies use torque coupling in rapid banked turns. *Science, 361*(6407), 1089–1094. doi: 10.1126/science.aat0350.

Khayyam, H., Javadi, B., Jalili, M., & Jazar, R. N., (2020). Artificial intelligence and the internet of things for autonomous vehicles. In: Jazar, R., & Dai, L., (eds.), *Nonlinear Approaches in Engineering Applications* (pp. 39–68). Springer Cham. doi: 10.1007/978-3-030-18963-1_2.

Kuntzman, M. L., & Neal, H., (2014). Sound source localization inspired by the ears of the *Ormia ochracea. Applied Physics Letters, 105*, 033701. https://doi.org/10.1063/1.4887370.

Leonard C. etal (2020). Plasmonic ommatidia for lensless compound-eye vision. *Nature Communications,* Volume 1, Article number: 1637 (2020).

Lonari, Y., Zhou, X., & Kundu, S. K., (2020). *IoT-Enabled AI Technologies for Self-Driving Connected Cars.* Industrial AI blog/. https://www.hitachi.com/rd/sc/aiblog/023/index.html (accessed on 17 August 2023).

Malkin, R., Thomas, M., Natasha, M., Thomas, S., & Daniel, R., (2014). Energy localization and frequency analysis in the locust ear. *J. Roy Soc. Interface, 11*(90), 20130857 DOI:10.1098/rsif.2013.0857.

McIntosh, A. C., & Beheshti, N., (2008). Insect inspiration. *Physics World, 21*(4), 29–31.

McIntosh, A. C., & Lawrence, J. (2018). The extraordinary design of the bombardier beetle – A classic example of biomimetics. *Proc. Intern. Conference on Creationism* (Vol. 8, pp. 268–276). J.H. Whitmore, Pittsburgh, Pennsylvania.

Mendoza-Galván, A., MunñozPineda, E., Järrendahl, K., & Arwin, H., (2018). Pitch profile across the cuticle of the scarab beetle *Cotinis mutabilis* determined by analysis of Mueller matrix measurements. *Roy. Soc. Open Sci., 5*, 181096. http://dx.doi.org/10.1098/rsos. 181096.

Miles, R. N., (1995). Mechanically coupled ears for directional hearing in the parasitoid fly *Ormia ochracea. J. Acoustical Soc. Amer., 98*, 3059. https://doi.org/10.1121/1.413830.

Press Trust of India, (2018). *Novel Insect-Inspired Flying Robot Developed.* Business-Standard London. https://www.business-standard.com/article/pti-stories/novel-insect-inspired-flying-robot-developed-118091400352_1.html (accessed on 17 August 2023).

PTI, (2016). *Insect-Inspired Sensor Can Detect Harmful Airborne Chemicals.*https://www.indiatoday.in/pti-feed/story/insect-inspired-sensor-can-detect-harmful-airborne-chemicals-689440-2016-09-19 (accessed on 17 August 2023).

Pye, J. D., (2010). The distribution of circularly polarized light reflection in the *Scarabaeoidea* (Coleoptera). *Biol. J. Linn. Soc., 100*, 585.

Rinaldi, A., (2007). Naturally better. Science and technology are looking to nature's successful designs for inspiration. *EMBO Rep., 8*(11), 995–999. doi: 10.1038/sj.embor.7401107.

Rob, M., (2016). *How Insects Are Advancing Hearing Aid Technology.* News week, https://www.newsweek.com/how-insects-can-advance-hearing-aid-technology-495291 (accessed on 17 August 2023).

Schmitt, O. H., & Bleckmann, H., (1998). The photomechanic infrared receptor for the detection of forest fires in the beetle *Melanophila acuminata* (Coleoptera: Buprestidae). *J. Compar. Physiol., 182*(5), 647–657.

Schmitz, H., Kahl, T., Soltner, H., & Bousack, H., (2011). Biomimetic infrared sensors based on the infrared receptors of pyrophilous insects. *Proc. SPIE–The Intern. Soc. Optical Engineering,* 7975. doi: 10.1117/12.879149.

Schneider, E. S., Schmitz, A., & Schmitz, H., (2015). Concept of an active amplification mechanism in the infrared organ of pyrophilous *Melanophila* beetles. *Front. Physiol., 6,* 391. https://doi.org/10.3389/fphys.2015.00391.

Shigong, Y., & Rind, F. C., (2006). Collison detection in complex dynamic scenes using an LGMD-based visual neural network with feature enhancement. IEE Transactions on Neural Networks, 17(3), 705–716.

Sugiura, S., & Sat, T., (2018). Successful escape of bombardier beetles from predator digestive systems. *Biology Letters*, *14*(2), https://doi.org/10.1098/rsbl.2017.0647.

Shinji Sugiura, Sugiura S. 2018. Anti-predator defences of a bombardier beetle: is bombing essential for successful escape from frogs? PeerJ 6:e5942 http://doi.org/10.7717/peerj.5942

Swartz, M. M., (2017). *The Fabrication and Characterization of Aluminum and Magnesium Nanoparticles*. PhD thesis, The University of Utah. https://www.proquest.com/openview/15158204a8d70449530f896540192084/1?pq-origsite=gscholar&cbl=18750 (accessed on 17 August 2023).

The Economic Times News, (2018). *Dutch Engineers Develop Robots, Flying as Agile as Flies*. https://economictimes.indiatimes.com/news/science/dutch-engineers-develop-robots-flying-as-agile-as-flies/articleshow/65802434.cms?from=mdr (accessed on 17 August 2023).

University of Toronto, (2017). *This Fly's Incredible Hearing is a Curiosity to Those Developing Better Hearing Aids.* Science Daily. www.sciencedaily.com/releases/2017/05/170515111136.htm (accessed on 17 August 2023).

Yue, S., & Rind, F. C. (2013). Visually stimulate motor control for a robot with a pair of LGMD visual neural networks. *Int. J. Adv. Mechanotronic Systems, 4*(5–6), 237–247. doi: 10.1504/IJAMECHS.2012.052219.

Yue, S., Rind, F. C., Keil, M. S., Cuadri, J., & Stafford, R., (2006). A bio-inspired visual collision detection mechanism for cars: Optimization of a model of a locust neuron to a novel environment. *Neurocomputing, 69*(13–15), 1591–1598. doi: 10.1016/j.neucom.2005.06.017.

Zheng, Y., Song, L., Huang, J., Zhang, H., & Fang, F. (2019). Detection of the three-dimensional trajectory of an object based on a curved bionic compound eye. *Optics Letters, 44*(17), 4143–4146.

CHAPTER 4

Gecko-Inspired Biomimetic Adhesive Materials

ABSTRACT

Scientists and engineers are fascinated by geckos' ability to scale high and overhanging walls. Numerous synthetic gadgets can be made from Gecko skin. They can be used to attach and detach a variety of daily items. Using this ability, televisions or computers can be attached to walls, as well as medical, industrial, apparel, and household applications, with ease. As an original adhesive, a Gecko's toe pads have a unique set of structures. In milliseconds, geckos may link and disconnect their adhesive toes. Without the use of chemicals, Gecko Bandage works. Synthetic gecko feet have been attempted by many researchers by combining rigid materials like carbon fiber with a soft polydimethylsiloxane (PDMS) elastomer resin-like material (PDMS). Using a variety of synthetic and natural materials, scientists were able to mimic gecko-like stickiness. People can safely walk on a glass wall thanks to human-created technologies based on gecko feet. Gecko feet have been studied and emulated in the development of special gloves that allow people to scale glass walls, for example. NASA's Jet Propulsion Laboratory (JPL) collaborated with researchers to develop robots with sticky attachments for catching space trash, such as defunct satellites. During spacewalks, astronauts use Gecko tape to keep their clothes together. However, research into how stickiness and friction are produced in nature, namely in lizards, has proven to be a useful tool in developing new adhesives and climbing robots. Gecko adhesion research is a good example of the necessity of fundamental, curiosity-driven research.

Bio-Inspired Technologies for the Modern World: Emerging Trends and Applications.
R. Ramakrishna Reddy & T. Pullaiah (Authors)

4.1 INTRODUCTION

Leg attachment pads enable insects, spiders, and lizards to move on a variety of surfaces. Geckos' hairy attachment processes are quite complicated, and they're capable of a variety of clever adhesion techniques. They can stick to flat or textured surfaces and be easily removed. Some 3 million microscale hairs (setae) and hundreds of nanoscale spatulae (approximately 1 billion spatulas) at a density of about 14,000/mm^2 are thought to be used by these animals (Figures 4.1 and 4.2). Biomimetic microstructure research has long been attracted to geckos because of their remarkable climbing ability in a variety of situations, particularly on smooth glass or ceilings (Koch et al., 2010; Bhushan, 2007). Engineers used gecko feet's sticky properties as a basis to manufacture rubbery material strips that can pick up and drop items. Every business makes use of adhesives, which are an essential part of our daily life. For their part, single-use adhesives like traditional ones often include strong glues that might remain adhered even after they have been washed off. New pressure-sensitive adhesives based on nature are called dry adhesives. As with gecko feet, adhesion does not require liquid-like adhesives or chemical processes to produce a strong bind between two surfaces. It has never before been used to make adhesives using an electrospinning technique that is already widely used in the filtration business. Adhesive directionality is the key to the ShearGrip dry platform's success. Customizable roll stock with applications in development for a wide range of industry areas, such as consumer/industrial goods, packaging, transportation, and electronics, will be the first commercial products.

Researchers have long been fascinated by geckos' ability to scale walls and skitter over ceilings, prompting them to investigate ways of harnessing this remarkable feat of strength and agility. Human-made gadgets based on gecko feet are just the beginning of the applications of gecko adhesion technology.

It's possible that geckos wouldn't have inspired the development of nanostructured adhesives if they hadn't developed. Geckos are able to scale vertical surfaces at speeds up to 1 meter per second thanks to the adhesive setae on their toes. An adhesive must be able to endure the rigors of climbing while simultaneously being able to be rapidly removed. As a fantastic example of a PSA (Pressure-Sensitive Adhesives) that is both strong and tough to remove, duct tape is a great choice (e.g., sticky notes). Gecko's glue is a long cry from the usual glue. Traditional PSA formulations frequently contain deteriorating, fouling, self-adhesive, and clinging soft viscoelastic polymers. A gecko's seta is a branching hair-like structure made of hydrophobic keratin with an elastic modulus comparable to that of PSAs, which is found on the toes of geckos. Since they clean themselves, setae can be utilized in dirty conditions for months without losing any of their function. Glue made by Gecko is a far cry from the typical glue used in construction. Polymers used in traditional PSA formulation include viscoelastic soft polymers that degrade stink, and self-adhere. An elastic modulus comparable to that of PSAs can be discovered in gecko seta, a branching hair-like structure comprised of hydrophobic keratin found on the toes of geckos. Setae will continue to perform at the same level, no matter how unclean the environment is. As an anisotropic 'frictional adhesive,' setae require the preservation of a proximally directed shear force in order to adhere, allowing for either a strong connection or spontaneous dissociation. Gecko-like synthetic adhesives may have a bright future when it comes to glue and screws. A schematic depicting the hierarchical hierarchy of the adhesion system is given in Figures 4.1–4.3. The Gecko Adhesion is a model system for integrative biology (Autumn et al., 2014)).

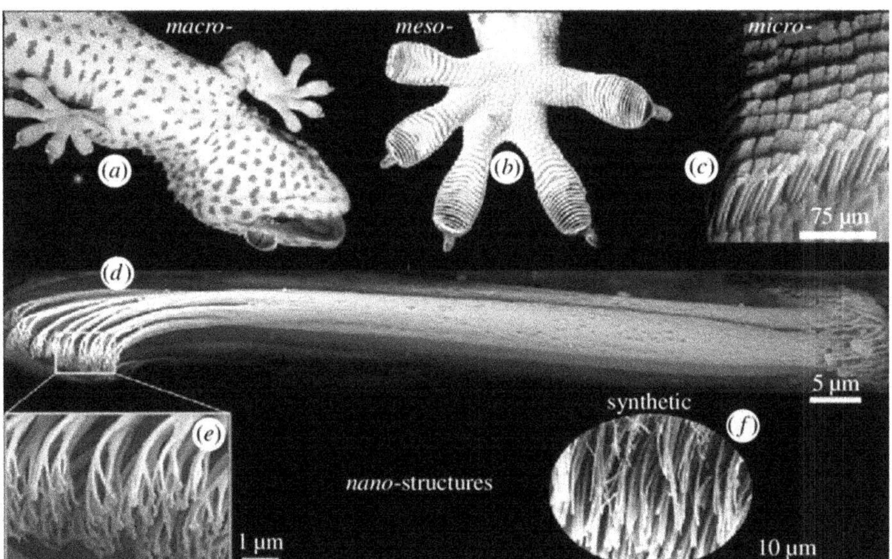

FIGURE 4.1 Gecko structural hierarchy of the adhesive system.

Source: Reprinted with permission from Autumn and Gravish, 2008.

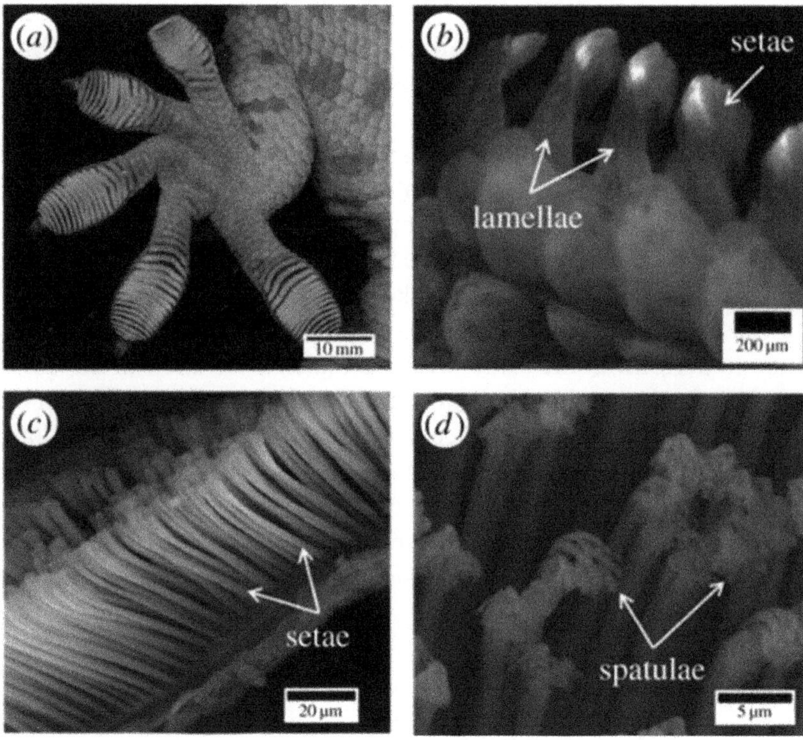

FIGURE 4.2 The gecko's adhesive system is structured in a hierarchical manner. Setae (thin hair-like structures) cover (a) flattened pads, (b) lamellae, (c) skin folds, and (d) the flattened tips of the setae that branch out are referred to as spatulae.

Source: Reprinted with permission from Stark et al. (2016) Copyright. 2016. Royal Society Publishing

A gecko's toepads are made up of a mixture of proteins, as shown in Figure 4.3 (Gamble et al., 2012).

- More than 5,00,000 hairs (spatulae) each toe; each hair contains hundreds of nanoprojections.
- One foot's adhesive force is 100 newtons.
- A 45-pound child may be lifted by a dime-sized area.

4.2 DESCRIPTION OF GECKO PAWS

Nanoscale features on the foot of the gecko allow it to cling to vertical or inverted surfaces, allowing it to cling to the wall. Overlapping adhesive lamellae covered in hundreds of spatula-shaped structures cover the soles of gecko feet, which are covered with setae or superfine hairs. These nanometer-thick pads may be bent to fit into the tiniest cracks on the surface. Combined with the stickiness of each of these pads, the gecko is held in place. Adhesive tapes, self-dissolving bandages, and high-friction polymers that can sustain heavy weights on smooth surfaces are all being produced by exploiting this unique ability.

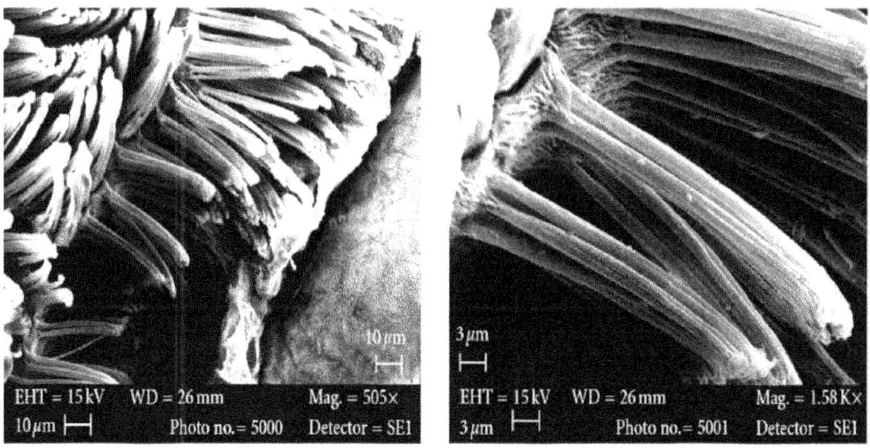

FIGURE 4.3 Scanning electron microscopy of the hierarchical structures of a gecko foot. Each toe contains hundreds of thousands of setae and each seta contains hundreds of spatulae.

Source: Reprinted with permission from Lepore et al., 2008. https://creativecommons.org/licenses/by/3.0/.

Toes have a series of ridges that are coated in hair-like setae, as shown in the images (Figures 4.1–4.3). Setae are further subdivided into hundreds of spatulas, each of which has a single split end. In order to catch their prey, geckos use Van der Waals forces, which are incredibly small interactions between molecules. On the soles of geckos' feet, there are roughly 14,000 setae per square millimeter of skin. Thus, a 70-gram (2.5-ounce) adult gecko's setae can bear a weight of 133 kilograms (293 lb).

Lizards' feet are cushioned by cushions. Below their fingers and toes, they have pads made of broad plates or scales. As a result, a lizard's toes curl and uncurl when walking or racing on horizontal or vertical walls, creating a suction pressure that allows it to cling to the walls. The suction pads on lizards' feet allow them to walk on walls. Low pressure is achieved by pressing out the air between the feet and the wall. Outside, the high-pressure air pushes feet against the wall, keeping them there.

4.3 TECHNOLOGY BASED ON THE GECKO

There are only a few animal species capable of climbing walls and walking on ceilings, such as geckos. Nanoscale attachment elements have been the subject of extensive past investigation (Bhushan, 2007; Autumn, 2006; Labonte et al., 2016). The soles of a gecko's toes contain roughly 1 billion miniscule sticky hairs, each measuring about 200 nm in width and length. With the help of these hairs, the gecko is able to directly touch the ground it is crawling over. Hairs with spatula-shaped ends have an advantage when it comes to adherence (Figure 4.4).

FIGURE 4.4 Gecko natural downwards position on a plexiglass surface. Preliminary In Vivo Experiments on Adhesion of Geckos.

Source: Reprinted with permission from Lepore et al., 2008. https://creativecommons.org/licenses/by/3.0/.

Climbing animals, like geckos, have the biggest mass and the most complicated hairy attachment mechanisms in order to cling and detach from a variety of smooth and rough surfaces. Geckos have no issue scaling slippery surfaces (Figure 4.4). About 3 billion spatulas (called sets) comprised of approximately 3 million tiny hairs (about 14,000/mm^2) cover each of these animals' two feet. The so-called partition of contacts can be used to create high dry adhesion. Gecko's multiple-level hierarchical surface building enables it to establish a wide region of actual contact with a wide range of surfaces.

Geckos have little issue scaling slick surfaces, such as windows and mirrors. Spider-Man gloves, for example, borrow their sticky toe design from the renowned footwear of the comic book superhero. Setae, the hair-like structures found on the toe pads of geckos, cover the pads of these animals' feet. By using van der Waals forces, the setae stick to any surface they come into contact with (Figures 4.5 and 4.6). To demonstrate the immense strength of these microscopic structures, consider the setae on a gecko's foot. They can support 20 times their own weight.

Adhesive-pad geckos can't be stopped even by the smoothest vertical walls. Because of their trotting gait, geckos can climb vertically at rates of up to 77 cm s^{-1} with a stride frequency of 15 Hz. Because of this, there were no forces on the animal's complete body when it attached or disengaged its toe pads. Fore-aft force at mid-step was two times the person's weight. As they ascended, geckos maintained a constant average speed while applying no braking forces to their center of mass. The mass-specific mechanical power needed to climb was 10 times more, even if the overall mechanical energy required was only 5–11% higher. If the fore- and rear legs draw in toward the midline, attachment mechanisms may be loaded. Around 15% and three-quarters, respectively, of the total time spent in stance was spent with feet attached or detached. Fore-aft force output may have been restricted when ascending speed rose because attachment and detachment periods remained constant (Stark et al., 2016; Mara, 2014; Bhushan & Sayer, 2007; Geim et al., 2003). A net pitching moment toward the top was created as a result of the forelegs drawing inward and the hindlegs pulling outward during ascent, which counteracted the pitch-back from the vertical surface. Leg functions appear to be necessary for both vertical and horizontal movement.

4.3.1 WHAT IS GECKO TAPE?

Lizards, such as geckos and skinks, are able to walk on their backs and scale walls. Special toe pads allow them to carry their weight. They have setae, which are small hairs or bristles that function like Velcro and allow their feet to grip surfaces. One mechanism proposed for geckos' stickiness was "thin-film capillary force," which is based on the power of hydrophilia (literally, "love of water"). Molecules can be hydrophilic–water-loving–or, the opposite, hydrophobic. Smaller points, known as septulae, can be found on each of the setae. They're all able to hold onto the surface. The setae and septulae of even the largest lizards can sustain the weight of a human adult, thus, they can walk on their backs. Lizards may hold on to a dry surface with their claws. They can't hold as well when the surface is damp and smooth (like wet glass). Lizards, on the other hand, are able to walk on tree branches, logs, and other surfaces with ease. SAR robots, chemical sensors, space-based positioning and positioning systems, and search and rescue robots are just a few of the uses for self-cleaning gecko seta arrays.

FIGURE 4.5 Gecko climbing on wall.

Image courtesy: Robert Michniewicz. https://creativecommons.org/licenses/by-sa/3.0/.

The "frictional adhesion" qualities and mechanisms of the gecko's foot, hairs, and seta have been studied in order to mimic the fast movement of a gecko on walls, ceilings, and other surfaces (Figure 4.5). Small size, variable articulation, self-cleaning characteristics, and ability to adapt to uneven surfaces are some of the benefits of wall-climbing robots that use adhesives inspired by geckoes (Figure 4.6). Since gecko-inspired adhesives and their attachment/detachment processes are so important for both theoretical as well as practical applications, studies into these processes are of great interest.

FIGURE 4.6 Higher power images of the Gecko foot parts showing the aligned rows of setae

Source: Reprinted with permission from Melrose, 2022. Copyright © 2022 by the author. Licensee MDPI, Basel, Switzerland. https:// creativecommons.org/licenses/by/ 4.0/.

Tape called Gecko Tape has nanoscopic hairs that look like those seen on the feet of geckos (Labonte et al., 2016; Stark et al., 2016) (Figure 4.7). Because of Van der Waals pressures exerted by these millions of flexible hairs (Autumn et al., 2002), there is a considerable adhesion impact (Figure 4.7). Only two of the many possible uses of this technology have been mentioned so far: underwater and aboard a space station. In order to get their hands on some gecko tape footwear and gloves, spider-man fanatics will have to put up with a long wait!

4.4 INTRICACIES OF GECKO ADHESION

Animals' adherence is aided by surface area, although it is not the sole determinant. University of Massachusetts in Amherst researchers examined the adhesive strength of gecko species. Nearly half of the increase in adhesion force with increasing size was attributed to the toe pad area (Autumn et al., 2000; Gilman et al., 2015). According to their findings, the most critical dry adhesive feature is the contact area to compliance ratio. The contact area to compliance ratio is the most crucial dry adhesive feature, according to the scientists who conducted the research. High stiffness is equivalent to low compliance, which means that the toe pads of larger geckos stretch less than those of smaller geckos at a given force.

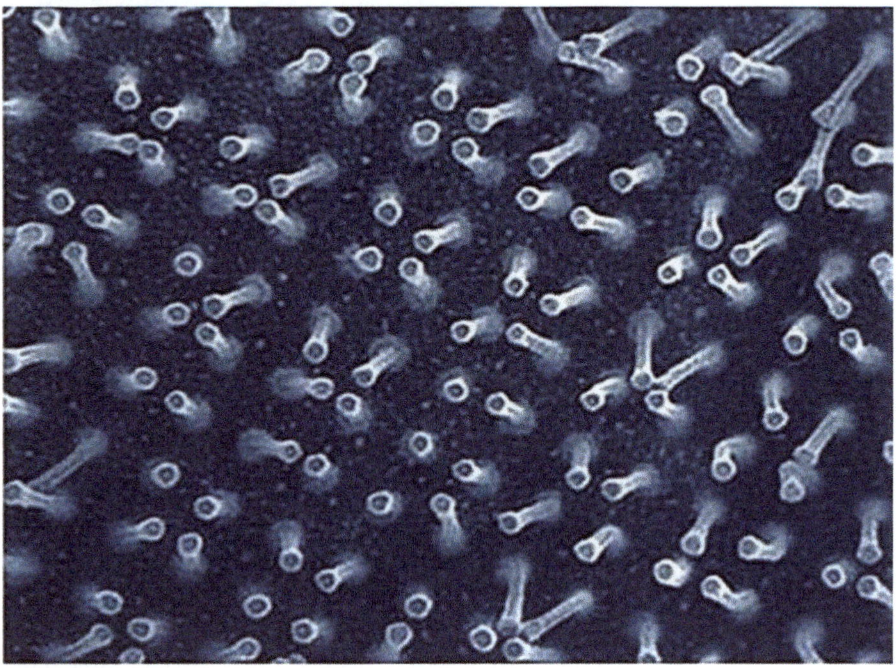

FIGURE 4.7 Micro view of nano tape.

Image courtesy: Douglasy from Geim et al, 2003. Wikipedia. https://creativecommons.org/licenses/by-sa/3.0/.

4.5 HUMANS CANNOT USE GECKO ADHESION

For a wide variety of species, it is well-established that the size of the adhesive toe pads grows in proportion to the body mass. According to UK researchers from the University of Cambridge, sticky toe pads must cover a larger and larger proportion of the body as weight rises (Labonte et al., 2016).

This means that in order for humans to be good climbers, the dried adhesive would have to cover nearly half of our bodies. Our hands and feet would need to be at least two feet wide and long in order to accomplish this task effectively.

4.5.1 GECKO TOE PADS WORK AFTER DEATH

Materials research is not for the faint-hearted. University of California, Riverside researchers have discovered that Gecko adhesion is passive, implying that no muscular effort is necessary to hold on (Stewart & Higham, 2014). In order to illustrate this, scientists compared the adhesive strength of dead and alive geckos.

4.6 IT'S POSSIBLE TO USE GECKO TAPES TO HOLD AND RELEASE SPECIFIC TIPS ON GECKO HAIRS

The gecko foot has long been a source of inspiration for scientists working on reusable, superstring adhesives. Adhesives that can be grabbed and released at will have been developed by Carnegie Mellon researchers (Figure 4.7). These structures mimic the hairy tips of geckos' toes, which are well-known for their climbing prowess.

As a surgical bonding agent, Gecko-like adhesives have already shown promising results. Climbing robots for reconnaissance and space exploration could benefit from the gecko's innate ability to climb, say researchers (Stark et al., 2016; Mara, 2014; Bhushan & Sayer, 2007; Geim et al., 2003). Geckos' toes are covered in millions of microscopic hairs that are densely packed together. Each hair follicle has hundreds of small spatula-like structures.

The Van der Waals forces hold each spatula to the object a gecko is trying to climb. Geckoes can stick to just about anything, thanks to the combined pressure of millions of spatulas. As far back as 2006, a group of mechanical engineering students at Carnegie Mellon University led by Metin Sitti developed flat mushroom tips that resemble spatulas (Sitti & Fearing, 2003). The tips had the ability to grip with a force comparable to that of a gecko, but there was no simple way to release them. Later, researchers discovered that altering the spatula's tilt was the key to managing stickiness (Sitti & Fearing, 2003) (Figure 4.8). Polymer fiber tips were slanted at a 28° angle to approximate the angle between a gecko's hair and a spatula before being placed on top of other polymer fiber tips, and the results were studied.

Increased contact area with the spatulas occurs when pressure is applied along the fibers. Pulling on the tape in the other way minimizes the contact area, giving it the nickname "gecko tape" by Sitti. The team's glue proved capable of supporting a kilogram of weight when applied at an angle. The tape's hold might be broken with as little as a 300-gram weight.

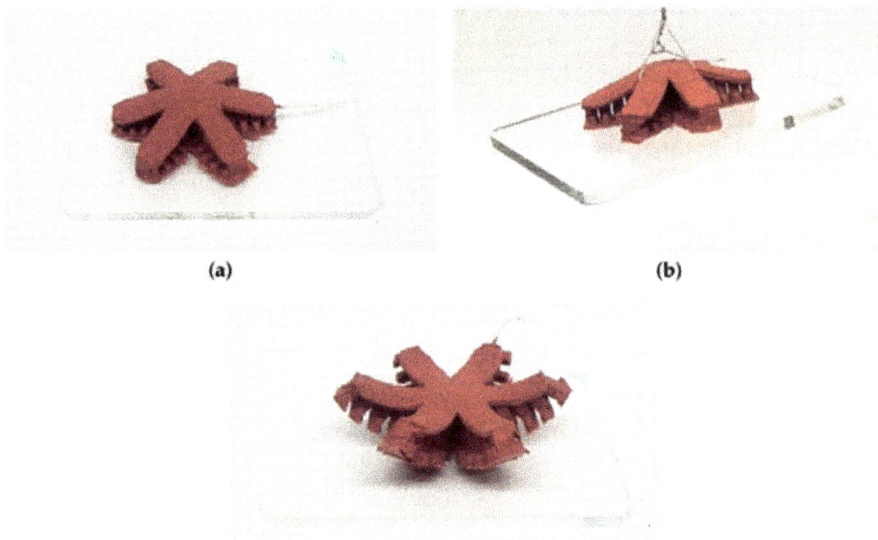

(a) (b) (c)

FIGURE 4.8 The presented gripper expands the range of existing soft grippers with gecko-inspired adhesives. The passive gripper can (theoretically) carry a load with a weight of up to 18 kg.

Source: Reprinted with permission from Arthur Seibel et al., 2020. https://creativecommons.org/licenses/by/4.0/.

Geckos may get their foot stuck in just about any hole or crevice in the earth. The gecko's capacity to cling and detach from objects has been mimicked by scientists for years, but mass manufacture for everyday usage has remained a pipe dream (Kim & Varenburg, 2020; Georgia Institute of Technology, 2022). Climbing robots and capsule robots will benefit from the adhesive, adds Sitti, a researcher who works on both in his lab. Carbon nanotubes have been used by other researchers to obtain sticking power several times greater than that of a gecko. However, Sitti points out that nanotube strands that are a little stiffer can stick to a wall but have a hard time hanging from the ceiling. That's what his glue was capable of holding. "Simply scaled in manufacturing and cost-effective," says Sitti of his polymer fibers. Because of the fibers, this can be a huge advantage. Using his company, nanoGriptech, Sitti sells his polymer fibers for use in athletic gear and skin adhesives.

His carbon nanotubes' tips are "intriguing," says Liming Dai, an engineering professor at Dayton University. CNTs were used to generate 10 times the force of gecko feet. Polymers have one advantage over metals in nanofabrication because they can be easily made. "Also, it's cheap." One-time-use, safe medical gecko tape was developed by bioengineer Jeffrey Karp of Harvard-MIT Division of Health Sciences. It will be fascinating to see if this method can be used in industrial settings or if the adhesives perform better in moist environments, which is a key limitation for many of the gecko-mimicking adhesives. Sitti's team is working on a covering to protect the mushroom-shaped tips. Sitti thinks this could be valuable in the medical profession, as it could keep medicinal patches from falling off the wearer's skin when they're perspiring, for example. Tasks of greater difficulty await. While this is still the case, the tips tend to come off quickly.

Professor Ali Dhinojwala of the University of Akron is a gecko-inspired adhesion researcher who expects that the adhesive can self-clean so that it can be reused (Dalvi et al., 2019). For gecko-inspired glues, polyimide, PVS, and PDMS are among the materials that have been examined. Polyurethane, polystyrene (PS), and polyethylene have also been considered. Using polyurethane for gecko-inspired adhesives may be possible because of the material's polar groups, according to a study.

4.6.1 DRY ADHESIVE PRODUCTS MODELED ON GECKOS

- Geckskin, a stiff, sticky fabric invented by UMass Amherst scientists, is now available for purchase (Grifantini, 2009). There is no residue left when the fabric is peeled away from a surface, and it can hold more than 300 kg/m^2 of weight. Hanging devices, including hooks and sealing tapes, are being developed by Felsuma, a UMass Amherst-based start-up company.
- There are biologically inspired climbing aids in development at DARPA (Department of Defense Advanced Research Projects Agency). These mounts allow warriors to carry heavy loads while being able to scale tall barriers. After receiving a grant in 2009 for Geckskin from the Z-Man program and conducting an experiment

in 2012 that demonstrated the fabric could support 299 kg of static load, Geckskin is now commercially available. Additionally, Z-Man has financed and will continue to support the Draper Laboratory, an American not-for-profit R&D organization, in the development of an adhesive product inspired by geckos.

- NanoGriptech, a business based in the United States, created a line of dry adhesives dubbed Setex after studying geckos. For example, it can be used to cover components during manufacturing operations without leaving behind a residue, and it can also be used to improve the grip of sports gloves for better ball control.
- University of Dayton researchers discovered a gecko adhesive in 2008 that can sustain 100 N/cm^2 and is reported to be 10 times stronger than the gecko's foot (https://scitechdaily.com/super-adhesive-geckskin-material-holds-700-pounds/, https://doi.org/10.1098/rsif.2014.0675).

4.6.2 THE DESIGN GECKO TAPE

For generations, philosophers and scientists alike have been fascinated by geckos' incredible ability to climb. In the last several years, scientists have made significant progress in understanding how geckos are able to sense their prey by using submicrometer keratin hairs on their soles. About 10 N cm^2 adhesion is strong enough to keep geckos upright and seated on a glass ceiling when hundreds of millions of hairs work together. A new sort of glue could be developed by mimicking the gecko mechanism. Gecko tape is a prototype manufactured by microfabricating dense arrays of flexible plastic pillars, whose design is tuned to ensure their collective adherence (Grifantini, 2009; Geim, 2003).

Geckos' tiny nanoscopic hairs could allow people to walk across ceilings and sheer surfaces, according to researchers who have developed a novel substance covered with these hairs. Manchester University's Andre Geim and colleagues claim that the material is strong enough to allow a person's hand to attach to the ceiling. The tape (Gecko Tape) was simply peeled away from the surface by slowly peeling it away from one side. New Scientist quotes Geim as saying, "spider-man is science fiction and will remain in comics." "Gecko-man" may, in the near future, go from a science fiction story to reality. Geckos can scale even the most slippery surfaces with just one toe on the ground. Setae, small keratin hairs on the foot's surface, are the key to this incredible ability to climb. Each of these hairs exerts a van der Waals force, an intermolecular phenomenon. Even though each hair has a negligible amount of force, the millions of hairs work together to create a tremendous sticky force (Grifantini, 2009; Geim et al., 2003).

4.6.3 WHAT IS IT?

- Gecko tape is an adhesive derived from the "sticky" qualities of gecko feet.
- You can hold 700 pounds on it with a chunk the size of an index card.
- Enables synthetic devices to attach and detach effortlessly everyday objects to walls to be developed.

4.6.4 WHY GECKO TAPE?

- Gecko adhesions were being studied by two separate teams of scientists.
- Gecko climbing and clinging adhesion have been the subject of biologist Duncan Irschick's research for over two decades.
- After working on polymer adhesion for years, Alfred Crosby became interested in creating a new sticky substance that could mimic the gecko's features.

4.6.5 HOW IT WAS DEVELOPED?

In the late 20th century, scientists discovered that geckos use Van der Waals forces to stick to almost any surface. When it comes to attracting molecules, Van der Waals forces (the transitory attraction between molecules due to dipole moment) are among the lowest known. The feet of geckos, on the other hand, are coated in millions of setae, which are like tiny hairs. The setae of a gecko's foot considerably enhance its contact area, allowing the gecko to walk with as little friction as possible on any surface. Since more of the gecko's foot is making contact with the surface, more molecules are making contact with the surface. The gecko's ability to stick to a surface is due to Van der Waals forces, the strength of which depends on the number of molecules that may interact with each other. Glues and tapes that borrow ideas from gecko feet's minuscule setae have been developed by engineers ever since the science underlying gecko feet was uncovered. Unfortunately, gecko glue doesn't work very well. On the other side, gecko tape has been very successful.

In the course of their study, the Crosby lab discovered that geckos possessed stiff tendons linked to their toepads, which functioned in a similar way to stiff fabric adhesives that they had been developing.

- In February 2012, Geckskin was unveiled to the public for the first time (https://geckskin.umass.edu/).
- A silicon wafer is first coated with a polyimide film substrate.
- This is done via electron beam lithography, which creates nanoscale patterns on a surface by using a stream of electrons. Using dry etching, an ion bombardment of the metal against the mask removes it, allowing only the substrate and projecting polyimide hairs to remain.
- Because of microscopic flaws, most surfaces are not flat. Finally, the silicon wafer material is removed from the flexible base and attached. An atomic force microscope (AFM) with a cantilever tip was used to assess the adherence of this polyimide hair array. A single gecko seta's adhesive force was assumed to be the same for all hairs in the animal.

4.7 REAL WORLD APPLICATIONS

- The Military is looking to apply this technology to allow their soldiers to climb vertical walls.
- "Real life spider-man"
- Human mobility in construction and inspection is being improved.
- Potential to replace conventional adhesives used today in medical, industrial, clothing, and home applications.

4.7.1 GECKO ADHESION IS CAUSED BY A VARIETY OF FACTORS

Researchers at Lewis & Clark College in Portland, Oregon, and the University of California in Berkeley researched this uncommon trait of gecko feet extensively. A connection was made between the foot and the surface of the gecko, allowing it to do this amazing feat. The toes of geckos are covered in setae, which are small hair-like structures. Each of these structures has a smattering of nanoscale points. Gecko hair molecules interact with the surface they're attached to via van der Wall interactions, thus, these so-called spatula structures need to be large in order for this to happen. Van der Waals forces, on the other hand, are among nature's weakest. Although spatulae provide a larger surface area, the grip remains modest. These miniscule hairs, in their millions, hold the gecko's foot firmly in place.

The "Gecko tape" was the first to make headlines (Figure 4.9). Millions of microscopic plastic fibers, less than a millimeter in diameter, make up the new adhesive (known as "gecko tape"). Gecko feet have hairs on their feet, and the fibers look like those hairs.

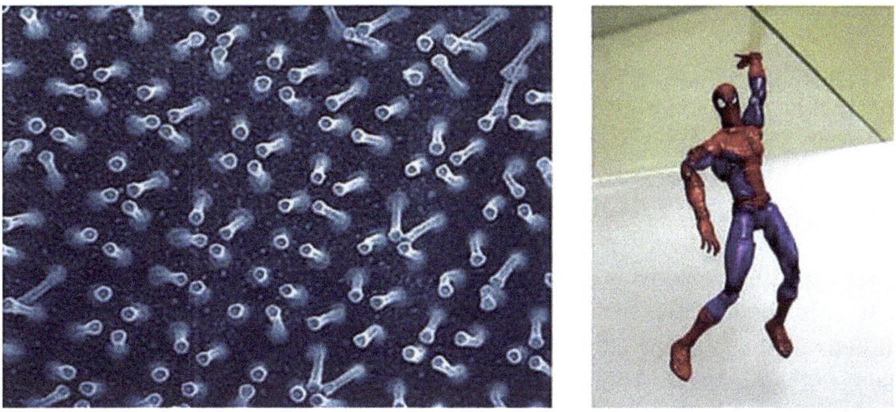

FIGURE 4.9 Micro view of nano tape Spider- Man test" of nano tape.

Image courtesy: Douglasy from Geim et al, 2003. Wikipedia. https://creativecommons.org/licenses/by-sa/3.0/.

The market for gecko-like surfaces is growing based on the increasing number of articles and patents on the subject. It is common practice for researchers to use materials such as polyimide, polypropylene, and polydimethylsiloxane (PDMS) in their studies (Kim & Varenberg, 2020). According to recent studies, carbon nanotubes are also often utilized in these systems.

4.7.2 SPIDER-MAN GLOVES INSPIRED BY GECKO CLIMB A WALL BY A SCIENTIST

There are now gloves in the form of Spider-Man's that allow you to climb up smooth, sheer surfaces. Skeptical? Elliot Hawkes, a mechanical engineer at Stanford, was the driving force behind the development of this method. "You can connect and detach [the gloves] with very little effort, and to make [the gloves] stay, all you have is hang your weight." Glass, varnished wood, polished stone, or metal are all good options for climbing surfaces. Scientists from all across the world have been working for the past decade to duplicate the gecko foot's amazing stickiness. It had been an apparently insurmountable problem to develop these materials to work on a human scale up until this point. Although the materials were sticky enough, no one could figure out how to properly balance the strain of a climbing human hand across a large surface of glue. This is the real problem. In the Journal of the Royal Society Interface, for example, Hawkes and his colleagues describe how they came up with a solution (Sitti & Fearing, 2003; Wang et al., 2020). This is a problem that scientists around the world have been grappling with for more than a decade now.

4.7.3 SKYSCRAPERS AND SPACE JUNK

However, this gecko technology has some drawbacks. For them to work, you'll need a somewhat flat surface, according to Hawkes. They lose some of their initial stickiness as the adhesive material becomes soiled from repeated exposure to the elements. There is a simple way to clean the microwedges with a piece of scotch tape. Scientists and engineers will have to work out how best to employ this new adhesive in the next years, which is exciting. It's encouraging to witness the researchers' leg-assisted scaffolding in action, and according to Autumn, we'll soon see kneepads and gloves with moldable shapes. In fact, NASA's JPL is working with Hawkes and his colleagues to develop robots that can connect the dry adhesives and launch space debris into orbit.

On a synthetic polyimide material, they employed photolithography, electron beam lithography, and plasma etching in an attempt to mimic the natural architecture of a gecko foot. Because of its distance from the gecko foot's 0.06 adhesion coefficient, Gecko tape was unable to match it (Anonymous, 2019; Bog, 2008; Dalvi et al., 2019; Geim et al., 2003; Grifantini, 2009; Kim & Varenberg, 2017, 2020; Wang et al., 2020). Already, some researchers have devised robots that can traverse vertical and horizontal surfaces while walking on their hind legs upside down. These robots move like geckos and have gecko skin on them. Several research groups around the world are investigating commercially viable, high-volume, easily scaled-up techniques for fabricating gecko surfaces. Many experts agree that gecko surfaces will be used in the real world within the next few years despite the numerous obstacles that must be solved (Figure 4.10).

For geckos with a mass of roughly 100 g, this is equal to 1,000 setae per 100 micro m^2 or 10 setae per m^2. Figures 4.11 and 4.12, which depict spatulae with a dimension of around 0.2 micro m, are completely in agreement with this.

The ideal attachment system configuration has a spherical tip. As a result, the smaller R must be as the mass of the biological system increases. Whereas the gecko's paws may generate adhesion forces of tens of N, synthetic devices with significantly higher adhesion forces are expected to be developed.

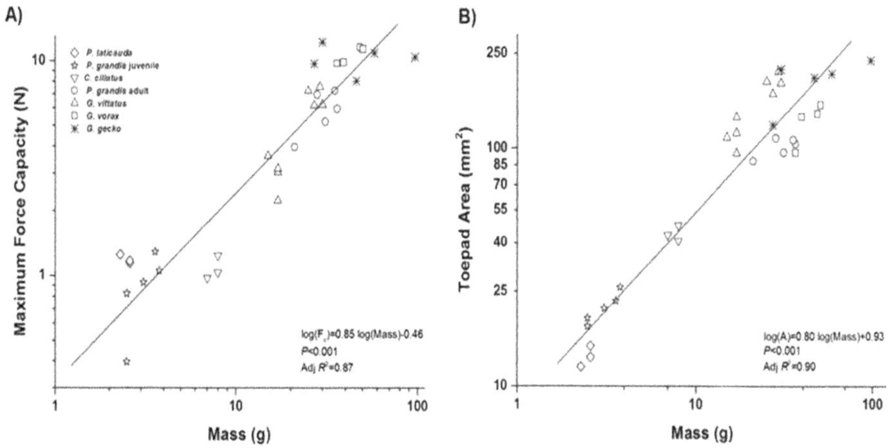

FIGURE 4.10 Mass vs. adhesion forces of tens of different species (Gilman et al., 2015).

Source: Reprinted with permission from Gilman et al., 2015. https://creativecommons.org/licenses/by/4.0/

Using Van der Waals forces, gecko feet are able to cling to or hang from surfaces because of the numerous setae and microhairs on their paws. In order to raise 127 kilograms, the millions of spatulae on a single gecko foot generate an adhesive force of about 10 N or 2.25 pounds. About 0.1 N/mm^2 of shear force is the highest that can be sustained for each seta (or approximately 10–5 N). Gecko's shear force is 40 N, multiplied by the coefficient of 0.1 N/mm^2 and multiplied by 400, according to the formula above. For a gecko, the weight is 0.245 N, and the shear force is 163/40 N/0.245 N = 163.

4.7.4 GECKO TAPES TODAY

For their project during Nordhausen's International Project Week (2016), a group of students focused on gecko tape. Their Product Design Award, one of the most prestigious design awards, was given to a gecko tape in 2011. Biomimetics researchers from the Christian-Albrechts-University in Kiel and Gottlieb Binder collaborated to create this tape. Non-glue-based products from Binder are well-known in the classic tape sector. Also, their tape is capable of adhering to the skin and may be used for medicinal reasons. For manufacture, it appears that they employ a casting or pouring method. A man can be carried on a 20 × 20 cm (8 x 8 inch) sheet that is affixed to a glass surface (Figure 4.11).

FIGURE 4.11 Gecko tape demonstration to hang household items.

Image courtesy: Peterwchen on Wikipedia. https://creativecommons.org/licenses/by-sa/4.0/deed.en

4.7.5 MOUNTS INSPIRED BY GECKOSKIN

Geckos' remarkable feet inspired the design of these wall mounts, which adhere to a variety of surfaces. It's not always easy to hang things on your walls. You may not be able to find a stud large enough to secure nails. Your screws may have been unable to secure themselves in their anchors. Felsuma, a manufacturer of building materials, may have a viable option. Felsuma's mounting devices, dubbed Geckskin, are said to adhere firmly to nearly any surface. In addition, they're reusable and can be peeled off to avoid harm. Plus, several of these adhesives are currently available for as little as $1.50 per (about £1.10 or AU $1.90).

It is possible that the phrase "geckskin" conjures up visions of science fiction for certain people. The method is based on natural phenomena. University of Massachusetts scientists came up with the concept. The long, coiled toes of geckos were particularly appealing to the group's members. Climbers like geckos are amazing to see. Glass, for example, is no match for many people when it comes to surface scaling. That these organisms aren't light-bodied insects is even more astonishing, as each one of them may weigh up to 100 grams (3.2 ounces) (Holloway, 2012; Hawkes et al., 2015; Reidel, 2016).

There are now 13 Geckskin devices listed on Felsuma's website. In total, there are four distinct subcategories (introductory, experimental, preorder, and concept). There are 10 of these things available for purchase at the time of this writing. The $5 grip hanger hook strap, the $6 grip seal strip (a weather strip for windows and doors), and the $1.50 Griphook FM are all examples of interesting products.

Though experimental, the Griphook (Figure 4.12) FM has a stunning sound. Only one can be used at a time. However, it's designed to grasp uneven surfaces, such as raw plywood. The Rectangle Bow Pad, a Geckskin adhesive, is sold in a four-pack for $3. Dual-sided Geckskin pads are also experimental and are designed for usage on smooth, solid surfaces. Sensors and thermostats, for example, can be mounted to these.

In spite of their experimental nature, several Geckskin gadgets can hold quite a bit of weight. For example, the Griphook FM has handled 10 pounds (4.5 kg) on very rough surfaces with no issues. Up to 5 pounds, Felsuma recommends the grip hanger hook strap (2.3 kg). Take a look at Felsuma's Geckskin hangers if you're interested.

4.7.6 HOW GECKO STICK TO A WALL LEWIS AND CLARK COLLEGE?

In 2012, scientists at the University of Massachusetts-Amherst led by Al Crosby and Duncan J. Irschick introduced Geckskin, a new superadhesive based on geckos' ability to attach to nearly any surface (Scientific American,

2002). Even NASA and the U.S. Military are eagerly anticipating future advancements in this discovery because of its wide range of applications (Reidel, 2016).

FIGURE 4.12 Geckskin technology was used to develop the grip hanger hook strap (Women hangs from Roof).

Source: https://www.cnet.com/home/smart-home/you-can-now-buy-geckskin-gecko-inspired-wall-mounts/.

4.7.7 *HOW DOES GECKSKIN WORK?*

A gecko's "setae," or microscopic hairs, aid in the animal's ability to grip smooth surfaces. Adhesion is formed by these small spatulae interacting with surfaces at the molecular level. When two molecules have the same electrical charge, they attract one another like hair to an inflated balloon. Van der Waals forces are a word used to describe intermolecular forces. Geckskin mimics the skin-pad-tendon connection rather than the setae on gecko feet. For the synthetic tendon, geckskin uses a combination of carbon fiber and soft, rubbery materials to mimic the skin (to represent the pad). To achieve "draping adherence," synthetic setae are forced to drape over surfaces like a tablecloth with a hard substance (Anonymous, 2019). This "gecko tape" was tested by attaching a sample to the hand of a 15 cm high plastic Spider-Man figure weighing 40 g, which enabled it to stick to a glass ceiling, as is shown in the figure.4.13. These materials are strong enough to hold up to 700 pounds, but they are also easy to dispose of.

FIGURE 4.13 Geckskin applications: Spider-man soldiers to home decor. Spider-Man test" of gecko tape.

Image courtesy: Douglasy from Geim et al, 2003. Wikipedia. https://creativecommons.org/licenses/by-sa/3.0/.

4.8 UTILIZATION OF MATERIALS WITH HIGH ADHESION

It was Al Crosby's 42-inch flat-panel television that was used to evaluate the robustness and longevity of his innovation. Using Geckskin, the television lasted for nearly three hours. For television mounting, the research team prepared a film confirming the substance can attach to almost anything. However, not enough testing has been done to commercialize the device.

4.8.1 *DECORATIVE GECKSKIN FOR THE HOME*

By eliminating the need for screws, tacks, tape, and other home-decorating staples, Geckskin could revolutionize the way people decorate their homes. Adhesives like this one could completely change the way we decorate and fix our homes. Renters, homeowners, and even college students living in dorms might benefit greatly from this.

4.8.2 *"GECKSKINSUPER-ADHESIVE" MATERIAL HOLDS 700 POUNDS*

On a flat surface like glass, a single index-card-sized piece of Geckskin™ can hold 700 pounds while being easily removed with no residue. According to CNN Money, it was one of the top five scientific breakthroughs of the year. Potential uses for Geckskin™, a synthetic adhesive and removable surface, include attaching and detaching ordinary household items such as televisions from walls.

4.8.3 *THE INVENTION*

Scientists and the general public have long been fascinated with gecko climbing on vertical and overhanging walls. With strong tendons implanted into their toepads, Gecko toepads can grip on smooth surfaces. Geckos' toepads, which are covered in millions of microscopic setae, offer the soft substrate required for adhesion to any surface. In spite of decades of study, scientists still don't know how geckos stick to each other.

Geckskin, a sticky polymer developed by Polymer Science and Engineering professor Alfred Crosby and Biology professor Duncan Irschick, was selected as one of the Top 10 Textile Innovations of 2013–2014 by the FabricLink Network.

An integrated adhesive with a soft pad woven into a stiff fabric was developed by a team of researchers at the University of Massachusetts Amherst because of the gecko's ability to generate sticky force. With 700 pounds of pressure, an index card-sized Geckskin (Figure 4.14) slice may stick to any flat surface and then be easily removed, leaving no residue.

FIGURE 4.14 A television be securely fixed using a Gecko skin cushion to a wall (Figure Redrawn).

4.8.4 SOLDIERS WITH GECKSKIN

Geckskin's military applications were seized upon promptly (King et al., 2014). DARPA, the US Department of Defense's advanced research department, has funded the Geckskin team's research. Fortress-climbing operations will benefit from this study's conclusions, which the military is already implementing. With a 218-pound researcher and an additional 50 pounds of gear, they've attempted to scale a 25-foot glass wall. Climbing aids like these could be useful for everyone, including first responders and construction workers.

4.8.5 GECKSKIN IN LABS

Geckskin could potentially be used in laboratories and hospitals. It's safe to use in a lab because it doesn't trigger any chemical reactions or leave any residue behind. Geckskin has yet to be tested for use on the skin, but wearables like patches and heart monitors could benefit from its properties.

4.8.6 ROBOTICS AND SPACE EXPLORATION: GECKSKIN

For a variety of purposes, including debris collection and minimizing collisions with other spacecraft, NASA is using the Geckskin material on its spacecraft (King et al., 2014).

4.9 GECKSKIN'S ENVIRONMENTAL SIGNIFICANCE?

According to Geckskin's chief scientist, Duncan Irschick, people will be able to acquire the product from a hardware store at a fair price once it is marketed (King et al., 2014). Geckskin is produced from common household items that are simple to build and mass-produce. For this project, the crew has switched to using only renewable resources like natural rubber and hemp. Because it can be recycled, Geckskin is more environmentally friendly than tape or Blu-Tack. Solar panels may even be detached and reattached using Geckskin's fabric, according to the company's founders. Geckskin has the potential to have a significant environmental impact by removing waste from outer space and reducing it on the planet's surface. One of the most cutting-edge materials is Geckskin. It could lead to new medical and scientific breakthroughs, as well as the development of new military equipment.

4.10 GECKO BANDAGE THAT WORKS WITHOUT CHEMICALS

In view of the excellent adhesive characteristics of the gecko-like dry adhesive surfaces, their potential as a medical adhesive tape has also been successively confirmed. It is expected to be used as a new type of tissue adhesive tape to assist medical procedures, such as replacing sutures to assist wound healing. The typical difference from the conventional gecko-like dry adhesive surfaces is that the adhesive enhancement for the soft deformable tissue is achieved by reducing the density of microstructure, namely reducing the ratio of the tip diameter to the pitch can enhance the interface contact (as shown in Figure 4.15). MIT scientists have created the first-ever surgical patch that resembles the way gecko feet might attach to vertical surfaces (MIT). The bandage can be used to treat wounds, prevent leaks, and replace sutures, among other things. Nanopillar-coated polymer would be used as a bandage, with the adhesive consisting of hundreds of thousands of nanopillars.

It is necessary to modify gecko-like wet adhesives and use medical-grade materials in order to ensure that they adhere to wet conditions. Flexible and biodegradable materials must be used to create products that can be easily molded to the body's tissues.

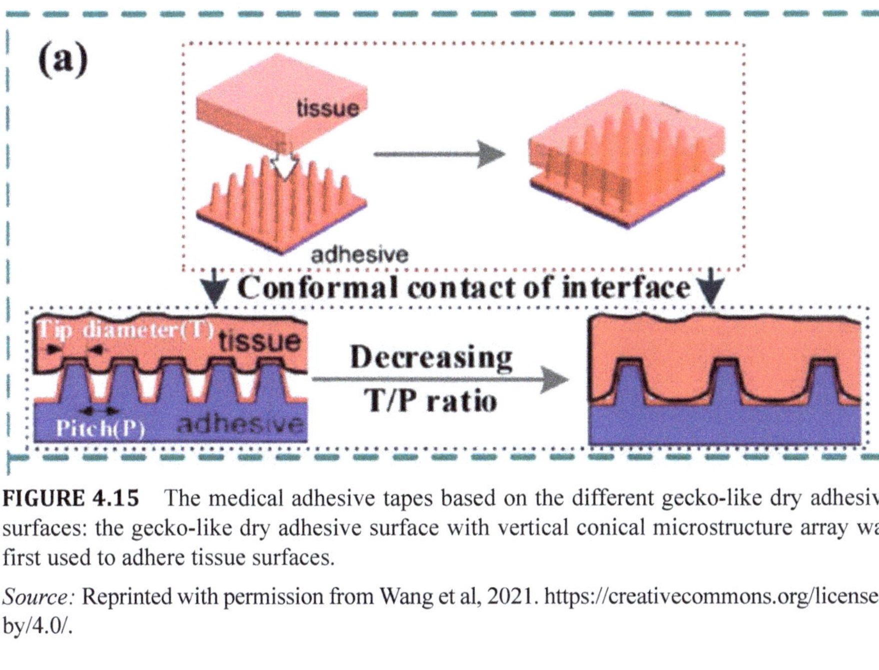

FIGURE 4.15 The medical adhesive tapes based on the different gecko-like dry adhesive surfaces: the gecko-like dry adhesive surface with vertical conical microstructure array was first used to adhere tissue surfaces.

Source: Reprinted with permission from Wang et al, 2021. https://creativecommons.org/licenses/by/4.0/.

Adhesive with nanoscale hills and valleys was inspired by the gecko (Dougherty, 2008). Other similar adhesives have addressed the issue of adhering to dry surfaces, and this one did as well. The MIT crew, on the other hand, went above and beyond. Wet strength was an important consideration for this new glue. For surgical purposes, this would be a tremendous benefit since it would allow it to augment internal sutures and staples. Is this the last step? Of course, we're talking about sugar here. A thin layer of sugar-based adhesive, to be exact. As a result, the bandage can stick to moist surfaces, and the biodegradable substance is ideal for surgery. An internal organ or vessel can be bandaged, and the surgeon can then walk away. The bandage, which is likewise non-inflammatory, will eventually dissolve.

4.11 GLASS? IS THIS POSSIBLE FOR A ROBOT TO DO?

Without employing suction cups, which are slow and inefficient, how can a robot scale a smooth surface like glass? A gecko's toes are the key to unlocking the mystery. An engineer at Stanford University named Mark Cutkosky has created the 'Stickybot' that uses the same type of sticky, dry adhesive that allows the lizards on the most unlikely of surfaces to stay put. For this 'directional adhesive,' geckos' split-end hairs on the ridges of their feet engage with the molecules of the climbing surface via millions of tiny hairs (nano-size) (King et al., 2014). Many gecko-like feet with particular movement mechanisms and dry adhesives have recently been developed and successfully used in climbing robots as shown in Figure 4.16.

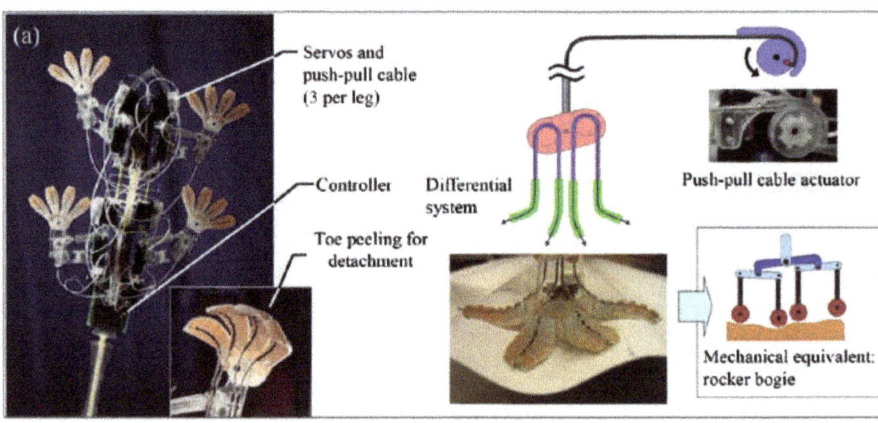

FIGURE 4.16 Multilegged climbing robots with dry adhesives: (a) Stickybot and its foot.

Source: Reprinted with permission from Xu et al., 2022. https://creativecommons.org/licenses/by/4.0/.

4.12 EXACTLY HOW DOES A GECKO GET ITS FEET OFF OF A THICK SURFACE?

Each time the lizards walk forward, their toes open up like a party favor whistle, and when they take a step back, they peel their toes up like a piece of tape.

4.13 ROBOTS' USAGE OF GECKO FEET

For climbing, a one-way adhesive is essential since it is easy to attach and remove a robot's foot. "You have to press the adhesive into the surface, and then you have to struggle to get it out. You can't walk around with chewing gum on your foot using other adhesives. This is similar to how directional adhesion works: you can hook on and unhook yourself from the surface as needed."

As soon as Cutkosky and his team realized that orientation was crucial, they inquired as to how to make phony materials for robots that might achieve the same results. A micro-scale mold was used to create a rubber-like substance containing tiny polymer hairs. Using a piece of glue the size of a child's hand, Stickybot's four tiny feet are affixed to its body. In the same way, as lizards climb walls, this mechanical lizard-like robot does so with ease.

New adhesives were developed in 2009 that resemble the gecko's lamellae and setae in their two-layer structure. Five times thinner than a human hair, the "hairs" are 20 micrometers wide compared to the originals. Stickybot can now scale surfaces like wood paneling, painted metal, and glass with these newer versions. No residue or harm is left behind by the material, which is robust and reusable. With vertical wall-climbing robots, access to risky or difficult-to-reach areas could be improved.

4.14 GECKO GRIPPER'S ADVENTURES: WHAT DO WE KNOW?

Defunct satellites and other space debris were first recovered for use on the International Space Station (ISS) using the gecko-grabbing technique. Using the Gecko Gripper, you can pick up objects with holes in their surface or smooth surfaces that a vacuum gripper can't. The Gecko Gripper attaches to a surface using millions of micro-scaled fibrillar stalks in the same way that geckos do when they climb. The Gecko Pick and Place device does not require compressed air or any other kind of power to attach or raise anything. The Gecko's capacity to lift without the use of compressed air or any other external power offers a more efficient and cost-effective alternative.

4.15 CLEAN-UP OPERATIONS IN ORBIT COULD BENEFIT FROM NASA'S GECKO-INSPIRED ROBOTIC GRIPPER

Nearly half a million bits of trash created by humans are hurtling around the planet at speeds up to 28,000 kilometers per hour. Satellites, spacecraft, and astronauts are all at risk from this garbage (The Indian Express, 2017). Cleanup is a hassle when suction cups are ineffective in a vacuum. Gecko-inspired adhesives combined with robotic grippers developed by NASA's JPL and Stanford University have been used to gather space garbage. Gecko Grip Microgravity Space Junk.

Traditional adhesives like tape cannot be used in such temperatures. Only magnetic objects can be used with magnets. Debris harpoons are one alternative, but any method that uses force to impact the debris runs the danger of moving it in unforeseen places. Engineers at Stanford University and NASA's JPL have developed a new type of robotic gripper that can collect and dispose of rubbish in an environmentally acceptable manner (Zhenqi et al., 2017). According to Stanford University Professor Mark Cutkosky, this adhesive gripper is inspired by a gecko's claw-like grip. Several zero-gravity conditions, including their own lab and even the ISS, were used to evaluate the gripper's performance (ISS). "There are other missions that would benefit from this, including rendezvous and docking and orbital debris abatement," Aaron Parness of NASA's JPL said.

According to a NASA official, a climbing robot assistant could be built in the future, which could crawl around the spaceship, make repairs, film, and inspect for defects. This technology has been used in climbing robots and even a system that allows humans to climb up certain surfaces. Gecko feet are capable of creating a Van der Waals force when they come into complete contact with the ground. This is how geckos can climb walls. To generate these forces, electrons must be placed in different positions on a molecular surface. Instead of being flat and stationary, it has a foldable adhesive grid on the gripper's front that extends outwards from each side, as if it were reaching out to hug you from all four sides at once. The grid may attach to flat surfaces, such as a solar panel, while the arms can grasp curved surfaces, such as the body of a rocket, for example. Researchers predict that in the future, they will be able to make more adhesives for less money. Gecko-inspired glue, according to him, could be as commonplace as Velcro in the near future. Even NASA is looking into the prospect of using Geckskin to catch space garbage and reduce collisions, as well as the possibility of robots grabbing spacecraft' exteriors to perform inspections and maintenance.

4.16 GECKO TOES INSPIRED PADS ENABLE TO WALK UP A GLASS WALL LIKE 'SPIDER-MAN

Scientists have finally figured out what makes a gecko's foot so sticky after a lot of effort. According to the BBC, scientists have developed gloves that allow people to climb walls like "Spider-Man" (https://www.theguardian.com/technology/2014/nov/19/spiderman-suit-us-military-tokyo-gecko-glass-climb). These tiny hair-like strands on the underside of geckos' feet allow them to do gravity-defying stunts. Researchers have discovered that setae are capable of producing a significant amount of friction when they come into touch with ascending surface molecules. In order to intensify the weak attraction, a gecko's foot is covered in hundreds of setae.

In addition to being able to hold on to practically any surface, gecko toes also have the ability to easily detach from it. For the sake of emulating the lizards, Stanford engineers have developed an adhesive solution that can attach to glass and support human weight (Figure 4.17). Due to the difficulty in creating new super-adhesives, scientists have attempted to mimic this architecture. A viable answer may have been developed by a group of Stanford University academics (Reidel, 2016; Jiang et al., 2017). For the first time, they discovered the root cause of their failures and devised a solution to the problem. A number of bioinspired dry adhesives have been developed, following the discovery of the gecko attachment mechanism. Some of them have been used in climbing robots and grippers. The morphology and potential mechanisms of biological dry adhesive organs, as well as the application of bioinspired dry adhesives to climbing robots and grippers. Human hands can be equipped with Gecko-like pads, which are capable of providing super-sticky climbing abilities, according to Hawkes and coworkers (2015). Humans are trying to climb the walls (Figure 4.17).

In tests, a 155-pound man was able to mount a 12-foot-high vertical glass wall utilizing the 140 square centimeter silicone pads in each of his palms efficiently. Hundreds of tests were carried out on the pads, and there were no issues. When they're attached to the pads using small springs, the microwedges work. With the help of springs, the tiny wedges can be used in the most efficient manner to maximize the stick force. In addition to allowing academics to live out their boyhood dreams of becoming "Spider-Man," the so-called "gecko gloves" could have enormous real-world repercussions. Window cleaners, for example, may make use of them to climb buildings' exteriors. The pads can be used to hold onto and operate objects by astronauts during a spacewalk. Even in the military, these gloves could prove handy. To date, the Gecko Gloves have a 200-pound (90-kilogram) weight limit (kg). However, if it is multiplied by ten, it can support up to 2,000 pounds (900 kilograms). A self-cleaning version based on how geckos care for their own sticky feet is being researched by a Stanford team.

FIGURE 4.17 A person is trying to climb a glass wall (Download this free HD photo of grey, human, person, and office building.

Image courtesy: Maxim Tolchinskiy on Unsplash. https://unsplash.com/photos/a-person-climbing-a-glass-wall-ZA6bQoplir0).

4.17 ROBOTS GRASP SPACE GARBAGE WITH GRIPPERS REMINISCENT OF GECKOS

The adhesive technique was tested in low gravity using a giant air hockey table. Find your way around. A new robotic instrument with low gravity can grab floating objects (Jiang et al., 2017). They could be employed to assist robots in the removal of dangerous space waste or in the ascent of space stations outside walls. There's a new technique to learn when there's no gravity: imitating gecko feet. On another planet, common adhesives like glue and tape will fail to hold things together. In space, most adhesives are ineffective, but a few are. Glues can't withstand such wide temperature swings, unfortunately. Because of the vacuum in outer space, suction also fails. A sticky substance alternative, on the other hand, maybe more effective. It is possible for geckos to rapidly scale walls. Scaly feet aren't a feature of their anatomy. In contrast, their ability to attach to surfaces is due to the Van der Waals forces. Mark Cutkosky has been making stickers like this for more than a decade. He is a mechanical engineering and robotics major at Stanford University in California. With the use of a gecko-inspired design, his team has created robotic gripper "hands." As long as they keep their hands on the object, they will not let go.

The crew put the grippers through their paces in the Robo-Dome. Nearby Pasadena, California's JPL is home to a massive air hockey table. With the use of Gecko grippers, an 815-pound robot was gently steered around. Aaron Parness works at the JPL. He and his colleagues tested the whole gripper hand on NASA's Weightless Wonder plane. A roller coaster ride, this plane flies in fast up-and-down arcs. A small amount of weight is lost each time the ship descends.

It is possible to connect the gripper hand to smooth surfaces by lightly pressing on the patches of gripping material on the hand. Gripper hands can be used to repair or transfer a dead satellite, according to Parness. This technology could also be used to attach smaller satellites, known as CubeSats, to larger spacecraft via gripping components.

By manipulating three different objects with just their hands, the crew managed to manipulate all three objects. This is only one of the many types of garbage that now clogs up our solar system. The majority of space debris is made up of discarded rocket parts and other discarded materials that have been launched into orbit. There may be hundreds of millions of them, but for the most part, they're just a collection of bits. If two pieces of this trash come into contact, they could shatter into smaller pieces. In addition, even the tiniest bits can represent a threat to spaceships and humans alike.

4.18 ASSEMBLING ADHESIVES INSPIRED BY GECKOES

Hurley (2020) emphasized adhesives for geckos in large quantities (https://www.techbriefs.com/component/content/article/tb/stories/blog/36937).

A gecko's large toes are coated with hundreds of minute hairs, allowing it to stick to any surface. Based on the lizard paradigm, scientists have spent decades inventing materials that reliably grab and release objects (Zhou et al., 2013). Because the gecko saw a way out, it climbed up. They can stick to nearly anything with their toes due to their capacity. Although gecko-like adhesive strips of rubbery material have long been known to exist, mass manufacture for everyday usage has remained beyond reach until now. Now, however, that has changed. Gecko-inspired adhesive materials may now be mass-produced at a significantly lower cost thanks to new research from the Georgia Institute of Technology (https://scitechdaily.com/reversible-gecko-adhesives-ready-for-easily-mass-production/; Bouten et al., 2014). As a result, flexible gripping strips might be mass-produced and made available to consumers and businesses alike.

Polymers with "gecko adhesion" surfaces can be employed on the same production line to make extremely adaptable grippers that can pick up a wide range of different things. With the ability to attach the picture at the same time, hanging photos became a breeze. Robotic vacuum cleaners using gecko-adhesive adhesives may one day be able to clean the faces of tall structures. It will adhere to just about anything, with the exception of Teflon. Each surface can be lifted without having to construct a gripper for each one. As a primary researcher and Assistant Professor at Georgia Tech's George W. Woodruff School of Mechanical Engineering, Michael Varenberg has developed Gecko-inspired adhesives that can lift curved objects like eggs and vegetables.

Clamps, magnets, and suction cups, for example, are only capable of lifting a limited number of items on assembly lines. In the future, it's feasible that gecko-inspired dry, glue-free grippers will replace or supplement many of the current gripping devices.

4.19 ADVANCES IN TISSUE ADHESIVES FOR CLINICAL MEDICINE

Tissue adhesives have gained importance in modern medicine since they were first developed 30 years ago and are now widely used in surgical procedures. Because of their ability to adhere tissue to tissue or tissue to non-tissue surfaces directly on the wound bed, they provide numerous advantages in the clinic, including reduced pain and suffering, a reduced need for stitches, improved cosmetic results, and targeted drug release. Non-invasive wound closure is becoming more popular with the use of tissue adhesives, according to Ge & Chen (2020).

Biodegradability, biocompatibility, sustainable hemostasis ability, forceful adhesion to wet tissues, noninterference with the processes of wound healing, simplicity of use, long duration, desirable cosmetic results, cost-effectiveness, and suitability for different procedures are some of the medical and surgical requirements for an exclusive tissue adhesive. To design new and improved adhesives for wound closure in living tissues, a thorough investigation of the constituents of tissue adhesives is required. Bonding durability, bonding efficiency, and bond strength are all influenced by the particular abilities of each component in the adhesive formulation. Effective formulations must incorporate compounds with certain properties (such as hemostasis, antimicrobial activity, and antiinflammation) in order to outperform new products. Natural polymers that are cross-linked through biological reactions by integrating certain chemical groups, according to scientific studies, show promise for the future creation of surgical adhesives.

Another type of biomimetic medical adhesive is based on gecko tissue adhesives. As previously mentioned, gecko feet have millions of nanoscaled fibrils that work together to generate an incredible stickiness (Geim et al., 2003). Tissue adhesives based on geckos are another biomimetic medicinal adhesive. By studying the gecko's foot, Geim and his colleagues (2003) came up with a way to bind polyimide hairs to a surface, resulting in an extremely strong adhesive. As a result, the production of large numbers of Gecko-inspired adhesives, as well as the removal of adhesions underwater, pose a number of obstacles.

4.20 INDUSTRIAL APPLICATIONS FOR GECKO'S ADHESIVE TECHNOLOGY ARE GETTING CLOSER AND CLOSER BY THE DAY

Scientists have long been fascinated by the gecko's ability to defy gravity, and they've tried to figure out a way to use that ability. It's not just spider-man 2 sequel shenanigans that could benefit from human-made equipment inspired by gecko feet: gecko-adhesion technology has many other potential uses as well.

Robot arms used in the production of computer chips are one possible use being investigated by the Georgia Institute of Technology researcher. For example, gecko adhesion might be used to handle fragile materials like silicon wafers used in the construction of computer processors, says Michael Varenberg (Kim & Varenberg, 2017; Dinte & Sylvester, 2017).

To employ gecko technology in robot arms and other devices, further research into the mechanical and physical characteristics of human-made sticky surfaces is needed. A study on this subject was published in the Journal of the Royal Society Interface ("Biomimetic wall-shaped adhesive microstructure for shear-induced attachment: the effects of pulling angle and preliminary displacement"). For example, Varenberg was able to reduce the range of angles at which the material would stick while also making it easier to release its grasp. Geckos' incredible ability to adhere to practically any surface is due to tiny, intermolecular hairs. The thin, film-like hairs are driven to the surface, where they are sheared off from the other end by the other side. As a result, they can either cling to the surface or be quickly dislodged.

Researchers must figure out how to impart weight to the robotic arm and silicon wafer in a factory that uses man-made adhesive technology in order to gain or loosen the grip between the two. Varenberg's team tested the adhesion of a polyvinylsiloxane wall-shaped microstructure surface that mimics the gecko's adhesive capabilities. The microstructure detaches with zero force when the pull-off angle reaches 140–160°, according to the researchers' tests. Wall-shaped microstructures provide a wide range of control over attachment and detachment, which will make building mechanical processes around that tolerance easier, said Kim & Varenberg (2017).

This could be used to replace the current method of processing and inspecting silicon wafers in the fabrication of computer processors. Ceramic chucks employ vacuum or electrostatic grippers to pick up and handle wafers. When the ceramic contact posts on a wafer begin to wear down from repeated loading, it is possible for defects to form on the front side, releasing particles that can damage the wafer's backside. Contrary to what Varenberg claims, cleaning requirements in the semiconductor industry are incompatible with this reality. Gecko adhesion microstructures are preferable because they don't destroy wafers and don't wear out over time.

Further steps, according to Varenberg, include optimizing the manufacturing process and using industrial-grade materials, as well as investigating the effects of environmental and surface geometry aspects.

KEYWORDS

- **adhesion**
- **gecko**
- **gecko bandage**
- **Gecko's adhesive technology**
- **geckskin**
- **nanostructures**
- **robotics**
- **space junk collection**

REFERENCES

Anonymous, (2019). *Geckskin Adhesive, Invented on Campus, is Now for Sale at the University Store*. UMass Amherst. https://www.umass.edu/archivenewsoffice/article/geckskin-adhesive-invented-campus-now-sale (accessed on 17 August 2023).

Autumn, K., (2006). How gecko toes stick. *Am. J. Sci., 94*, 124–132.

Autumn, K., & Gravish, N. (2008) Gecko adhesion: evolutionary nanotechnology. *Phil. Trans. R. Soc. A. 366:*1575-1590, doi: 10.1098/rsta.2007.2173366.

Autumn, K., Liang, Y. A., Hsieh, S. T., Zesch, W., Chan, W. P., Kenny, T. W., Fearing, R., & Full, R. J., (2000). Adhesive force of a single gecko foot-hair. *Nature, 405*, 681–685.

Autumn, K., Niewiarowski, P. H., & Puthoff, J. B., (2014). Gecko adhesion as a model system for integrative biology, interdisciplinary science, and bio-inspired engineering. *Annual Review of Ecology Evolution and Systematics, 45*, 445–470. doi: 10.1146/annurev-ecolsys-120213-091839.

Autumn, K., Sitti, M., Liang, Y. A., Peattie, A. M., Hansen, W. R., Sponberg, S., Kenny, T. W., et al., (2002). Evidence for van der Waals adhesion in gecko setae. *Proc. Natl. Acad. Sci., 99*(19), 12252–12256. https://www.pnas.org/doi/full/10.1073/pnas.192252799 (accessed on 17 August 2023).

Bhushan, B., & Sayer, R. A., (2007). Surface characterization and friction of a bio-inspired reversible adhesive tape. *Microsystems. Technol., 13*, 71–78.

Bhushan, B., (2007). Adhesion of multi-level hierarchical attachment systems in gecko feet. *J. Adhes. Sci. Technol., 21*, 1213–1258.

Bogue, R., (2008). Biomimetic adhesives: A review of recent developments. *Assembly Automation, 28*(4), 282–288. https://doi.org/10.1108/01445150810904422.

Bouten, P. J. M., Zonjee, M., Bender, J., Yauw, S. T. K., Goor, H., Van, H. J. C. M., & Hoogenbook, R., (2014). The chemistry of tissue adhesive materials. *Prog. Polymer Sci., 39*, 1375–1405.

Dalvi, S., Gujrati, A., Khanal, S. R., Pastewka, L., Dhinojwala, A., & Jacobs, T. D. B., (2019). Linking energy loss in soft adhesion to surface roughness. *Proc. Nat. Acad. Sci., 116*(51), 25484-25490.

Dinte, E., & Sylvester, B., (2017). *Adhesives: Applications and Recent Advances.* Intech Open. doi: 10.5772/intechopen.71854.

Dougherty, E., (2008). *Harvard-MIT Division of Health Sciences and Technology*. MIT News, https://news.mit.edu/2008/adhesive-0218 (accessed on 17 August 2023).

Gamble, T., Greenbaum, E., Jackman, T. R., Russel, A. P., & Bauer, A. M., (2012). Repeated origin and loss of adhesive toe pads in Geckos. *PLoS One, 7*(6), e39429. doi: 10.1371/journal.pone.0039429.

Ge, L., & Chen, S., (2020). Recent advances in tissue adhesives for clinical medicine. *Polymers, 12* (4), 939. doi: 10.3390/polym12040939.

Geim, A. K., Dubonos, S. V., Grigorieva, I. V., Novoselov, K. S., Zhukov, A. A., & Yu Shapoval, S., (2003). Microfabricated adhesive mimicking gecko foot hair. *Nature Materials*, *2*(7), 461–463.

Geim, A.K., Dubonos, S.V., Grigorieva, I.V., Novoselov, K.S., Zhukov, A.A. and Shapoval, S.Y. (2003) Source: "Microfabricated adhesive mimicking gecko foot-hair."

Georgia Institute of Technology, (2020). *Surfaces that Grip Like Gecko Feet Could be Easily Mass-Produced. Science Daily.* Retrieved from www.sciencedaily.com/releases/2020/05/200507094751.htm (accessed on 17 August 2023).

Georgia Institute of Technology. (2020). Reversible "Gecko" Adhesives Ready for Easy Mass-Production. SciTech daily. https://scitechdaily.com/reversible-gecko-adhesives-ready-for-easy-mass-production/ (accessed on 17 August 2023).

Gilman, C. A., Imburgia, M. J., Bartlett, M. D., King, D. R., Crosby, A. J., & Irschick, D. J., (2015). Geckos as springs: Mechanics explain across-species scaling of adhesion. *PLoS One*, *10*(9), e0134604. doi: 10.1371/journal.pone.0134604.

Grifantini, K., (2009). *Gecko Tape That Lets Go, Special Tips on Gecko Hairs Can Grip and Release.* MIT Technology Review. https://www.technologyreview.com/2009/01/22/126009/gecko-tape-that-lets-go/ (accessed on 17 August 2023).

Hawkes, E. W., Eason, E. V., Christensen, D. L., & Cutkosky, M. R. (2015). Human climbing with efficiently scaled gecko-inspired dry adhesives. *J. Roy. Soc. Interface*, *12*(102), 20140675. https://doi.org/10.1098/rsif.2014.0675.

Holloway, J., (2021). Gecko-inspired mess-free super-adhesive can stick a TV to the wall. *Science, 2012.* https://newatlas.com/gecksin-adhesive/21504/ (accessed on 17 August 2023).

http://news.mit.edu/2008/adhesive-0218 (accessed on 17 August 2023).

https://news.cgtn.com/news/3d55444d3255444e/share.html (accessed on 17 August 2023).

https://news.cgtn.com/news/3d55444d3255444e/share.html (accessed on 17 August 2023).

https://www.berkeley.edu/news/media/releases/2008/01/29_gecko.shtml (accessed on 17 August 2023).

https://www.cnet.com/home/smart-home/you-can-now-buy-geckskin-gecko-inspired-wall-mounts/ (accessed on 17 August 2023).

https://www.prescouter.com/2016/12/geckskin-applications/ (accessed on 17 August 2023).

https://www.space.com/37335-robotic-gecko-gripper-microgravity-space-junk.html (accessed on 17 August 2023).

https://www.technologyreview.com/2009/01/22/126009/gecko-tape-that-lets-go/ (accessed on 17 August 2023).

Hurley, B., (2020). *Gecko-Inspired: Researchers Find Easier Way to Mass-Produce Adhesives.* Digital Editorial Manager, Tech Briefs. Manufacturing & Prototyping Materials https://www.techbriefs.com/component/content/article/tb/stories/blog/36937 (accessed on 17 August 2023).

Jiang, H., Hawkes, E. W., Fuller, C., & Estrada, M. A. (2017). A robotic device using gecko-inspired adhesives can grasp and manipulate large objects in microgravity. *Science Robotics, 2*(7), eaan4545. doi: 10.1126/scirobotics.aan454.

Kim, J. K., & Varenberg, M., (2017). Biomimetic wall-shaped adhesive microstructure for shear-induced attachment: The effects of pulling angle and preliminary displacement. *J. Roy. Soc. Interface, 14*(137). http://dx.doi.org/10.1098/rsif.2017.0832.

Kim, J. K., & Varenberg, M., (2020). Drawing-based manufacturing of shear-activated reversible adhesives. *ACS Appl. Mater. Interfaces, 12*(17), 20075–20083 https://doi.org/10.1021/acsami.0c01812.

King, D. R., Bartlett, M. D., Gilman, C. A., Irschick, D. J., & Crosby, A. J. (2014). Creating gecko-like adhesives for "real world" surfaces. *Advanced Materials, 26*(25), 4345–4351. doi: 10.1002/adma.201306259.

Koch, K., Bhushan, B., & Barthlott, W., (2010). Multifunctional plant surfaces and smart materials. In: Bhushan, B., (ed.), *Springer Handbook of Nanotechnology*. Springer, Berlin, Heidelberg. https://doi.org/10.1007/978-3-642-02525-9_41.

Labonte, D., Clemente, C. J., Dittrich, A., Kuo, C. Y., Crosby, A. J., Irschick, D. J., & Federle, W., (2016). Extreme positive allometry of animal adhesive pads and the size limits of adhesion-based climbing. *Proc. Natl. Acad. Sci. USA, 113*(5), 1297–1302.

Lepore, E., Brianza, S., Antoniolli, F., Buono, M., Carpinteri, A., & Pugno, N. (2008) Preliminary in vivo experiments on adhesion of geckos. *Journal of Nanomaterials 2008*, Article ID 194524, 5 pages, doi:10.1155/2008/194524.

Mara, W., (2014). *From Gecko Feet to Adhesive Tape.* Cherry Lake Publishing, Ann Arbor, MI.

Melrose, J. High Performance Marine and Terrestrial Bioadhesives and the Biomedical Applications They Have Inspired. Molecules 2022, 27, 8982. https://doi.org/10.3390/molecules27248982.

Reidel, H., (2016). *Geckskin Applications: Spider-Man Soldiers to Home Décor.* Prescouter.

Science News, (2002). *Scientists Prove How Geckos Stick, Unlock Secrets to Making Artificial Gecko Glue*. Lewis & Clark College, https://www.sciencedaily.com/releases/2002/08/020828063412.htm (accessed on 17 August 2023).

Seibel, A., Yildiz M. & Zorlubas, B. (2020). A Gecko-Inspired Soft Passive Gripper. *Biomimetics* (Basel). *5*(2):12. Doi: 10.3390/biomimetics5020012.

Sitti, M., & Fearing, R. S., (2003). Synthetic gecko foot-hair micro/nano-structures as dry adhesives, *J. Adhesion Sci. Technol., 17*(8), 1055–1073.

Stark, A. Y., Subarajan, S., Jain, D., Niewiarowski, P. H., & Dhinojwala, A., (2016). Superhydrophobicity of the gecko toe pad: Biological optimization versus laboratory maximization. *Phil. Trans. R. Soc. A, 374*, 20160184. http://dx.doi.org/10.1098/rsta.2016.0184.

Stewart, W. J., & Higham, T. E., (2014). Passively stuck: Death does not affect gecko adhesion strength. *Biol. Lett., 10*, 20140701. http://dx.doi.org/10.1098/rsbl.2014.0701.

The Indian Express, (2017). *NASA's Gecko-Inspired Robotic Gripper May Help Clean up Space Debris.* PTI, Boston. https://indianexpress.com/article/technology/science/nasas-gecko-inspired-robotic-gripper-may-help-clean-up-space-debris-4728687/ (accessed on 17 August 2023).

Wang, L., Hui, Y., Fu, C., Wang, Z., Zhang, M., & Zhang, T., (2020). Recent advances in gecko-inspired adhesive materials. *J. Adhesion Sci. Technol., 34*(21), 2275–2291. https://doi.org/10.1080/01694243.2020.1760478.

Wang, W., Liu, Y. & Xie, Z. (2021). Gecko-Like Dry Adhesive Surfaces and Their Applications: A Review *Journal of Bionic Engineering 18*:1011–1044: https://doi.org/10.1007/s42235-021-00088-7.

Xu, K., Zi, P. & Ding, X. (2022) Learning from biological attachment devices: Applications of bioinspired reversible adhesive methods in robotics. *Front. Mech. Eng. 17*(3), 43. https://doi.org/10.1007/s11465-022-0699-x

Zhenqi, X., (2017). *Stanford's 'Gecko' Robot Could Help Clean up Space Trash.* NASA's Gecko-inspired robotic gripper may help clean up space debris, PTI, Boston, Tech & Sci., 13, 37. https://news.cgtn.com/news/3d55444d3255444e/share_p.html (accessed on 17 August 2023).

Zhou, M., Pesika, N., Zeng, H., et al., (2013). Recent advances in gecko adhesion and friction mechanisms and development of gecko-inspired dry adhesive surfaces. *Friction, 1*, 114–129. https://doi.org/10.1007/s40544-013-0011-5.

CHAPTER 5

Bats Inspired Scientific Advancements

ABSTRACT

Bats-inspired scientific investigations helping humans are noteworthy. They are used in the treatment of blood clots, in low-temperature surgery, and in military sonar to aid the visually impaired. The new technology inspired by bats using ultrasonic waves can help blind people detect obstacles. Technologists have come up with a stick ('UltraCane') that gives vibrations when approaching near objects. By sensing vibrations, blind people become alert, take care of themselves, and escape from accidents. The bat-inspired robot uses ultrasound to find its way around new environments. Scientists are analyzing the study done on bat echolocation in an effort to find biological solutions that could help in the development of new sonar tracking and navigation systems. Sonar technology is also adopted in self-driving cars with park-assist detection. A group of scientists studied how bats travel in groups so effortlessly. This investigation could help the concept of driverless cars and drones. Drones developed based on bat technology could be applied to control weather or fertilize crops and sonar tracking technologies. Using membrane wings inspired by the wings of the bat, scientists have developed new MAVs (micro-air vehicles) that have superior aerodynamic qualities and can go long distances while also being more efficient when running. A system based on bat technology can also be used to monitor water availability in fields. If there were no video cameras, the bat-like sensors might keep buildings safe from attackers, track the movements of unwell patients in nursing homes, or notify nurses and doctors. Support for the recent inquiry on SARS-CoV-2-related coronaviruses in bats and pangolins is substantial.

Bio-Inspired Technologies for the Modern World: Emerging Trends and Applications.
R. Ramakrishna Reddy & T. Pullaiah (Authors)

5.1 INTRODUCTION

*"Engineering is the art and science of harnessing the vast power of nature for the benefit of mankind."—**Thomas Tredgold.***

Mother Nature is like a huge classroom. Those who use it as an opportunity to ponder and grow will find plenty to take away from it. We can get many things widely used today that inspire how devices run in this universe. Nature has provided people with numerous ways to meet their basic necessities. There are many ways in which it can be a source of inspiration and motivation. Bats have inspired and enriched people in a variety of ways. Surgical procedures requiring low temperatures, blood clot therapy, and military sonar all make use of these technologies. They are useful for the visually impaired. These inventions offered by Bats can help blind people 'see' their surroundings using echolocation. Bats are talented in echolocation and driving under challenging atmospheres like forest cover in a simple way. Compared to bat skills, human inventions like drones are considered not superior. Lessons from bats can be used in both sonar and radar systems because of their breadth.

Dracula, an anticoagulant derived from the saliva of vampire bats, is yet another lesson learned. Stroke and heart attack victims can benefit from this drug's ability to dissolve blood clots and thin the blood. Only three of the 1,300 different species of bats are known to feed on human blood. Research on bat immunity and endurance is another promising topic for study. Despite their diminutive stature, bats are able to survive for a long period of time. The mouse will live for a year or two. It reveals the difference between life spread and frame mass. A particular type of Bat called insectivorous Brandt's Bat of Eurasia weighs only about 4–8 grams. This body mass and the life span difference between the two is extreme for all mammals. The North American Little Brown Bat has a 30-year life expectancy.

Exactly how do bats manage to survive for so many years? They can live a long time because of their unique biology and adaptations, including hibernation and a low reproductive rate. Bats' ability to fly is the key to their lengthy life spans. Around 1,000 times per minute is the rate at which the heart of a bat beats when it's flying. After just a few short hours, its metabolism jumps 15-fold. For short runs, a running rodent's metabolic rate is seven times higher than that of a sedentary one. Free radicals are formed as a result of an increased metabolic rate. If it weren't for the bat's molecular repair team, it would enhance the bat's DNA. Bats' lengthy lifespans can be attributed to their ability to withstand DNA damage. Bats are extraordinarily resilient to viruses and other infections, and they don't appear to get cancer.

5.2 FROM BATS, TECHNOLOGICAL INNOVATIONS

There is a great deal of progress in the field of biomimicry because of bats (https://discoveries.nsf.gov/?cntnid=125883). Cutting-edge technologies take their cues from the latest advances in science. When it comes to nature, it relies on what is available. In part, echolocation in bats is responsible for the development of shipboard sonar and ultrasound. To track their fast-moving prey at night, most bats utilize echolocation to avoid colliding with trees, buildings, or other obstacles while engaging in daring aerial dogfights and high-speed chases. As an example of how bat echolocation works, consider the following example: An echo is created when the bat's sound travels through the air and is reflected off of adjacent objects. A bat may generate a three-dimensional map of its environment by comparing the delay of reflected sound and the structure of echoes of the original sound to infer its distance from the objects and the size and shape of those items. Bat echolocation is so sensitive that a bat can fly at a speed of 25 mph even though its brain is only the size of a peanut. In full darkness, bats would be able to identify changes in echo delays of less than a microsecond, according to Seth S. Horowitz's "Universal Sense: How Hearing Shapes the Mind."

How do bats locate their prey using sonar echoes? Will they be able to concentrate without being annihilated by the echoes of other objects? The above-mentioned inquiry about the most recent bat echolocation study has an answer, thanks to an NSF video. In addition, the ability of bats to fly provides promise for future use. The only mammals capable of flying are these creatures. In just three wing beats, the bats' aerodynamic collection switches flight direction 180° while flying at full tilt. Bats are graceful fliers thanks to their wings' dexterity. A fighter pilot would be envious of the flexibility of their skin and the muscle power of their wings, said Horowitz. Aeroplane maneuverability may be improved in the future thanks to ongoing research into the structure and physics of bat wings.

5.3 ECHOLOCATION

Acoustic pulses are sent to a target, and the echoes it returns are picked up by echolocation (Griffin et al., 1960), an auditory method for locating and identifying objects. Many mammals, including dolphins, whales, and bats, use echolocation, according to Fullard & Dawson (1997). An echolocation is a form of sound communication that is similar to a sonar or radar. Ultra-high frequency noises are used to navigate and locate prey by echolocation. Bats and marine mammals are able to "see" using their ears. The animal's "picture" of its surroundings is formed by the echoes that return. Species-specific "Auditory Imaging Systems" are available. The Bats' echolocation is superb. Using the echoes of sounds, vocalizations are able to create 3-dimensional information. By expelling air from their nasal cavities, dolphins generate sounds. Afterwards, they focus on making sounds with their melon. The dolphin's melon has a fatty patch on its front. You can use it to focus on specific sound waves. The dolphin's lower jaw is then able to receive the sound waves once they are reflected off of the objects. Prey can be found at considerable depths with little or no light when echolocation is used. During their dives, whales employ a variety of safety measures. The clicks get more frequent as the depth climbs. The whales' clicks are close together and sound like a continual buzz as they approach their target. The average distance from the beginning of the clicking to the first buzz is roughly 390 meters.

In order to navigate, bats use echolocation (Fullard & Dawson, 1997; Moss & Surlykke, 2001) (flying around at night). As a result of their extraordinary ability to fly in total darkness, bats' echolocation has been the subject of several scientific studies. Bats are able to catch their prey thanks to echolocation (Figure 5.1). Bio-sonar signals are the means through which they communicate. Bat echolocation cries, for example, are awe-inspiring examples of "great design." The bats' ability to locate prey (such as small insects) will improve thanks to the development of echolocation-assisted vision. When prey comes close, each bat adapts its sounds to best serve the task at hand, changing its shape and frequency accordingly. In 1960, Griffin witnessed the first recorded echolocation of a bat. Researchers studying bats' biology and physiology have also been intrigued by their capacity to fly. Engineers are studying bat echolocation activity in order to adapt what they learn to sonar technology.

When an animal uses echolocation, its sounds are typically in the ultrasonic range of 20 to 200 kilohertz (Griffin et al., 1960). The human ear can detect sounds up to a frequency of 20 kHz. These calls can be categorized based on their frequency, intensity (dB), and duration (ms). Emission of high frequency sound (ultrasonic sound, roughly 20 kilohertz) is used to detect the presence of things (including food) by their echoes. Those species, such as bats, which must hunt in the dark, can use it as a substitute for their eyesight. The larynx (voice box) of insectivorous bats of the suborder Microchiroptera produces these high-frequency sounds.

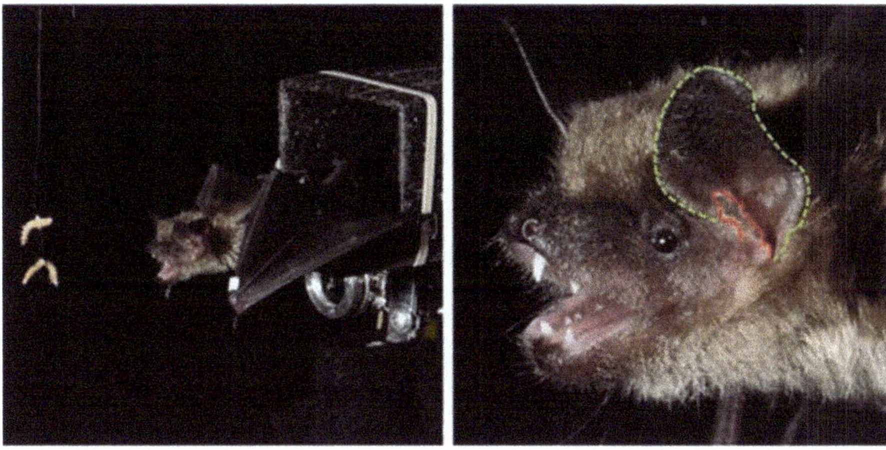

FIGURE 5.1 Bats are trained to perch on a platform and produce echolocation calls to track and intercept approaching targets.

Source: Reprinted with permission from Diebold et al, 2020. Copyright © 2020 by the authors. http://creativecommons.org/licenses/by/4.0/.

5.3.1 WHAT IS BAT ECHOLOCATION?

In the dark, bats are able to see. Mouth or nose-produced sound waves. The bat's hearing receives echoes from the item, which helps it form a picture of its surroundings (Kue, 2013). Humans cannot hear most of these shrill, high-pitched sounds.

5.3.2 WHO IS THE FIRST BAT DETECTOR MAKER?

A heterodyne detector was first created in the early 1900s by George Pierce. In 1944, a New York-based scientist by the name of Donald Griffin researched bats. Echolocation is a phrase he coined. In the beginning, he was unable to distinguish bat echolocation calls. A lack of Pierce's support would have made it impossible for him to analyze the sounds of crickets and birds in the ultrasonic range. Sonic amplifier was the first heterodyne detector with an audio output capable of detecting birds' calls. In the Netherlands, a researcher had already discovered that bats use sound to distinguish in the dark. On the other hand, Donald Griffin and Robert Galambos used the heterodyne detector to discover the first echolocation noises. Even while Pierce's detector couldn't record sound length, it could record repetition rates, and better microphones would be developed later. In the mid-1960s, the first commercially available detectors were sent.

5.3.3 THE RESEARCH ON BAT ECHOLOCATION

When it came to studying bat echolocation between 1940 and 1960, only a few scientists were interested. Griffin (1960) experimented with this idea in great detail. In addition to the Constant Frequency (CF) cries made by several bat species, he was familiar with the frequency-modulated (FM) sound wave structure. Further investigations by Robert Galambos and his team have demonstrated bat sonar or echolocation. Experiments were conducted in order to see how well the bat could avoid obstacles when its other senses were impaired. Additionally, Griffin and Pierce discovered that bat cries have directional cues. The bat's echolocation call may be picked up by placing the detector immediately in front of the animal. In addition, it was discovered that various species echolocate in distinct manners.

Echolocation sounds are made by most microbats using the vocal cords located in their voice boxes or larynxes. Some fruit bat species use their tongues to produce noises. Oral communicators, such as the pipistrelle bat, fly with their lips open to provide echolocation. Horseshoe bats fly while their lips are closed. Bats have a unique ability to support and echolocate. People's attitudes toward bats shifted dramatically as a result of these discoveries, from one of dread to one of interest. Because they lack ultrasonic orientation, the mega-chiropterans are unable to find their way. Fruit, flower nectar, and other sweet substances are all acceptable food sources for them.

It is likely that *Rousettus* and *Epomophorus wahlbergi* will make noise. Ultrasonic parts are designed by clicking the tongue rather than the larynx. Only bats have ultrasonic direction noises, making them unique among mammals. Bats are born with the ability to judge the speed of sound. In the dark, they avoid colliding with trees by using echolocation. Bats seem to be born with the ability to echolocate, unlike birds and lions, which both must learn to do so (Grinnell, 2018; Lewis, 2014; Brigham et al., 2004; Yin, 2019) and also detect the speeds of sound? Experiments on adult Kuhl's pipistrelle bats were carried out by Tel Aviv University scientists (Eran Amichai and Yossi Yovel). Bats (*Pipistrelle kuhlii*) may fly into a chamber filled with a mixture of Helium and Oxygen, where they can rest. When compared to the other gases in our atmosphere, Helium has a lower density. As a result, the sound here travels more quickly. The bats' echolocation was thrown off by the Helium's interference, which caused them to miss the perch entirely.

It is possible for bats to produce supersonic noises using a highly complex system. These reverberated when they struck an object, causing a sound effect. At the time of flying, bats create ultrasonic sounds with a frequency of 1,00,000 cycles per second. It's like having a radar gun in your hand. Hearing ranges from 20 hertz to 20,000 hertz, which is common for mammals. To allow social activities like mother-infant bonding, recognizing each other, and warning calls, all bats possess and use low-frequency noises (below 20 kHz – within the usual auditory range of humans). In 1940, Jones and Telling looked at how bats use echolocation (Jones & Telling, 2006). The concept of echolocation is found in a variety of birds. Research on bats has been bolstered by the discovery of this gene. From 1938 until 1942, Griffin and his Harvard University classmate Robert Galambos did groundbreaking research. Researchers have shown that echolocation may be found in a wide range of animals ranging from bats to whales to porpoises to oilbirds and swiftlets.

5.3.4 HOW BATS INSPIRED ECHOLOCATION ULTRACANE WORKS?

The GPS was inspired by animals like dolphins and bats (Sekar, 2013, https://news.mongabay.com/2013/11/dolphins-bats-and-the-evolution-of-echolocation/).

The unique ability of dolphins and bats is based on the use of Echolocation. They can hear and taste high-frequency sound vibrations. How did this happen? When a bat uses echolocation, it makes a sound that is reflected back to it. To get around barriers and get their preferred meal, these noises are made by animals sliding along the ground. Bats use a technique known as echolocation, in which they produce sounds that reverberate back to them. It aids them in learning more about the location and figuring out any problems they may run against along the way. Sonar helps dolphins travel, hunt, and gather food (Figure 5.2).

There is a plethora of new technologies thanks to the pioneering work of scientists (Sekar, 2013; Brigham et al., 2004). UltraCane is a significant ultrasonic wave-based invention that allows one to identify impediments. The sonar system is being used in underwater surveys.

It possesses an ultrasonic cry that detects ripples created by the dorsal fin of a small fish on the water's surface.

To catch the fish, the bat uses its long nails and legs. When the cane sense detects an obstruction, it vibrates, allowing the blind to "see" where they are going. Biosonar signals will give multiple pieces of information. Most bats are nighttime creatures and hunt during the nighttime hours. For the first time in history, a biology student at Harvard discovered that bats utilize echolocation to find and catch their prey. In bat echolocation, ultrasonic noises are emitted in order to form an echo.

FIGURE 5.2 GPS borrowed from Dolphin and Bats (Redrawn this Figure). Echolocation is a technique used by Bats and Dolphins.

5.3.5 WHAT IS THE ULTRACANE?

The UltraCane (Yin, 2019; Wahab et al., 2011) manages as "Bat cane." Bats navigate in total darkness using "echolocation," and Bats inspired it. It became "UltraCane" in honor of the ultrasound technology it utilizes.

UltraCane is a radical step-forward invention in first aid for the vision-impaired user. A regular white cane uses the latest electronic technology to do something honest to identify its surroundings (Figure 5.3).

The UltraCane was developed by hundreds of vision-impaired testers who wanted a user-friendly device based on the white cane. Listed below are a few noteworthy deviations from the norm: Detection of items in the path of the user is made possible using an ultrasonic echo device. It resembles the side mirrors of parked cars by holding objects at a set height. In order to avoid conflicting with other risk information, it provides precise information, but not in the audible range. The cane's ergonomic design has a variety of buttons for indicating the location of potential obstacles. The strength of the vibration determines the obstacle's distance. You can utilize this information to determine the exact location in your brain where the problem is. To do this, you must immediately construct a mental map of your immediate surroundings.

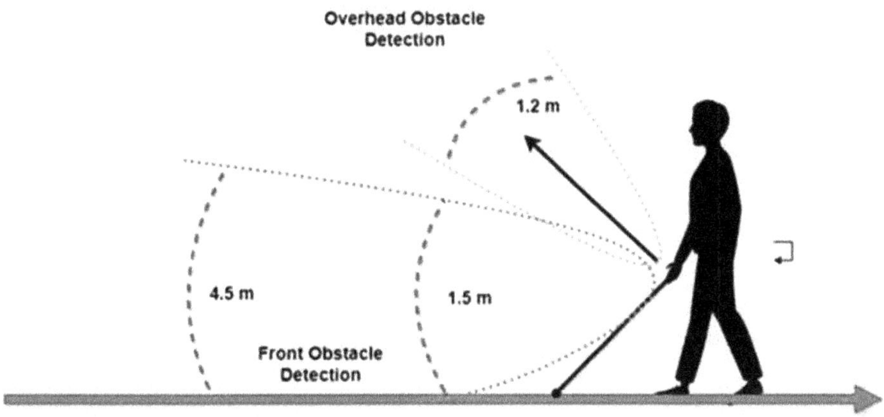

FIGURE 5.3 Ultrasonic sensors' placement on the smart stick.

Source: Reprinted with permission from Muhammad Siddique Farooq et al., 2022. Copyright © 2022 by the authors. Licensee MDPI, Basel, Switzerland. http://creativecommons.org/licenses/by/4.0/.

University of Leeds (in the UK) created the echolocation technique used by bats. They adjusted their results to develop a cane for the Blind. The UltraCane manufactures, marketed, and sold by Sound Foresight Ltd. (Figure 5.4).

In order to use UltraCane, it is necessary to have a thorough understanding of how bats behave when it is pitch black. Bats, on the other hand, are able to "see" in the dark because of ultrasonic echoes. An UltraCane alerts the visually impaired to obstacles ahead. It identifies the location of potential problems. The UltraCane has several sensors that allow the user to detect items that are higher than their head.

5.3.6 BATS INSPIRE SMART CANE FOR THE BLIND

Today, blind people rely on walking intelligent canes. But new technologies have covered how they can claim their independence—these bright canes borrow the bat's echolocation technique. 'UltraCane' from Britain and

'SmartCane' from India use echolocation to help blind people walk. They help them how to avoid more obstacles.

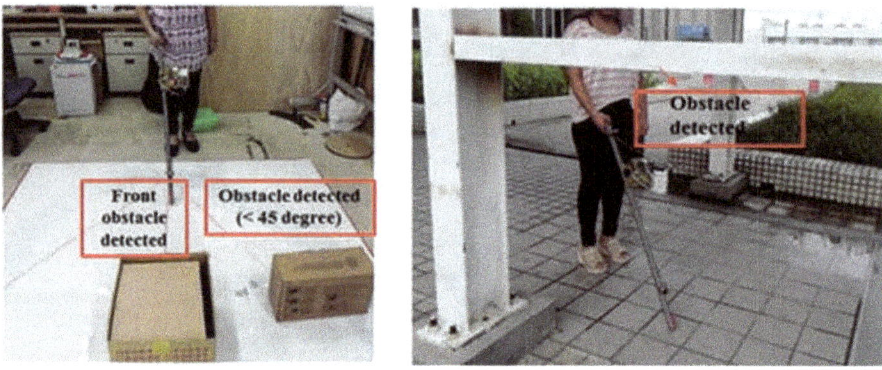

(a) (b)

FIGURE 5.4 Experiment on Visually challenged person using ultrasonic sensor cane and it could detect the front obstacle: (a) Indoor testing; (b) outdoor testing (Open access Journal).

Source: Reprinted with permission from Nilima Sahoo et al., 2019. Copyright © 2019 by the authors. Licensee MDPI, Basel, Switzerland. http://creativecommons.org/licenses/by/4.0/

Transmission and reception of information is carried out by UltraCane and Smart canes by means of ultrasonic waves (Yin, 2019; Kim & Cho, 2013; Wahab et al., 2011). These canes use ultrasonic waves to deliver and receive sound waves. Meenakshi Balakrishnan, a professor and lead engineer at the Indian Institute of Technology, is the inventor of this new technology. Changes in the environment can be detected by the handle's vibrating motions. The closer the user is to the obstacle, the faster the vibration signal. The sound waves in these canes can detect obstructions above the knee, which makes them superior to traditional walking canes. UltraCane costs approximately €735 (as of 2011), but SmartCane is only about €36. Most of India's blind people can afford it. Walking canes and seeing-eye dogs are essential tools for blind people on their journeys across the world (Figure 5.5).

5.3.7 ULTRACANE–AN AWARD-WINNING PRIMARY MOBILITY AID, IS NOW AVAILABLE

Blind people's first line of defense is the UltraCane, a product protected by a patent. Assistive technology that allows safe obstacle avoidance has taken a huge step forward.

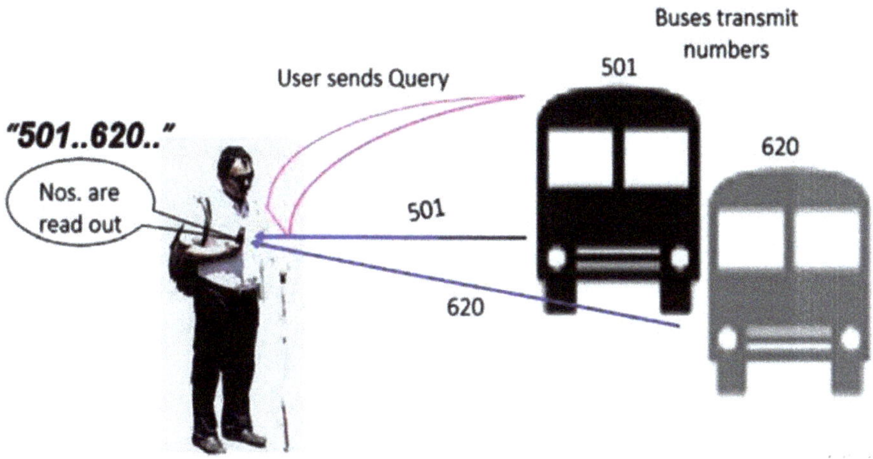

FIGURE 5.5 User presses the query button (user module): (1) RF query is sent to all buses in the vicinity and buses respond with their route numbers. (2) User module reads out all the route numbers one by one.

Source: Reprinted with permission from Balakrishnan, 2021. http://creativecommons.org/licenses/by/4.0/.

Blind people may now navigate their way around buildings, pavements, and even mountain bike trails with a new BBC television show (https://www.leeds.ac.uk/news/article/3329/bat-inspired technology showcased). 'Miracles of Nature' on the BBC featured a blind mountain biker who used the same bat-inspired technology to successfully navigate the course. At the University of Leeds, researchers are working on a similar concept.

The patented invention is Blind people's major electronic mobility assistance, which is the UltraCane. As a "step ahead" in support technology, it helps users to safely avoid obstacles. This technology enables the user to navigate around impediments in their forward route, as well as protecting the user's head and chest (Anand, 2014). The UltraCane uses state-of-the-art narrow beam technology to detect traffic barriers within 2 or 4 meters. Ultrasonic waves are sent from two sensors in order to detect objects. It can also catch up to 1.5 meters in front at chest/head height. In order to obtain information, the user can employ this strategy while maintaining control of his thumb. Vibrations from two buttons trigger a mental "mind map" of the barrier's location and direction. The user is able to get data while still avoiding potential hazards.

5.3.8 PEOPLE BEHIND SMART CANE

Statistics in the first World Vision Report] released by the World Health Organization (WHO) in 2019 show that more than 2.2 billion people are visually impaired or blind worldwide. Among many guide equipment for the blind, the intelligent cane, also called smart blind cane, or smart cane, is highly valued by the visually impaired due to its stable structure and high practicability. Laser sensing and vision sensing smart canes can improve the convenience of travel for the visually impaired. Aya Dernayka etal (2021) with their method used and checked whether or not the cane can be used to detect under standard walk conditions at different distances (Figure 5.6).

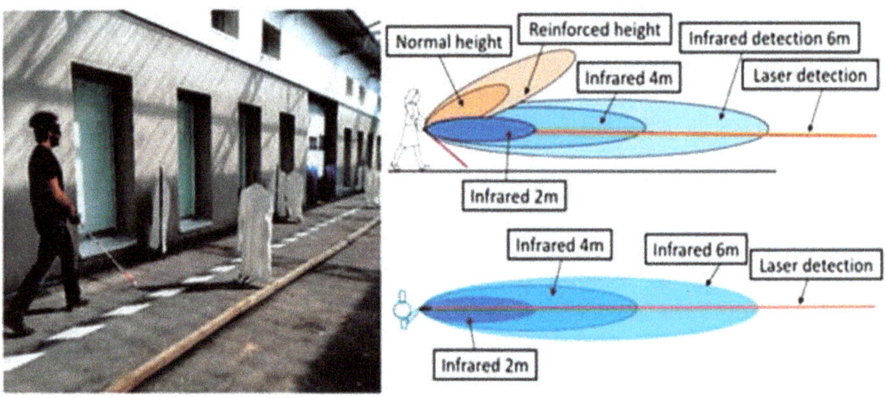

(a) (b)

FIGURE 5.6 (a) A visually impaired holding the Tom Pouce III laser smart cane; (b) the detection range of the smart cane.

Source: Reprinted with permission from Dernayka et al. 2021. Copyright © 2021 by the authors. Licensee MDPI, Basel, Switzerland. https://creativecommons.org/licenses/by/4.0/.

Technology for the blind is the focus of Assistech (IIT-Delhi). A new technology has been used by SmartCaneTM. An IIT and Rhodes scholar, Rohan Paul, developed SmartCaneTM for the blind, a gadget that allowed them to travel freely and safely. From a distance of 3 meters, it is able to identify any obstacles above the knee. Rohan Paul, a man from India, was the first to implement this technology. He wants to get the product out of the lab and into the hands of the public. Research and marketing help was provided through a network of organizations by Saksham director Mr. Dipendra Manocha and the president of the National Association of the

Blind in Delhi. Phoenix Medical Systems, the industrial partner of Smart-CaneTM, has completed the R&D and construction of the product, as part of the R&D for Affordable Healthcare in India, the Wellcome Trust has given over $4,50,000 to IIT Delhi to further develop SmartCane™.

The unique device features are as follows:

- Variability of holding grips;
- Adjustment for angles;
- Detachability of the cane;
- Robust design;
- Small form factor;
- Affordability.

A novel method of locating objects is being developed by a researcher based on the design of bats. Rolf Mueller, a mechanical engineering professor at Virginia Tech, invented bio-inspired technology by studying the inner workings of a bat's ear. The source of a sound can be pinpointed with this technology. Using this technology, Mueller has come up with a simpler and more accurate concept of where a good site is. For this technique, a human ear serves as a model. For the first time, he's doing an investigation to discover how to locate sounds. These findings can be found in Nature Machine Intelligence. "Bats have always had a special place in my heart because of their extraordinary ability to navigate difficult natural environments using ultrasound. It's possible that the animal's ears are to blame for this," Mueller chimed in.

5.3.9 CAN BAT SONAR INSPIRE BETTER SELF-DRIVING CARS?

Animal subjects of identification and return give better biological answers, from which we can encourage new technologies primarily for services in critical and loud atmospheres. By analyzing target identification and interception research by echolocating bats, Diebold et al. (2020) focused on biological solutions. It directs new ways to substitute sonar tracking and travel ways. The bat species adapt wideband echolocation signals locate thick forests and seek false insects in the dark. The bats show progressive changes in flight trajectory. A variety of assignments like tracking and recovery reveal biological solutions. By reviewing research on target following and interception by echolocating bats, the investigators (Diebold et al., 2020) aim is to specify biological solutions that could find new ways to substitute sonar tracking and shipping systems. Bats employ wideband echolocation sounds to navigate dense woodlands and hunt for elusive insects in the dark. The design of the sonar beam and the echolocation signal will be crucial to the success of these creatures.

Many creatures use a unique travel system called echolocation, out of which whales, dolphins, porpoises, and birds, bats are noteworthy. The aforementioned creatures and birds rely on their hearing to navigate and track down objects in the dark using their senses of sight and hearing. Sonar technology is the product of this unique ability. Sonar is also used in parking assistance and rear bumper detection systems. In huge groupings, bats move without slamming against each other. This observation initiated the scientists to invent self-driving cars. How do bats travel in groups so effortlessly? This topic was investigated by researchers at Saint Mary's College of Indiana. Inspired by the pattern of driverless automobiles and drones, it encourages new ways of thinking. "Before, people believed the bat was not adept at echolocation and only made these simple clicks," says Wu-Jung Lee (Scientist at the Applied Physics Laboratory of the University of Washington) (Lee et al., 2017) (Figure 5.7).

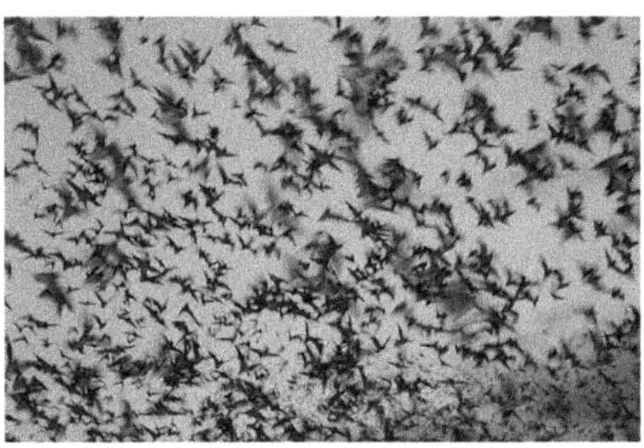

(A)

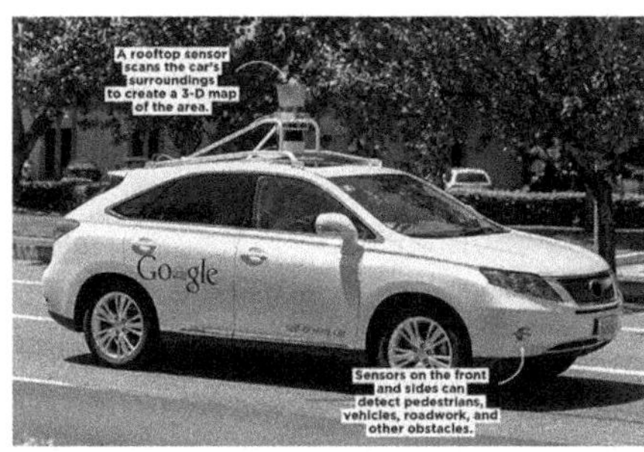

(B)

FIGURE 5.7 (A) Bats are traveling without hitting each other; and (B) a driverless car.

Image courtesy: A) Dolovis at Wikimedia Commons. https://creativecommons.org/licenses/by/2.0/deed.en

Image courtesy: B) Eric Risberg/AP Images from https://junior.scholastic.com/issues/2017-18/112017/are-driverless-cars-a-good-idea.html?language=english.

When self-driving cars first appeared in science fiction, they were a rarity. Recently, they have become a reality in many American cities and can be found there. Hundreds of self-driving cars are being tested by Uber on public roads in Arizona, California, Michigan, Pennsylvania, and elsewhere. Laura Kloepper, an Assistant Professor of biology at Saint Mary's, says it's an "impossible undertaking" to keep bats from colliding with one another (Duke, 2021). When bats cluster in large quantities, their signals are said to interfere with each other and produce confusion. Observing how bats obtain their resources may help humans improve self-driving automobiles. Using sonar to gather data will be advantageous to researchers. Radio waves are used by radar, which is a piece of technology.

With various ideas, Kloepper and her team (Duke, 2021) tried to know the information when they entered a group of bats into a colony. In their attempt, they gathered relevant details. They depend on drones and arrange recorders on zip lines through the cloud. The scientists also used full sterile suits and respirators to enter bat caves and conducted experiments. They wish to know how bats can fly nearby without throwing each other off with overlapping sonar and colliding. Self-driving cars begin with a broad spectrum of technologies. These technologies include a variety of devices like cameras, lasers, sonar, and radar. LIDAR (Light Detection and Ranging) mounted on the car's roof is used by Google's self-driving car. Much as bats send out signals, the LIDAR sends out beams. They bounce back and then detect. Google employs a laser that rotates 360°. It captures over 1 million messages in a second. They are accurate up to 100 meters (about 328 feet). Google's cars also use cameras with overlapping fields of vision and bumper sensors.

5.3.10 WEATHER AND CROP POLLINATION ARE TWO OF THE MANY USES FOR BAT-LIKE DRONES

An engineer at Northeastern University, Ramezani, has conducted some research on bats' capacity to stand on their hind legs. An inquiry into the capacity of a robot to hang its feet up and preserve energy conduct like bats. Bats spend their entire lives hanging upside down, including birth and reproduction. They spend most of their time doing this. The Egyptian fruit bat is at the heart of autonomous vehicles and unmanned aerial vehicles (UAVs). For example, this bat employs modern-day surveillance tactics in its navigation. New research reveals that this bat uses navigational tactics similar to those used by modern military and civil surveillance. The fruit bat's high-pitched squeals are distinct from those of other bats. It makes clicks with its tongue, similar to those made by a dolphin. When it comes to vision, fruit bats are able to move between bright and dark environments with ease. Bats perform some of the most impressive flying maneuvers in the animal kingdom in order to arrive at that dangling point. A bat's perch can last for hours after a quick midair turn. When it's time to fly away, they merely have to let go and flap their wings.

Scientists have been copying nature to develop pollination drones that can assist real bees and human farmers.at can assist and human farmers. Very small drones are developed in labs around the world to tackle this global issue by replicating bees' role in pollination. Bat-inspired drones could be used to monitor weather or pollinate crops. Scientists are creating a new generation of ultralight drones that are inspired by the aerodynamics of birds and insects Plant protection unmanned aerial vehicles (UAVs) play a crucial role in agricultural aviation services (Nahina Islam *et al.,* (2021)).

Drones based on bats can be used to control the pollination of crops (Ramezani et al., 2017). A new generation of lightweight drones has been developed by scientists. They are inspired by birds and insects' aerodynamics. Traffic, weather, and even product pollination could all be controlled by tiny drones in the near future.

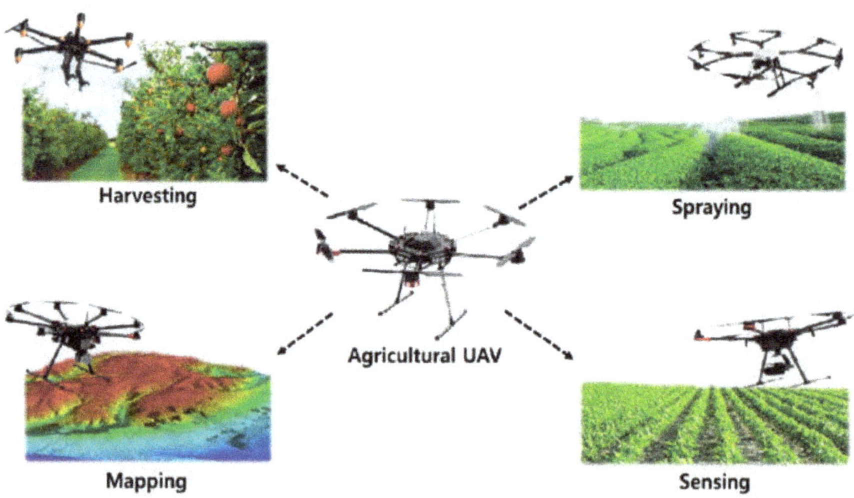

FIGURE 5.8 Different types of agricultural UAVs (Harvesting UAV, Spraying UAV, Mapping UAV, Sensing UAV)].

Source: Reprinted with permission from Nahina Islam *et al., 2021.* Copyright © 2021 by the authors. Licensee MDPI, Basel, Switzerland. http://creativecommons.org/licenses/by/4.0/.

A drone that mimics the flying experience of bats was proposed by Alireza Ramezani from Northeastern University. Professor Ramezani said that bats "have arguably the most advanced flight mechanisms of any animal." Alireza Ramezani saw this unique point of the bats and came up with an idea to design drones that can mimic complex aerobatics and free energy. Bat landings are a lot like freestyle swimming flip turns. One of its wings is tucked in, propelling its weight upward, and the belly is turned out. The bat's locomotive activity during landing was examined by Alireza Ramezani, who then developed a robotic mechanism to mimic it. The bat's mechanical structure can be seen in this device. Harpoon is the name of this technology, which can be installed in drones to improve the eyesight of the drones. Drones now use a lot of energy to fly for lengthy periods of time, and they are ineffective in windy and confined environments.

Bat Bot is the name given to Professor Ramezani and his team's free drone. Its wings can beat, skip, and make rapid twists, yet it weighs just over three ounces. Unlike larger drones, this one dives to get into tight spots that would otherwise be inaccessible. Drones based on bats could be used to control the pollination of crops that need them. Even though it's barely over three ounces, it has wings that can beat, skip, and perform quick turns, all at the same time. This drone dives into narrow spaces that are otherwise unreachable, unlike larger models. This technology could be used to control pollination of crops that require it. According to Professor Ramezani, a drone of this type would be useful for surveillance purposes if it flew over city streets. He made the point that these trendy drones look a lot like quadcopters. They're built of strong materials, have rigid frames, fast-rotating blades, and propellers, and shouldn't be operated near people. Because of its carbon fiber skeleton and flexible membrane, the Bat Bot would not injure people or other things, even if it crashed with them. Moreover, the bio-inspired drone is far less bothersome. The Northeastern team is building a more advanced version of the Bat Bot that includes cameras and sensors. It is used to store and organize many types of information.

5.3.11 BAT-INSPIRED ROBOT TO DRIVE UNFAMILIAR ENVIRONMENTS

Many animals are capable of planning to move for the first time in a new environment. Researchers simulated this experience by producing a robot that shoots sound like a bat. Bats can send sound waves and collect the reflected echo information from things nearby. In order to create a map of space, they studied the returning echoes (The Robot Report Staff, 2018). The developed Robots focus on the demand for new distinct ways to help targets such as barrier clearance. These robots are able to distinguish the object and arrange their route.

Creating a map of an unknown environment is a difficult problem for many robots to solve. While also figuring out how to get around in these new surroundings. Echolocating bats will be used to find a solution to this problem. Bats are aware of their surroundings on an acoustic level (other animals also solve this task daily using a range of sensory modalities). It is through the use of sound signals and echoes that bats can navigate and map new habitats. In response to this capability, the 'Robat,' a fully autonomous terrestrial robot that completely relies on bat-like SONAR to navigate and map a new environment, was built (Eliakim et al., 2018). The robot was able to navigate a wide outside new environment and map it in real time utilizing frequency-modulated chirps at a regular bat rate with two ears and a single emitter (mouth).

Bats use echolocation to find food in the dark, sending ultrasonic waves from their lips and analyzing the echoes that return to them to plan their surroundings. Tel Aviv University researchers (Eliakim et al., 2018) have made use of this outstanding new robot capacity. Using a sonar inspired by nature, this robot is able to navigate its environment on its own. They utilize their mouths to make noises that bounce off cave walls and their ears to pick up the echoing sounds and interpret the information. Ultrasonic speakers and ultrasonic microphones are employed by this robot. Bat-like sounds can be produced with a lower volume using this configuration. It models its surroundings in great detail using an inbuilt artificial neural network. As it moves around barriers, it can determine whether it should avoid a wall or a plant, and it can even work out whether it should do so.

Before designing new sonar systems, researchers use the science of echolocation. As far as we can tell, it's the first time that technology has been integrated into a working robot. Team members hope that their work will pave the way for an entirely new category of robots. In order to navigate through varied settings, it employs this strategy. Tel Aviv University graduate student Itamar Eliakim believes that his biorobot is "the first fully autonomous, bat-like biorobot that goes across a novel area while mapping it purely based on echo information." "The boundaries of things and the free paths that connect them are the subject of this report. This study demonstrates how sound can be used in robotics in the future."

5.3.12 SONAR TRACKING SYSTEMS INSPIRED BY BAT ECHOLOCATION

The echolocation and flying execution of bats are responsible for a slew of modern technological advances (Diebold et al., 2020). Even while non-natural technologies have improved a few critical elements, many aspects of bat sonar tracks remain unaffected. Bats rapidly alter the spectro-temporal characteristics of echolocation sounds, as previously described. It is important for sonar systems to be able to alter their behavior in order to drive complicated surroundings and track and catch objects. Adaptive sonar with wideband waves and fully realized behaviors opens up a plethora of new technological options. Examples of how bats have had an impact on technology in the past are provided herein.

Using a metaheuristic algorithm is possible when using a normal Bat Algorithon (BA). There are only a limited number of parameters that can be used by the conventional BA. It also makes use of echolocating bats for this purpose (Jenson et al., 2005; Yang & Gandomi, 2012). The following are some of the idealized behaviors.

Using echolocation, bats are able to determine how far away their prey is from them, as well as define the environment they are operating in.

It is possible for bats to adjust their sonar parameters depending on their distance from a target so that they are able to catch prey at the right frequency and intensity. Call volume fluctuates between a high absolute value and a low fixed value.

5.3.13 ROBOTS CAN NAVIGATE OUTSIDE THANKS TO THE GOOD LOCATION INSPIRED BY BAT EARS

Inspired by the ears of Bats, Virginia Tech researchers Yin & Müller (2021) have developed a new method for discovering the source of sound. A synthetic ear inspired by horseshoes and leaf-nosed bats was created by Virginia Tech researchers. They hooked it to a rope and a motor that flutters the ear when it detects an approaching sound. The incoming audio waves are analyzed by a massive neural network, which is constantly being trained to determine the direction of each received audio input. According to Mueller's statement, "We intend to provide intelligent freedom to a variety of outdoor settings." Agriculture and forestry, environmental surveillance, biodiversity monitoring, and defense and security-related issues are all part of the system. According to Mueller, the old access to this wonderful position was inspired by the way the human ear works. With the ability to either gather numerous frequencies or employ multiple receivers, pressure targets like microphones are used. Based on his previous studies with bats, Mueller was aware that their ears were far more flexible than the human ear at picking up sounds. Rather than having several receivers or frequencies, his team decided to keep working toward a single frequency and distant receiver.

5.3.14 BUILDING THE EAR

Yin & Müller (2021) attempted to mimic the ability of a bat's ears by using a single receiver and a single frequency. They developed a soft synthetic ear after getting inspiration from a horseshoe. He set up a simple motor that was timed to create the ear flutter of the Old-World leaf-nosed bats. It connects to receive incoming sound signals. These particular bats' ears can receive a problematic change of sound waves. Nature's plan appears to be a good selection. This change begins with the pattern of the outer ear, announced the pinna. For the party, it makes use of the ear's motion as it hears noises to generate a variety of forms. To receive the signal frequency, Yin and Mueller had to overcome the most significant obstacle: single-receiver frequency. The incoming sound waves are converted into information by the receiving computer. They received the Ear for developing a bat-related device. Doppler shift traces were formed by the fluttering pinna's rapid motions. These signatures are from the Source and aren't easy to interpret the models.

To test the system's performance, which includes the ear and machine function, one should mount the ear on a rotating rig. It includes a laser pointer. Like the ear, the sounds released by a loudspeaker are received in many different ways. Rotating the rig is controlled by the computer once the noise direction has been determined. This time, the laser pointer hit a target linked to the loudspeaker, which pinpointed the exact location. With our ears, we normally define a location within 9° of that position. Within 7.5°, the technique produces the best results. Mueller replied, "All of this can be achieved with far less work. Environmental monitoring, and different outdoor environments, including precision in agriculture and forestry, are all made easier with this knowledge." Biodiversity monitoring, defense, and security are just a few of the many uses for this technology. When the system flushes the outer ear, Doppler shift fingerprints linked to the sound's source can be seen. Because the patterns were too complicated to solve, a neural network was employed to build the reference region for each installed echo.

Doppler shift fingerprints relating to the sound's source are observed when the system flushes the outer ear. The neural network was used to create the reference region for each installed echo since the patterns were too difficult to solve. The human-inspired systems require multiple receivers, each tuned to a different frequency. There is a lot of potential for the scientists' bat-based technology to assist robots in navigating "complex outdoor areas" (Fingas,@jonfingas, 2020) with little effort. It is possible to witness farm robots that harvest with precision, biodiversity monitors with remarkable accuracy, and military robots that traverse rough terrain with little input from humans. We can take inspiration from the most successful biomimicry in our biology, even if no tangible results are produced.

5.3.15 BAT-INSPIRED ROBOTS NAVIGATE LIKE A REAL ONE

A favorite tree foraging can be identified by a bat's echoes, which are used to categorize and identify items. Nearly the same can be accomplished by a robot. A 68% success rate in distinguishing between plants and was reported non-plants (Allen, 2018). In order to gather data, the robot must pause every half-meter for around 30 seconds. The researchers, on the other hand, believe that robots should be able to echolocate with a few adjustments, such as a speaker with a wider beam that can move. The robot could come in helpful in cases where visual systems might be unable to do so effectively. At night, when bats are most active, robots can conduct search and rescue missions in smoke-filled buildings and other hazardous environments.

5.3.16 MICRO-AIR VEHICLE WINGS INSPIRED BY BAT WINGS

Scientists have constructed new membrane wings inspired by the bat wing structure at Southampton University. Improved aerodynamics, greater range, and cheaper operating costs are just a few of the advantages that micro-air vehicles (MAVs) provide over regular aircraft. They move in reaction to external pressures, acting as if they were real muscles. Because there are no moving parts, the MAVs that connect them may be serviced more quickly. Because of the use of electroactive polymers in the wing's design, it hardens. When a voltage is applied, they contract, and the form of the electroactive membrane changes. Since the aerodynamic qualities can be altered while in flight, this is an option. The prototype wing will be able to fly for longer distances in the future than is currently possible. Developing the right wing requires a new series of experiments. They used a new set of practical tests to investigate in-flight. At Imperial College London, researchers from the University of Southampton and the Engineering and Physical Sciences Research Council worked together to conduct these studies. Air Force European Aerospace Research and Development (EAER) contributes to the effort (EOARD). MAVs have a circumference of 15 centimeters and are used for a range of civil and military purposes. The design of MAVs is increasingly being influenced by natural elements. In comparison to small drones, these vehicles can fly farther and have more maneuverability because of their wings.

An emphasis has been placed on mimicking bat biology by the Southampton-Imperial team. To speed up the planning process, the Imperial team devised new computational rules. Some of these choices and tests are part of the Southampton team. In their design, they are intended to glide over the ocean's surface. A test MAV, complete with groundbreaking "bat wings," was built with their assistance. The car is put through its paces in a nearby coastal spot as part of extensive testing before it goes on sale. In the next step, the above team incorporated active wings into the standard MAV designs. In the next few years, they could be put to use in real-world scenarios.

5.3.17 BAT-INSPIRED SPY PLANE

A bat-shaped, six-inch robotic spy plane would collect information on urban combat zones by analyzing the sights, sounds, and scents there. University of Michigan College of Engineering are developing a six-inch robotic spy plane, modeled after a bat (Figure 5.9). The US Army thinks so too, and they've given $10 million over five years to the College of Engineering at U-M to make it happen. This contract, which has an option to be extended for another five years and $12.5 million (Yoneda, 2008; Defense INSIDER, 2016), would fund research into microelectronics and Biomimetic Advanced Technology at a university in the United States.

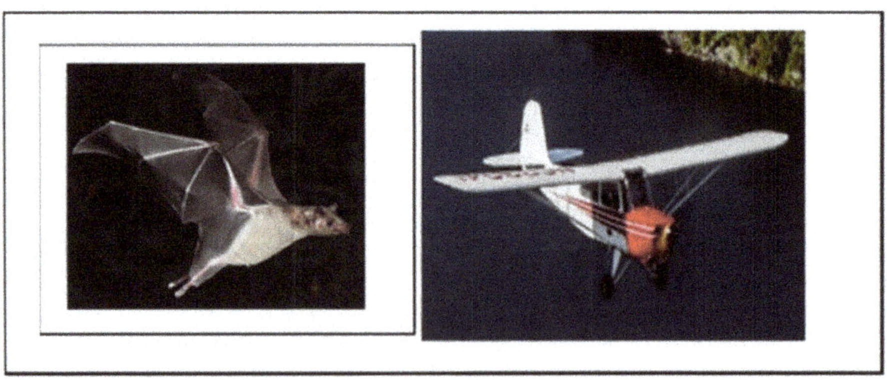

FIGURE 5.9 Bat Inspired Spy Plane. (Redrawn from Kristina Delp on Unsplash.

Source: https://unsplash.com/photos/white-and-red-airplane-flying-over-green-trees-during-daytime-BeoPmdh9Fmo.

The United States uses the U-2, a single-seat, high-altitude jet aircraft, for espionage, surveillance, and reconnaissance missions. The U-2, also known as the Dragon Lady, has been in use by the CIA since 1956 and is often considered the most iconic spy plane ever built.

Sights and sounds would be gathered and transmitted to a soldier via a six-inch robotic spy plane that appeared after a bat. Together, they created a small, autonomous aircraft that could detect and communicate with its surroundings. Planned short-term missions for robotic bats will aid advanced troops on battlefields. As the director of COMBAT (Center for Objective Microelectronics and Biomimetic Advanced Technology) stated, their team is pushing our technologies to their limits in order to reach functionality that was previously unimaginable for them to do. The COM-BAT project included the University of New Mexico and the University of California, Berkeley. There are a number of subsystems in the robotic bat that each university is responsible for creating.

Researchers at the University of Michigan are focusing on microelectronics (Yoneda, 2008). All of these components will be made by them for the new "Bat" micro-aerial vehicle. In the future, engineers hope to create miniature stereo cameras, miniature microphone arrays, and miniature radiation and poison gas detectors. Low-power miniaturized radar and a vulnerable navigation system will allow the robot bat to navigate in the dark. Radio signals and sun power will renew the bat's lithium-ion battery, and it will communicate with soldiers via radio waves. Our sensor must be able to provide high-resolution navigation and sensing even in the dark, exactly like bats do, Sarabandi said in an interview. In order to find their way around, bats use echolocation, a method in which they make noises and listen for echoes. The problems of systems dependent on visual signals are no longer an issue for science. The robot will have a maximum length of six inches and a maximum weight of about a quarter pound. It's estimated that this device will utilize 1 watt of power.

University of Michigan researchers (Yoneda, 2008) are focusing on improving current technology. Solar cells with quantum dot technology will have two times the output of current solar cells. For the robot's autonomous navigation system, which would guide its motions, it needs a system that is 1,000 times smaller and more energy efficient. A communication system that is 10 times smaller and lighter will be developed if the planned upgrades are successful. It's more energy-efficient than anything that's currently available. Research on tiny air vehicles has recently been reviewed by Ward et al. (2017). The goal of most scientists is to come up with new technical advancements. An expert in the field, Sarabandi, suggested that it might be used to solve a wide range of technical problems—ranging from medical to industrial. Researchers at Harvard University have created a micro-robotic fly. Both the Lemur and the rat robot, both designed by NASA for use in rescue operations, have also been featured on TFOT recently.

5.3.18 BATS-INSPIRED TECHNOLOGY TO LOCATE WATER AVAILABILITY IN FIELDS

Nature provides a wealth of instances of active sensing that are both fascinating and unusual. Echolocation, for example, has evolved over millions of years to allow some creatures to actively sense their surroundings. Synthetic systems are often envious of their unique in-built sensing processes. Instead of relying solely on their eyesight, bats employ a variety of techniques that are similar to those used by other animals. Sound waves allow them to see in the dark.

It's common knowledge that bats don't like anyone and have a skewed public image. Fruit on branches and flying insect prey ricochet back to the bats' ears (EARSC, 2021). People find bats disturbing rather than inspirational because of their bloodthirsty nature. However, Dynacrop's soil moisture monitoring service would be unable to operate if they weren't there (EARSC, 2021). In a variety of conditions, neither humans nor animals are able to discern barriers in front of them. Active remote sensing techniques such as echolocation, sonar, and radar can be used in a variety of contexts, such as a giant tree, a stone wall, or an impending drought. The underlying concept is straightforward. A signal or pulse of sound can be sent if you can't see. Pay attention to what comes back. Modern radar uses the same technique bats use to locate prey. During World War II, this approach was developed in secrecy for the military. The employment of radar for friendly purposes across international borders is now commonplace. While flying, bats are able to track their prey with pinpoint accuracy. Radar is also used by meteorologists to keep tabs on rain and wind. Radar is increasingly employed in modern cars to avoid collisions with other vehicles.

The idea that satellites are spying on us from space is no longer science fiction. For the most part, you use satellite photos on a daily basis to create maps for your mobile devices. There are a variety of satellites out there, some of which produce what we may call photos, while others use entirely different methods. One of these is bat-like technology, which may come as a surprise to you. We can now monitor natural occurrences all across the world using a satellite-mounted radar sensor. It's all due to the same enchantment. There is a constant exchange of electromagnetic impulses between the satellite and Earth. Copernicus Sentinel 1A and 1B, operated by the EU and the European Space Agency, are two examples of such satellites. Everybody who wishes to access Sentinel 1's radar data has been able to do so for free since 2014. In the world of radar satellites, things are anything but straightforward. The wavelength, polarization, angle, and other characteristics of a satellite signal are all important considerations. When it comes to analyzing these kinds of satellites data, you can imagine how difficult it is. Even after numerous rounds of pre-processing, it still seems this way.

5.3.18.1 DYNCORP SERVICE IN MEASURING SOIL MOISTURE

Technologists can be a godsend to farmers in dire situations like drought, irrigation, or precise fertilizing and seeding tactics. They started a new set of monitoring tools to anticipate how much water would be available for crops in the field ahead of time. The soil moisture index (SMI) is a new service provided by Dynacrop to farmers. Because dry terrain reflects radar satellite signals differently than wet soil, scientists can use this to their advantage. We can determine the present and actual moisture levels of any field in the world using radar data from Sentinel 1 satellites (EARSC, 2021). By analyzing this data, it is possible to identify areas where crops have a greater chance of obtaining additional water. Changes in seeding and fertilization methods are implemented as a result of this information by farmers to boost yields. Even in overcast weather, the service is available. Radar satellites, like optical satellites, can make judgments even if there are clouds in the sky. You may always send a signal and see what comes back if you design technology that is tested on bats.

5.3.19 LESSONS LEARNED IN ECHOLOCATION FROM BATS

BIAS (The Basic Technology program for Biologically Inspired Acoustic Systems) plans to ask for that call. It included scientists and engineers from Leicester, Southampton, Edinburgh, Strathclyde, and Leeds. Other disciplines were also involved, including animal acousticians and mathematicians; signal processing experts; sound engineers; geological and medical scientists; and acoustic engineers, all of whom collaborated. Animals alter their echo locator calls and release them in order to assist them in finding their way around and catching prey. Their communication methods are extremely diverse, and they are all based on sound pulses with set and varying frequencies, many lasting less than a millisecond (Rees, 2011). These calls were recorded and evaluated by the BIAS team, which has preserved them in a call database. Rees (2011) aimed to replace the signals recorded by microphones as bats fly past with recorded bat echoes. As a result, one can create a small, lightweight recorder that can be fastened to the backs of adult Egyptian fruit bats and record their movements. Biosonars used by whales and bats were developed independently and differ greatly in many ways. Prestin-coded protein sequences from the bottlenose dolphin and microbats show that the two groups are closely related, even though they are distinct species.

5.3.20 BAT PACKS

Using a microphone to record the noises bats make as they fly has revealed some of the methods they use to grab prey, such as the utilization of relevant frequencies. These kinds of tests led us to recognize that we needed to pay extra attention to the complicated changes in sound waves as they alter over time. Rees (2011) built their signals and new methods to help humans understand such alterations in animal signals because it isn't straightforward to analyze them. In terms of accuracy and success, these signals are superior. Early distance measurement devices and the most noteworthy ones. They resemble the organisms' emitted light in appearance. Some remarkable results have been achieved in a variety of sectors thanks to the usage of these bio-inspired signals by researchers (Rees, 2011). A good example of this is the use of high-frequency sound to transmit and understand signals resembling those of bats and dolphins in human tissues for cardiac imaging measurements. Solid and granular geological materials were studied using lower frequency signals by Rees (2011). In the short term, the research has resulted in the creation of many innovative transducers, which are devices that transmit and receive bio-inspired signals. Standard competitors can't match their increased sensitivity and bandwidth.

The scheme has made substantial progress in the investigation. This new information is for real-world applications. It gives objects a sense of identity and a sense of purpose. The NDE (non-destructive assessment) of elements is a well-known example. To see the predicted content of cargoes, select the cargo's content from the drop-down menu. Additionally, robotic positioning vehicles working in noisy environments, such as nuclear reactors, could benefit from this technology. Deaf persons could benefit from research into better hearing aids or cochlear implants that can recognize objects. Furthermore, the BIAS team is currently investigating environmental science applications, including the characterization of underwater silt.

5.3.21 SMARTPHONE LOCATOR WITH A BAT-INSPIRED INTERFACE

To follow and intercept erratic prey from interrupted sensory input, bats use echolocation to compute their three-dimensional (3D) position. This raises questions regarding the tactics they use to track and intercept prey from interrupted information (Wilkins, 2021). By making noises and listening for the echoes that follow, bats are renowned for locating objects. Humans, on the other hand, can mimic bats in many ways. Glasgow University researchers have developed a machine learning method that can be used with cellphones and PCs to do the same procedure. Investigators create a virtual representation of their immediate surroundings by sending out notes and measuring the echoes that return. Without the use of security cameras, the bat-followed device could assist in keeping buildings safe from intruders. Using a bat-followed device, nurses and doctors can keep tabs on the whereabouts of elderly or critically ill patients in nursing homes and be alerted when assistance is on the way.

In order to create 3D visuals from a single source of sound, the technology relies on advanced computer processing methods; the method might work on any device with a speaker and microphone. "Echolocation in animals is a fantastic capacity, and science has succeeded in reproducing the ability to generate three-dimensional images from reflected echoes in various different ways, like Radar and LiDAR," stated Dr. Alex Turpin. In order to produce three-dimensional images, this research relies solely on information obtained from a single input, namely, data obtained from the microphone or antenna. First of all, we believe that if you have either of these pieces of kit, you can use our method to turn any gadget into an echolocation device.

Because of this, the price of 3D imaging might be drastically decreased, allowing for a wide range of new uses. Clearly, there is a lot of potential for sensing the world in new ways. There is a lot of excitement about the prospect of generating additional high-resolution pictures in the future. There is a lot of money to be made in being able to precisely and economically create 3D models of locations. In the past, sophisticated cameras, radar, and LiDAR—which utilize lasers or radio waves to map the environment—were needed. Even Boston Dynamics' robot dog, Spot, is equipped with LiDAR sensors for mapping its surroundings. However, it would be a huge step forward in mobility if anyone could map areas with only a smartphone.

5.3.22 USING AI-GUIDED DESIGN AND CONTROL, AERIAL ROBOTS CAN INSTRUCT A BAT

The scientific and technological basis for the development of bat-shaped UAVs with racing capabilities is highlighted by Sihite et al. (2020). Aerial drones that look like bats will cause the least amount of disruption possible. Humans can safely work in shared locations in order to gather situational data. According to Sihite et al. (2020), the bat-inspired design would also have to be collision-resistant in order to market a crowded product. Conditions in the physical world are hard to foresee. With this technology, smart cities' public safety and dynamic traffic control, as well as environmental calamity monitoring, can be considerably improved. In addition, large-scale systems, such as residential structures (such as construction zones, roadways, and state parks), can benefit from the UAV's communications, processing, and sensing capabilities. Unmanned Aircraft Systems (UAVs) Transdisciplinary testing at the junction of theory and experiment will be possible for the next generation of scientists and engineers. The project's curriculum and training tools will include bio-inspired robotics, machine learning, artificial intelligence (AI), and nonlinear control theory. With these skills, you should be able to have a substantial impact on the world and earn a decent salary.

An AI-guided framework is used by Sihite et al. (2020) to examine bats' various flight techniques, including zero-path flight, hovering, and perching (i.e., upside-down landing). Using the bat's wing action as a low-dimensional subspace is a good way to visualize it. In order to extract low dimensional information from the flight data of a real animal, researchers will use an auto-encoding shifting thought approach. Sihite et al. (2020) have streamlined the engineering process for building bio-inspired aerial co-robots that mimic the flight behavior of a target animal. As a result, it was able to plan out its flight path because it was aware of its location. A data-driven approach will help us understand more about the bat's sensory feedback system. Sihite et al. (2020) propose to investigate and simulate the flying patterns of bats using AI and then build a soft and collision-resistant airborne agent based on bats that can land on structures. Because of the intricate interdependencies among the various components, these two stages will be performed by a multidisciplinary team. Our goal is to lower the learning threshold for bio-inspired multimodal co-robot control and design concepts.

5.3.23 SARS-COV-2 ANTIVIRAL DRUG PREDICTION OPTIMIZER USING RECURRENT NEURAL NETWORKS

The most efficient methods of optimization are those that draw inspiration from nature. Yang and Gandomi (2012) developed a new nature-inspired metaheuristic optimization method called the bat algorithm (BA) to solve engineering problems. The echolocation of bats inspired the development of the BA. An antibody may be able to stop the virus from spreading by attaching itself to the virus and blocking it from connecting with human cell receptors.

HIV inhibitors Chloroquine and Remdesivir are both available. They are considered to be good prospects for eradicating the infection. Brilacidin and leron limb are also being studied as possible treatments for the virus by Innovation Pharmaceuticals and CytoDyn, respectively. Many additional institutions are investigating a viral treatment. Elansary et al. (2020) developed a collection of candidate medications for further study. To find the best binding affinity, scientists first employed the coronavirus as a target and attached all of the sample candidate medications to it. The virus may be cured with the medicine that has the highest binding affinity. Global health officials are concerned about the source of COVID-19 at this time. The zoonotic source of the infection in Wuhan City's wet market was COVID-19. COVID-19 has no specific treatment, according to scientists. The coronavirus is then employed as a source for optimizing the model parameters of RNN using BA. The virus may be cured if the medicine has the best possible kinship to the virus. The 6,77,044 SMILES dataset is used to train and assess the suggested technique. Repetitions, salts, and chemical information are all removed from this dataset. Preprocessing nucleic acids and peptides falls outside of the chemical area in which we intended to get samples. When compared to RNN, the suggested model's promising experimental results revealed a high level of perversion and performance.

Coronavirus 2 (SARS-CoV-2) is currently causing illness and death in people all over the world (SARS-CoV-2). This coronavirus, COVID-19, has no recognized treatment or vaccination for the serious disease it causes. The focus of Elansary et al. (2020) was on viruses and hosts as potential subjects of research. The study of medicinal chemistry can utilize it as a guide to finding novel remedies for this fatal virus. Theoretically, all of the CoV enzymes and proteins participate in viral replication. The name of the host cell machinery is anticipated to be a druggable target for the therapy of SARS-CoV-2. Structure-based analysis by Gil et al. (2020) identified therapeutic targets and reported medicinal compounds that have been demonstrated to be effective against SARS-CoV-2 and several other circulating viruses. Coronavirus infection can be treated using an innate immune response and its therapeutic potential.

5.3.24 SOUTHEAST ASIAN BATS AND PANGOLINS HAVE BEEN IDENTIFIED TO CARRY SARS-COV-2-LIKE CORONAVIRUSES

The COVID-19 pandemic was caused by a previously unknown coronavirus (severe acute respiratory syndrome coronavirus 2) (SARS-CoV-2). More than 75 million people were infected with SARS-CoV-2 in a short amount of time after being infected with the COVID-19 pandemic. Because of the issues with COVID-19, a number of critical scientific topics remain unanswered. These researchers claim that COVID-19 is the source of SARS-CoV-2 and that it was the first transmission event from animal to human, according to Guo et al. (2020).

The coronavirus disease 19 (COVID-19) is a highly transmittable and pathogenic viral infection caused by SARS-CoV-2, which emerged in Wuhan, China, and spread around the world. Many researchers (Guo et al., 2020) have been able to look at the genomic features of SARS-CoV-2 and have found that evidence does not support that SARS-CoV-2 is a laboratory construct. If it were a constructed virus, its genomic sequence would show a mix of known elements. This is not the case. COVID-19 was declared a global pandemic on March 11, 2020. As of July 24, 2022, over 567 million cases, including over 6.3 million deaths, have been reported worldwide.

There are still a lot of questions about the COVID-19 outbreak. They transmit SARS-CoV-2. Animal hosts may play an important role in the early stages of animal-to-human transmission. Evidence of SARS-CoV-2 (SC2r-CoVs)-related coronaviruses have been found in molecular and serological studies, according to experts (Callawy et al., 2020). It was discovered that the RaTG13 bat coronavirus was transmitted by bats and other mammals. Right now, it's making the rounds across Southeast Asia. Cave-dwelling bats (*Rhinolophus acuminus*) from Thailand provided the only genomic order isolation (RacCS203). The most relevant *Rhinolophus malayanus* RmYN02 and RmYN02 isolates are found in Yunnan, China. An animal checkpoint in Southern Thailand found SARS-CoV-2 neutralizing antibodies in two bats from the same colony and one pangolin (Whitman et al., 2020). Taxonomically speaking, SARS-CoV-2 belongs to the SARS genus (Figure 5.10).

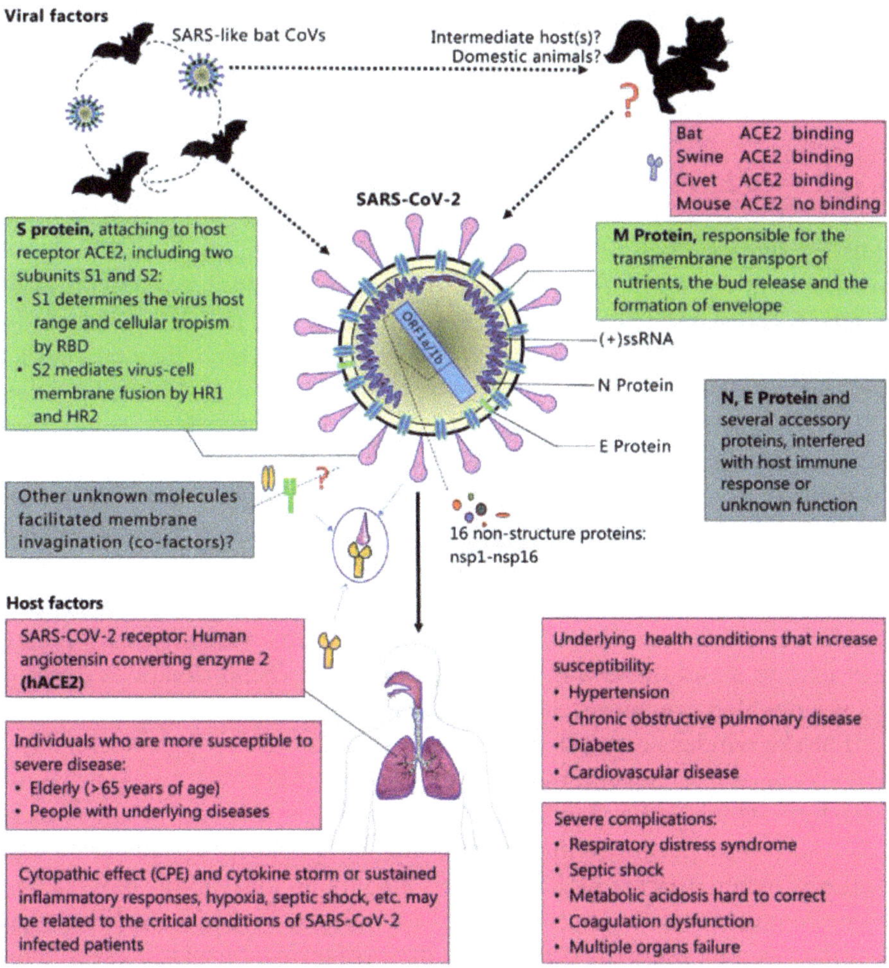

FIGURE 5.10 COVID-19 origin, transmission, and clinical therapies.

Source: Reprinted with permission from Guo et al. (2020) http://creativecommons.org/licenses/by/4.0/

SARSr-CoV is a coronavirus in the Betacoronavirus genus of the Coronaviridae family. RaTG13, a similarly related SARSr-CoV genome sequence, was discovered by Zhou et al. (2020). As a whole, it had a 96% genomic match. SARS-CoV-2 genomes from bats from Eastern China and Japan and pangolins in China have been considered since then. However, the virus's immediate animal parent or progenitor is the equivalent of the >99% identical SARS-CoV sequences that have been discovered. Determining the virus's origin and the viruses that came before it is crucial in the academic world. SARS-CoV-2 and other closely similar virus epidemics must be avoided at all costs in public health (Callawy et al., 2020).

5.3.25 DO BATS PRACTICE SOCIAL DISTANCING?

Bats are hard to study because they're small, difficult, and the only flying mammals of the night. Investigators, on the other hand, have found a way to attach better tracking systems to bats by modifying technology normally reserved for sophisticated devices. As part of their research, they take walks with the animals to see how well they can handle physical distance. Silicon Labs, an Austin-based chipmaker, applauded the research. It could help advance the field of animal tracking. Simon Ripperger, an animal biologist, and Niklas Duda, an engineer, collaborated to develop a wireless sensor network for bat teaching. It's a game changer for animal research around the world. An Ohio State University biologist and a Friedrich-Alexander-Universität Erlangen-Nuremberg electrical engineer, Ripperger, studied this research together with Duda (Ripperger et al., 2020). Vampire bats who are sick tend to avoid those who are close to them in social situations (https://www.eurekalert.org/news-releases/772210).

For the wireless network and tagging system, Silicon Labs technology plays a major role. Several fresh findings and studies on bat social behavior have been developed and published by experts in 2019. As a result of their actions, it was determined that animals are better at keeping a lower "social distance" when they are ill. Researchers also found that captive bats used in labs maintain social bonds when released into the wild. When it comes to feeding time, mother bats don't guide their offspring very far. Research on bats was conducted in Germany, Belize, Costa Rica, and Panama. Silicon Labs' sensor system is a substantial improvement over earlier equipment. In comparison to prior systems, this one is simpler, more adaptable, and has fewer components. Silicon Lab's technology has been used by scientists since 2017. In order to improve the reliability of their investigations, this group's technology makes use of Flex Gecko chips. Sabolcik stated the tracking system is customized for the Internet of Things (IoT) using Silicon Labs' technology. Internet of Things (IoT) refers to the use of non-computer items connected to the Internet to enhance or expand their capabilities.

5.3.26 SIGNIFICANT IMPACT ON WILDLIFE ANIMALS TO MAINTAIN SOCIAL DISTANCE

Researchers found that sick vampire bats spend less time with their fellow bats in the area. Their neighborhood begins to get infected with the disease gradually but certainly. A field experiment was conducted by the researchers (Ripperger et al., 2020) in order to verify their findings. A pandemic can make it harder to maintain one's social distance. However, this is a common occurrence in nature. In order to minimize the spread of disease, animals will flock to the path of least resistance. The preservation of social separation in animals is something that many individuals are unwilling to undertake, as evidenced by a study published in the journal Behavioral Ecology. The sick wild vampire bats in their colony have been proven to spend less time with their fellow bats. As a result, the disease has a lower risk of spreading throughout the neighborhood. According to US experts, vampire bats (*Desmodus rotundus*) kept their distance from other bats in lab settings. They undertook this experiment to see if the bats would act in this way in the wild.

Around 31 adult female vampire bats were discovered roosting in a hollow tree in Lamanai, Belize, by the team led by Ripperger et al. (2020). To make the bats appear ill, an immune-suppressing drug was given to half of them at random (lipopolysaccharide, or LPS). Bats in the control group receive saline injections, while those in the experiment receive a placebo. Before returning to their native home, the bats were fitted with proximity sensors. Researchers observed the behavior of bats and their relationships with one another. In the group, the number of sick bats was lower, and those who spent time with other bats and were socially attached to healthy group members spent less time together as well. Four fewer bats interacted with "ill" bats than with healthy control bats. As a result, the interaction time between "sick" bats was reduced by 25 minutes for each pair. Healthy bats had a 49% chance of being in the same group as each control bat, but "sick" bats had a 35% chance of being in the same group as each control bat. This distance is greater than 400 meters (437 yards) for wading birds and songbirds, according to a study conducted by this team. Species had different threshold distances for mammal encounters. Squirrels can live up to 1,000 meters (1,093 yards) away from larger animals like moose and elk. Animals and wildlife populations can be negatively affected when people engage in outdoor recreational activities. Those in charge of spatial planning and natural resource management should be aware of this conclusion.

KEYWORDS

- **AI-guided design and control**
- **bats**
- **DynCorp service**
- **echolocation**
- **robots**
- **SARS-CoV-2**
- **self-driving cars**
- **social distance**

REFERENCES

Aerospace and Defense Technology. https://www.aerodefensetech.com/component/content/article/adt/insiders/defense/stories/24189 (accessed on 17 August 2023).

Allen, M., (2018). Watch this bat-inspired robot navigate like a real one. *Science*. https://www.science.org/content/article/watch-robot-bat-navigate-real-one (accessed on 17 August 2023).

Anand, R., (2014). *This Brilliant Invention from IIT Delhi May Just Change the Way Blind People Move Around Forever*! Blindness Community Media. https://www.thebetterindia.com/11954/invention-blind-smartcane-mobility-visually-impaired-innovation/ (accessed on 17 August 2023).

Balakrishnan, M. (2021), ASSISTECH: An Accidental Journey into Assistive Technology. In *A Journey of Embedded and Cyber-Physical Systems*. Springer Cham, pp.57-77. DOI:10.1007/978-3-030- 47487-4_5

Brigham, R. M., Jones, G., Kalko, E., & Parsons, S., (2004). *Bat Echolocation Research Tools, Techniques, and Analysis*. Bat Conservation International, Austin, Texas. https://www.researchgate.net/publication/281210205_Bat_Echolocation_Research_Tools_Techniques_and_Analysis, https://www.researchgate.net/publication/281210205 (accessed on 30 September 2023).

Callaway, E., Ledford, H., & Mallapaty, S. (2020). Six months of coronavirus: The mysteries scientists are still racing to solve. *Nature, 583*, 178, 179.

Defense INSIDER, (2016). *Researchers Design Bat-Inspired Wings for Micro Air Vehicles*.

Dernayka, A.; Amorim, M.A.; Leroux, R.; Bogaert, L.; Farcy, R. Tom Pouce III, an electronic white cane for blind people: Ability to detect obstacles and mobility performances. *Sensors* **2021,** *21,* 6854.

Diebold, C. A., Salles, A., & Moss, C. F., (2020). Adaptive echolocation and flight behaviors in bats can inspire technology innovations for sonar tracking and interception. *Sensors (Basel), 20*(10), 2958. https://doi.org/10.3390/s20102958.

Duke, C., (2021). *Bats Don't Have to Learn the Speed of Sound – They're Born Knowing it*. New Scientist News Letters. https://www.newscientist.com/article/2276404-bats-dont-have-to-learn-the-speed-of-sound-theyre-born-knowing-it/ (accessed on 17 August 2023).

EARSC. (2021). *How We Use Technology Inspired by Bats to See Water Availability in Fields.* European EO Industry, Members News, EARSC. https://dynacrop.space/en/how-we-use-technology-inspired-by-bats-to-see-water-availability-in-fields-2/ (accessed on 17 August 2023).

Elansary, I., Hamdy, W., Darwish, A., & Hassanien, A. E., (2020). Bat-inspired optimizer for prediction of anti-viral cure drug of SARS-CoV-2 based on recurrent neural network. *J. System Management Sci., 10*(3), 20–34. doi: 10.33168/JSMS.2020.0302.

Eliakim, I., Cohen, Z., Kosa, G., & Yovel, Y., (2018). A fully autonomous terrestrial bat-like acoustic robot. *PLoS Comput. Biol., 14*(9), e1006406. https://doi.org/10.1371/journal.pcbi.1006406.

Farooq, M.S., Shafi, I., Khan, H., Diez, I.D.L.T., Brenosa, J., Espinosa, J.C.M., & Ashraf, I. (2022). IoT Enabled Intelligent Stick for Visually Impaired People for Obstacle Recognition. *Sensors 22*(22), 8914; https://doi.org/10.3390/ s22228914

Fingas,@jonfingas, (2021). *Sound Location Inspired by Bat Ears Could Help Robots Navigate Outdoors.* April 16th, Engadget, Virginia Tech Daily. https://www.engadget.com/bat-inspired-ai-sound-location-140117971.html (accessed on 17 August 2023).

Fullard, J. H., & Dawson, J. W., (1997). The echolocation calls of the spotted bat *Euderma maculatum* are relatively inaudible to moths. *J. Exp. Biol., 200*, 129–137.

Gil, C., Ginex, T., Maestro, I., Nozal, V., Barrado-Gil, L., Cuesta-Geijo, M. Á., Urquiza, J., et al., (2020). COVID-19: Drug targets and potential treatments. *J. Med. Chem., 63*(21), 12359–12386. doi: 10.1021/acs.jmedchem.0c00606.

Griffin, D. R., Webster, F. A., & Michael, C. R., (1960). The echolocation of flying insects by bats. *Animal Behavior, 8*(3, 4), 141–154.

Grinnell, A. D. (2018). Early milestones in the understanding of echolocation in bats. *J. Comparative Physiol., 204*, 519–536.

Guo, Y. R., Cao, Q. D., & Hong, Z. S., (2020). The origin, transmission, and clinical therapies on coronavirus disease, 2019 (COVID-19) outbreak – an update on the status. *Military Med. Res., 7*, 11. https://doi.org/10.1186/s40779-020-00240-0.

https://www.leeds.ac.uk/news/article/3329/bat-inspired_technology_showcased (accessed on 17 August 2023).

https://www.nsf.gov/discoveries/disc_summ.jsp?cntn_id=125883 (accessed on 17 August 2023).

Islam, N., Rashid, M.M., Pasandideh, F., Ray, B., Moore, S., & Kadel, R. (2021). A review of applications and communication technologies for internet of things (iot) and unmanned aerial vehicle (uav) based sustainable smart farming. *Sustainability 13*(4), 1821; https://doi.org/10.3390/su13041821

Jensen, M. E., Moss, C. F., & Surlykke, A., (2005). Echolocating bats can use acoustic landmarks for spatial orientation. *J. Exp. Biol., 208*, 4399–4410. doi: 10.1242/jeb.01901.

Jones, G., & Telling, E. C., (2006). The evolution of echolocation in bats. *Trends Ecol. Evol., 21*(3), 149–156. doi: 10.1016/j.tree.2006.01.001.

Kim, S. Y., & Cho, K., (2013). Usability and design guidelines of smart canes for users with visual impairments. *Intern. J. Design, 7*(1), 99–110.

Kue, B., (2013). *Echolocation Allows Dolphins, and Bats to Navigate Without Sight.* Scitech. http://thetartan.org/2013/10/28/scitech/howthingswork (accessed on 30 September 2023).

Lee, W. J., Falk, B., Chiu, C., Krishnan, A., Arbour, J. H., & Moss, C. F., (2006). Tongue-driven sonar beam steering by a lingual-echolocating fruit bat. *PLoS Biol., 15*(12), e2003148., https://doi.org/10.1371/journal.pbio.2003148.

Lewis, T., (2014). *Dolphins May Use Magnetic Sense as GPS*. Scientific American, Live Science. https://www.scientificamerican.com/article/dolphins-may-use-magnetic-sense-as-gps/ (accessed on 17 August 2023).

Li, Y., Liu, Z., Shi, P., & Zhang, J., (2010). The hearing gene Prestin unites echolocating bats and whales. *Curr. Biol., 20*(2), R55, 56. doi: 10.1016/j.cub.2009.11.042.

Moss, C. F., & Surlykke, A., (2001). Auditory scene analysis by echolocation in bats. *J. Acoust. Soc. Am.*, 110, 2207–2226. https://doi.org/10.1121/1.1398051

Pesce, N. L., (2020). *Even Vampire Bats Practice Social Distancing When They Get Sick*. MarketWatch.

Ramezani, A., Chung, S. O., & Hutchinson, S. (2017). A biomimetic robotic platform to study flight specializations of bats. *Science Robotics, 2*(3) doi: 10.1126/scirobotics.aal2505.

Rees, J., (2011). *Lessons in Echolocation from Bats and Dolphins*. Planet Earth. https://www.discoverwildlife.com/animal-facts/what-is-echolocation (accessed on 30 September 2023).

Ripperger, S. P., Stockmaier, S., & Carter, G. G., (2020). Tracking sickness effects on social encounters via continuous proximity sensing in wild vampire bats, *Behavioral Ecology, 31*(6), 1296–1302. https://doi.org/10.1093/beheco/araa111.

Sahoo, N., Lin, H.-W., & Chang, Y.-H. (2019). Design and implementation of a walking stick aid for visually challenged people. *Sensors 19*(1), 130; https://doi.org/10.3390/s19010130

Sekar, S., (2013). *Dolphins, Bats, and the Evolution of Echolocation*. https://news.mongabay.com/2013/11/dolphins-bats-and-the-evolution-of-echolocation/ (accessed on 17 August 2023).

Sihite, E., Kelly, P., & Ramezani, A., (2020). Computational structure design of a bio-inspired armwing mechanism. *IEEE Robotics and Automation Letters*, *5*, https://doi.org/10.1109/LRA.2020.3010217.

The Robot Report Staff, (2018). *Robat, a Bat-Inspired Robot, Uses Echolocation to Map, Navigate Environment*. https://www.therobotreport.com/robat-echolocation-map-navigate/ (accessed on 17 August 2023).

Turpin, A., Kapitany, V., Radford, J., Rovelli, D., Mitchell, K., Lyons, A., Starshynov, I., & Faccio (2021). 3D Imaging from multipath temporal echoes. *Phys. Rev. Lett., 126*, 174301–174330.

Wacharapluesadee, S., Tan, C. W., Maneeorn, P., et al., (2021). Evidence for SARS-CoV-2 related coronaviruses circulating in bats and pangolins in Southeast Asia. *Nat. Commun., 12*, 972. https://doi.org/10.1038/s41467-021-21240-1.

Wahab, M. H. A., Talib, A. A., Kadir, H. A., Johari, A., Noraziah, A., Sidek, R. M., & Mutalib, A. A., (2011). Smart cane: Assistive cane for visually impaired people. *Intern. J. Computer Sci.*, *8*(4), 21–27.

Ward, T. A., Fearday, C. J., Salami, E., & Soin, N. B. (2017). A bibliometric review of progress in micro air vehicle research. *Intern. J. Micro Air Vehicles*, *9*(2), 146–165 https://journals.sagepub.com/doi/pdf/10.1177/1756829316670671.

Whitman, J. D., Hiatt, J., Mowery, C. T., et al., (2020). Evaluation of SARS-CoV-2 serology assays reveals a range of test performance. *Nat. Biotechnol., 38*, 1174–1183. https://doi.org/10.1038/s41587-020-0659-0.

Wilkins, A., (2021). *Scientists Invent Bat-inspired Tool to Find Your Location with a Smartphone*. https://metro.co.uk/2021/04/30/scientists-invent-bat-like-tool-to-find-your-location-with-smartphone-14492966/ (accessed on 30 September 2023).

Yang, X., & Gandomi, A. H., (2012). Bat algorithm: A novel approach for global engineering optimization. *Engineering Computations, 29*(5), 464–483. https://doi.org/10.1108/02644401211235834.

Yin, S., (2019). *Why Is Creating Electronic Canes for the Blind so Hard?* Why, PPBS, The Pulse. https://whyy.org/segments/why-is-creating-electronic-canes-for-the-blind-so-hard/ (accessed on 17 August 2023).

Yin, X., & Müller, R., (2021). Integration of deep learning and soft robotics for a biomimetic approach to nonlinear sensing. *Nat. Mach. Intell.*, *3*, 507–512. https://doi.org/10.1038/s42256-021-00330-1.

Yoneda, Y., (2008). *The Solar Powered COM-BAT Spy Plane.* INHABITAT, Design Innovations, University of Michigan, https://inhabitat.com/com-bat-solar-powered-spy-plane/ (accessed on 17 August 2023).

Zhou, P., Yang, X. L., Wang, X. G., et al., (2020). A pneumonia outbreak associated with a new coronavirus of probable bat origin. *Nature, 579*, 270–273. https://doi.org/10.1038/s41586-020-2012-7.

CHAPTER 6

Spider Innovative Technologies

ABSTRACT

Spider silk is tougher than steel and more powerful per unit weight. Spiders use a variety of silks, each of which has a distinct function in their lives. Despite being bulletproof, the material is extremely supple. The simple spider has inspired the development of a slew of new materials, including construction tools and medicines. Fibers made from spider silk are produced by Kraig Biocraft Laboratories, Inc. (OTCQB: KBLB). These fibers have their inherent elasticity and strength, and the material was developed from genetically modified silkworms and created body armor. The latest armor under consideration of U.S. Army chemical composition is new kevlar. It is a significant invention and new feature. US-based Kraig Biocraft Laboratories is making advanced substances employing biotechnology research and development companies. Spider-silk microcapsules have been produced that can transport vaccines directly to the core of the immune system's cells. In the words of experts, tiny spider-silk microparticles may be able to shield vaccine particles. In cancer treatment, silk particles are utilized as carriers of therapeutic compounds because they can enter free cells and stimulate T-cell immunity without causing harm. With the aid of double-sided tape, organs such as the liver or heart can be implanted with medical devices without risking injury or puncturing the tissue. New biomaterial applications have been opened up thanks to the advancement of large-scale biotechnological spider silk proteins. When it comes to biotech silk materials, only one company can compete with AMSilk.

6.1 SPIDER SILK'S PROPERTIES AND APPLICATIONS

Silk is the fine thread generated from a silkworm that spins its cocoon. In its cocoon, the silkworm becomes a pupa. A silk protein fiber has a lot in

Bio-Inspired Technologies for the Modern World: Emerging Trends and Applications.
R. Ramakrishna Reddy & T. Pullaiah (Authors)

common with human skin in terms of its makeup. As a result, silk is also resistant to the formation of mildew, fungi, and mold, all of which are common issues for humans. A broad name for any protein fiber spun by insects, silk, is also commonly used to describe silk. A silk material's qualities might vary greatly depending on where it was sourced. Nature's small packages contain sheets and silk fibroin polymers. With regards to silk webs, cocoons, and other tricks, polymers utilized in this process can be a great resource. A range of biomaterials are now being developed using these proteins and their mechanical characteristics. Their capacities for strength and resiliency vary widely. Additionally, silk is non-immunogenic, bioresorbable, biocompatible, biodegradable, and morphologically flexible. It is also biocompatible and environmentally friendly.

Each spider species has a unique amino acid chain and unique behavior that determines the sort of silk protein they produce. An upsurge in the need to obtain spider silk protein for diverse biological purposes has been attributed to new research. When compared to a silkworm's 600–900 meters of silk, a spider's silk can range from 12 to 137 meters in length. Using DNA technology, scientists have been able to artificially increase the production of spider silk. Spider silk offers a wide range of qualities, including high elasticity, tensile strength, and light weight.

The prominent applications of silks are in soaps, cream, and nail varnish, etc. The recombinant spider silk surfaces change the bio-molecular corona. Researchers have created spider silk using DNA technologies to artificially improve production rates. Many visible effect changes reveal this study. Significant potential future uses of recombinant spider silk biomaterials are expecting from this research (Aigner et al., 2018). The materials are used to find in bulletproof vests, rugged clothing, parachutes, and the construction of buildings. Bullet-proof material, remarkably flexible yet robust building tools, breakthroughs in medicine, all inspired by the humble spider! (Vollroth, 2000; Gerristen, 2000).

Spider silk is more powerful by weight than steel, making it a formidable weapon. In comparison to steel, a single pencil-thick strand of spider silk can hold a Boeing 747 aloft. Stronger than steel, silk from spiders can be. Silk glands contain a liquid that turns into a solid that is more durable than Kevlar, according to scientists (Figure 6.1). When cooled to–40°C, its strength remains unaffected even though it is more sensitive than hair (the threads are often only several microns in diameter). The Golden Orb-Weaving Spider's (*Nephila clavipes*) wire silk is particularly noteworthy.

FIGURE 6.1 Spider silk: A garden spider spinning its web.
Photo credit: Duk on Wikipedia. https://creativecommons.org/licenses/by-sa/3.0/.

Although spider silk may be stretched 2–4 times its original length without breaking, capture silk (sticky silk) cannot. Silkworm silk is pliable and absorbent, whereas spider silk is tougher, more elastic, and more waterproof. An atomic force microscope (AFM) was used to study the silk. The brown recluse spider spins a silk unlike any other produced by known arachnids or insects. Instead of being round, the recluse silk fibers are flat and extremely thin, like a silky nanoribbon. And they're spotted with tiny spherical dots, a team of scientists reported in *Advanced Materials*. When it comes to building ground webs and storing eggs, spider silk outperforms steel according to research. Thousands of nanostands, each measuring barely 20 millionths of a millimeter in diameter, were found to be thinner than a human hair, according to the researchers.

The incredible variety in chemistry, structure, and function that can be found in spider silks is what sets them apart. Adhesives and cocoons are also included in this category. There are several uses for spider silk, from textiles to biological materials, which have been domesticated for over 5,000 years. Researchers are exploring the possible uses of spider silk that are comparable to silkworms. New materials and technologies are being developed by scientists thanks to the genetic engineering of spider silk. Research and development in the future is expected on spider-silk proteid self-assembly processes and the selective study of material applications (Basu, 2015).

- Spider silk is a fiber secreted by spiders.
- Spider silk is a solid material, the strongest fiber known to us.
- When it comes to building structures, spiders often use their silk, either for the protection of their children, or for other animals.
- Many tiny spiders use silk threads for ballooning.
- They project several threads and let themselves become taken away with upward winds.

6.2 TYPES OF SPIDER SILK

Silk is an absolute classic, sought after for its beauty and exquisite softness, shunned for its price tag and supposed frailty. Silk is a fine, strong, soft, lustrous fiber produced by silkworms, particularly mulberry silkworms, in making cocoons and collected to make thread and fabric. Silk is biocompatible, biodegradable, easy to functionalize and has excellent mechanical properties, in addition to optical transparency. Figure 6.2. shows seven types of spider silks.

There are currently more than 34,000 known spider species, each with a unique silk and mechanical qualities. Silk may be used sparingly by some spiders. Most spiders construct large nests, traps, and cocoons, all of which are made possible by a variety of glands found in their bodies (Hinman et al., 2000; The Sciences of Spider Silk, 1869; The Conversation, 2019). The silk of some spiders may be used sparingly.

Spiders have seven silk glands, each of which can produce a distinct type of silk. There are at least three males (dragline attachment and swathing silk) or four females (egg sac silk) in each spider, however they don't have all seven glands.

The six types of silk and their uses are:

- Egg sac silk is produced by a cylindrical gland.
- Producing silk to encase and bind prey is the job of the achniform gland.
- A spider's non-sticky dragline silk is produced by its ampullate glands (major and minor), which also serve as safety lines in case a spider falls or moves outside of the web. Due of its function, dragline silk is the strongest type of spider silk.

- Attachment threads and discs are manufactured by the pyriform gland, which also produces silk threads.
- The sticky silk at the center of a web is produced by a gland called a flagellum, which is responsible for trapping animals.
- The adhesives that are placed along the threads of sticky silk are the product of a gland that collects and releases droplets.

In addition to its attractiveness, silk's remarkable physical characteristics made it a popular choice for humans. However, once the history of silk is revealed, it is transformed into a high-tech fabric (Hinman et al., 2000; The Sciences of Spider Silk, 1869; The Conversation, 2019). Silk's structure and function have been studied in the past to obtain a better knowledge of it.

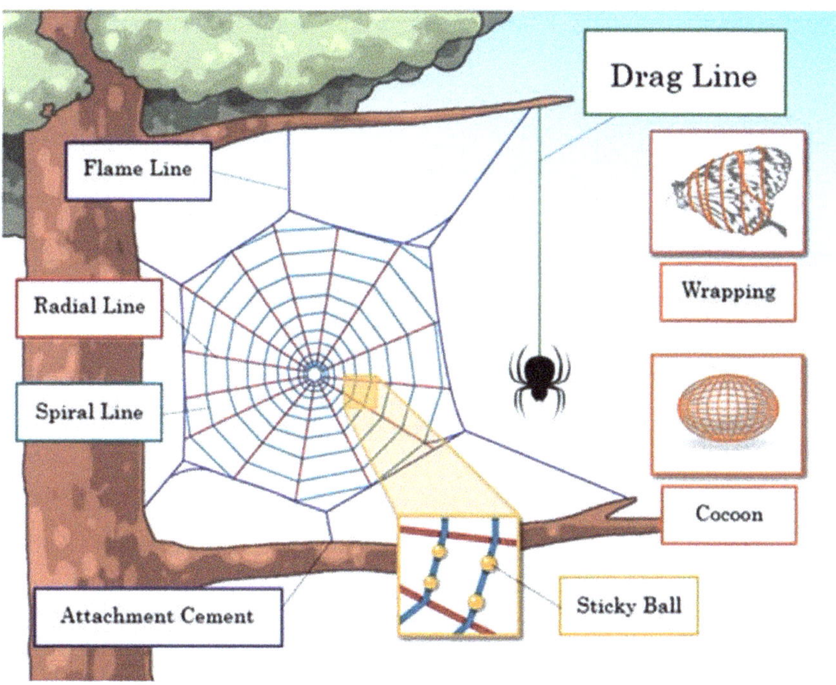

FIGURE 6.2 Highlights seven types of spider silks

Source: Reprinted with permission from Asakura (2020). https://creativecommons.org/licenses/by/4.0/

The biocompatibility and degradation studies are primarily focused on human silk performance. Sewn sutures, surgical mesh, and textiles are only a few examples of the many modern medical applications for silk. Clinical experiments involving silk are critically important. The great majority of research to far has focused on silk-based biomaterials for bone tissue engineering (Sofia et al., 2001). Tissue engineering and regenerative medicine frequently use silk scaffolds. The cells require the scaffold's mechanical stability in order to adhere, grow, divide, proliferate, and differentiate. The connective tissues are the tendon, ligament, bone, and cartilage. Spider silk provides outstanding mechanical and biocompatible properties while inorganic nanomaterials may offer functional features to harness high-performance hybrid materials (Figure 6.3).

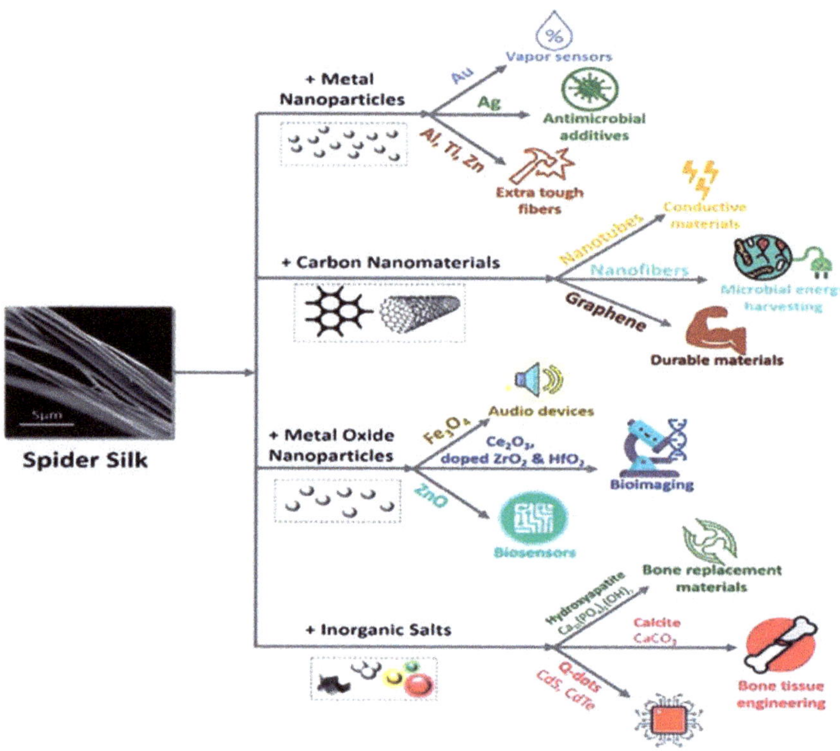

FIGURE 6.3 shows spider silk outstanding mechanical and biocompatible properties while inorganic nanomaterials may offer functional features to harness high-performance hybrid material.

Source: Reprinted with permission from Kiseleva et al., 2020. https://creativecommons.org/licenses/by/4.0/.

About 0.003 millimeters is the diameter of one spider web thread. However, the strength of spider silk in comparison to steel may be demonstrated (The

Conversation, 2019). Metals are mixed together to create steel, which is an alloy. Iron is a component of steel. For its make, one has to mix iron and chromium. Strong steel is used in the buildings and cranes. Various materials are synthesized by mixing iron with different metals. For example, stainless steel is employed in knives and forks so that they don't rust. The Bike was built by handling steel and iron, nickel, cobalt, molybdenum, titanium, and aluminum.

6.3 SPIDER SILK PROPERTIES

Silk fibers' most important qualities must be directly linked to their structure. In terms of solidity, spider silk is the strongest naturally produced fabric (Stilwell, 2020). Spider silk's tensile strength is on par with that of premium steel. For the most part, spiders use their silk to create structures, either to shield their young or to prey on other animals. Now, scientists have figured out how these silken threads are able to withstand so much stress. Spider silk is more powerful than steel in most cases.

It's possible to use an AFM to examine the silk used by deadly brown recluse spiders to construct their underground networks and to hold their eggs. There are thousands of nano-stands in each strand, making each strand several times as strong as previously thought. On a nanoscale, nano-stands are at least 50 times as long as these fibers are wide—and scientists believe they may stretch even further. Fibers with these qualities can be used to build a variety of different materials. In biomedicine, satellite technology, textiles, planes, and automobiles, these may be used in various ways.

Compared to other tough polymer materials (e.g., Kevlar®), dragline spider silk (shown in Table 6.1) is stronger than it (de Souza e Silva et al., 2020). The brown-colored silk was employed by scientists in this study. It is a ribbon-shaped fiber that is flat on the surface. Using a strong microscope makes it easier to see. The flat ribbons and looping technique do not apply to all spiders, according to scientists. They're often used in technical sectors. The efforts of scientists will be of use to materialists.

TABLE 6.1 Toughness, Tensile Strength, and Densities of Different Materials (de Souza e Silva et al., 2020)

Toughness of the Material (J/g)		Tensile Strength (M Pa)	Density (g/cm^3)
Dragline Spider Silk	120–160	1,100–29,001	18–136
Kevlar	30–50	2,600–4,100	144
Steel	2–6	300–2,000	784

6.4 UTILIZATION OF SPIDER SILK

6.4.1 BODY AMOUR IS NOW MADE USING SPIDER SILK

Kraig Biocraft also employed patented genetic proteins to make silkworms, just like the researchers at Utah State. Kraig Biocraft may be the first to create body armor that is flexible, lightweight, and comfortable. Spider silk body armor (Nguen, 2016) is a big and interesting development. In 10 years, the army will be procuring this very light body armor for its soldiers. It's made up of spiders' webs. It's not a new trend to incorporate spiders and their webs into fashion. Silk from the Golden Orb Spider of Madagascar was used to create the vest (Figure 6.4). Using a million spiders over the course of eight years, the designers created this garment.

FIGURE 6.4 Golden orb-weaver spider (Nephila inaurata madagascariensis) female.
Image courtesy: Charles J Sharp. https://creativecommons.org/licenses/by-sa/4.0/.

A kind of jumping spider that secretes milk to nourish its young was previously found by a team of Chinese scientists. Compared to cow's milk, the milk has four times as much protein. Dried-out form could be shaped to meet various needs, for example. Unlike steel, spider silk is very flexible and can be stretched to tens of thousands of times its length. A bullet could be held in place by a spider's silk. Nobody had figured out a way to make spider silk impervious to bullets before now. The DNA of silkworms can now be altered by researchers at Utah State University. When compared to typical silkworm silk, the new silk has a far higher strength-to-elasticity ratio. A slow-moving bullet was successfully deflected by the resulting substance. In contrast, today's conventional body armor requires a minimum of 33 layers of Aramid (Carlson, 2018).

In a recent announcement, Kraig Biocraft Labs said that it would be producing a significant number of panels for the United States military. There won't be any need for a million spiders to make it happen. Genetically engineered Kraig Biocraft silkworms are identical to those created by Utah State University researchers using their patented genetic proteins. Typical body armor was manufactured by Kraig Biocraft. In order to truly defend oneself, it was the first to design and manufacture body armor that was all three of those things at once. The Researchers (Jon Rice and Kraig Biocraft COO) mention that body armor produced gives exciting thrill and impressive. This unique breakthrough technology was created by those Scientists using the U.S. Army. It is allowing protecting the people from putting their careers protected and considered enormous honor.

6.4.1.1 KRAIG BIOCRAFT LABORATORIES CREATED POLYMERS

Polymers' mark applications are noteworthy and use many industries to combine different molecules' characteristics to create a more robust chained network. The two primary goals of the polymer-creation approach are the manufacturing of more essential commodities at a lower cost on a big scale. The next generation of biomaterials is currently being produced by biotechnology businesses. Founded in the United States, Kraig Biocraft is dedicated to producing high-quality products. Polymers incorporating the business's unique spider silk protein sequences are being developed by the company. The natural method of making spider silk takes a long time and produces a small amount of silk. In order for the product technology to be widely adopted, this solution is necessary (https://textilelearner.net/innovation-and-developments-of-spider-silk/).

One of the most innovative biotech companies in the world today is Kraig Biocraft Laboratories, Inc. Spider silk is the primary focus of this company. Because of this, it's capable of producing silk of a high quality. Kraig was founded by Kim Thompson, who is also the developer of the technology. Kraig is able to make spider silk and a variety of proteins, textiles, and materials at a reasonable cost. Continued collaboration with the University of Notre Dame and strengthened the relationship between Kraig and his research team. KBLB (OTCQB: KBLB) is one of the best places to work in the field of spider silk-based materials. In order to produce the first textile models of Monster Silk™, the Company's product catalog has successfully produced a necessary number of cocoons. Kraig completed a pilot manufacturing program for its Monster Silk™ spider silk product in 2013 and is now scaling up commercial-scale production. It's a crucial milestone in the commercialization plan's progress. Additionally, Kraig is in the final stages of developing a more advanced spider silk product called "SpiderPillar." SpiderPillar's spider silk will be pretty natural. In the technical textile sector, this material has the clout to make substantial advances.

6.4.1.2 ARE SPIDER WEBS BULLETPROOF WEBS?

It is anticipated that soldiers in the future will be outfitted in Bulletproof Spider Silk. Spider silk is one of the most stiff fibers available to the general population. Soldiers will be safe from gunshots and other dangers when this technology is fully implemented. Spider silk is more powerful than high-grade steel despite its lighter weight. In the future, the Army will use "spider silk" body armor that is stronger than Kevlar. As a bulletproof vest of the future, it will use spider silk instead of super-strong polymers (Carlson, 2018). Body armor and underwear made of synthetic spider silk are increasingly popular among the American military. When compared to Kevlar's strength and flexibility, what has the genetic make-up of a spider but is produced from a worm? (Tucker, 2017).

6.4.1.3 SPIDER SILK-INSPIRED MATERIALS AND SUSTAINABILITY

Under the biomimetic category, Spider silk considers as one of the best examples. The complex hierarchical structure of this fiber shows well-known properties, incredibly unique tensile characteristics. Spinning in an aqueous solution under particular conditions yields silk strands. Because of its inefficient physicochemical requirements, it uses the least amount of energy to manufacture the thread. The silks spun by spiders exhibit a wide range of physical and biological characteristics. Structure-function relationships can be used to influence the development of custom materials. In addition to silk's performance and spinning process efficiency, silk proteins (spidroins) may be a renewable and biodegradable material source with good biocompatibility properties. The use of materials inspired by spider-silk could help humans while also reducing society's impact on the environment. Nevertheless, creating materials that are both environmentally friendly and highly efficient is a difficult task, and there is still more to be done. When it comes to large-scale production of spider silk-inspired materials, questions concerning sustainability, the suitability of using spiderroin, and other more general difficulties arise.

6.4.1.4 A BULLETPROOF JACKET MADE OF SPIDER WEB IS GOING TO USE FOR MILITARY PROTECTION

It's imperative that soldiers wear bulletproof vests, which use spider webs to deflect enemy bullets before they reach the body. A spider silk jacket was created by Craig Biocraft Laboratories of the United States. As long as you've done your homework, the spider webs, so the theory goes, can stop rapid-fire shooting. Experts believe some spiders make a unique fiber that can work better if used for military protection. But what is this, and how will it work, let us tell you: It is Kraig Biocraft Laboratories that create the Spider-Man suit and the new ballistic shoot pack panels. However, it isn't as strong as steel by a large margin. It has a greater tensile strength than Kevlar. Military and police officers frequently use this type of bulletproof material. Using genetic engineering, Kraig managed to make silkworms produce spider silk (Figures 6.5 and 6.6) by manipulating their DNA. This cloth, made in the style of Superman, is useful to the U.S. Army. Something out of the Spider-Man film.

As a result of Kraig Biocraft Laboratories' advancement in the commercialization of its unique Dragon Silk material, the U.S. Army purchased ballistic shoot pack panels from Kraig Biocraft. Ballistic shoot pack panels were made by them as an alternative to traditional protective clothing in the United States. The panel evaluates whether or not the bullet-proof vests give Warfighters a lighter and more comfortable experience. All military troops who despise the heaviness and bulk of their bulletproof pads will rejoice.

FIGURE 6.5 The improved outer tactical Vest (IOTV) in Multicam, as issued to United states Army soliders.

Source: https://en.wikipedia.org/wiki/Improved_Outer_Tactical_Vest

FIGURE 6.6 Testing a bulletproof vest in Washington, D.C., September 1923.

Source: https://commons.wikimedia.org/wiki/File:Testing_bulletproof_vest_1923.jpg

Experts enlightened that some particular spiders make silk protein, resilient, and solid compared to fine-quality silk. Surgical outfits for doctors can be made from this silk because it is so high-quality. So far, gloves have been made which are under testing. Its specialty makes it quite important. It is mentioned that this silk is much more robust in strength used in the current jacket. One of the most fascinating natural fibers ever discovered is this incredibly strong spider silk. It has a unique ability to protect soldiers from gunfire if worn as a jacket by the military. Experts say that by wearing this silk jacket, the army will feel light and comfortable. With this, the army will also be very active. Their running and fighting abilities will also increase.

6.4.1.5 GENES THAT ARE UNIQUE TO A NEW GOAT BREAD

Spider silk comes in a wide variety of forms. Various sources of silk are available. Some spiders have as many as seven distinct glands. Despite its small weight, spider silk looks to be more durable and more resilient than steel in contrast to cotton. In comparison to human hair, spider silk is extremely thin. They can carry hundreds of times as much weight as it does. Arthropods of the smallest size create it, turning it into complex structures. Silken threads are investigated in the domains of chemistry, material science, and engineering. Silk threads are used by spiders to capture prey. Eventually, a spider silk parachute may fall on a spider silk wire-draped suspension bridge and be utilized to fend off thieves with light spider silk armor. A spider silk spacesuit would be ideal for anyone who desires to venture into outer space.

A novel breed of goats with golden orb spider genes has had its genetic makeup effectively altered by University of Wyoming scientists, enabling the goats to produce milk rich in spider silk proteins (https://www.techquila.co.in/scientists-breed-super-goats-that-produce-spider-silk/). Molecular biology professor Dr. Randy Lewis has shown a great deal of interest in this topic. A farm at Utah State University is home to Lewis' spider goats. The silk protein gene would be found in three of these individuals. A bigger amount of spider silk proteins will be extracted from the milk of the transgenic goats after they have matured.

Industries are interested in employing silk protein for a variety of purposes because of its natural strength. Tendons and ligaments and coat implants are important medical uses for artificial eye sutures. Additionally, the protein's antibacterial properties allow it to be used to produce wound patches, safety, and sensitive equipment used in the military, sports, and automobile industries. For example, silk protein is utilized in the production of bullet-proof jackets, tyres, and airbags for automobiles.

6.4.2 SPIDER SILK-BASED CAPSULE TO DELIVER VACCINES

Crosshairs made from spider silk are used in microscopes, telescopes, firearms, and bomb-guiding systems. It has now been discovered to have a wide range of unique features. Small nanocrystals immersed in an amorphous matrix of molecular nano-springs are formed during silk spinning by amino acids. Elasticity is provided by the amorphous regions, whereas strength is maximized by the nanocrystalline domains.

Spider silk microcapsules that can transport vaccinations straight to the center of immune cells have been produced by scientists (Lucke et al., 2018; Rosen, 2020). In the fight against cancer, vaccines are being developed that activate the immune system, allowing it to recognize and eliminate cancerous cells. However, it is not always possible to get the desired immunological response. Researchers discovered these new spider silk microcapsules in order to boost the efficiency of vaccines on the immune system and, in particular, on T lymphocytes trained in detecting cancer cells. Silk-based nanoparticles have also been extensively studied to deliver anti-cancer agents like paclitaxel, doxorubicin, floxuridine, methotrexate, etc (Figure 6.7).

Immune cells can now be activated by spider silk microparticles that have been successfully introduced into the body without harming healthy cells. B lymphocytes are essential to the human system. Immune protection against a wide range of pathogens is provided by their antibodies and T cells. Carole Bourquin, a researcher at the University of Geneva, said that the development of immune healing drugs is effective against cancer. For T cells to have any effect, they must be activated, and this activation process is more involved than that of B lymphocytes. T-cells are a major target of modern vaccinations. There is a cation in T lymphocytes that stimulates them.

A synthetic spider silk biopolymer was used by scientists. They're made of materials that are light, biocompatible, and nontoxic, and are also extremely resistant to heat and light. When the body is left with only a limited amount of protein, it quickly depletes it before it reaches its destination. A peptide with vaccination characteristics was inserted into this particular silk by researchers in Geneva. Peptides in vaccines can be preserved in the body by transporting them on silk microparticles (Kennedy, 2018). In order to boost T lymphocyte responses, they deliver the peptide to the core of lymph node cells. They also say that the protein chains then disperse to produce injectable microparticles by the aforementioned investigators. According to them, the validity of their research has been proven. Their demonstration demonstrated the effectiveness of a new vaccination program that is both long-lasting and simple to implement.

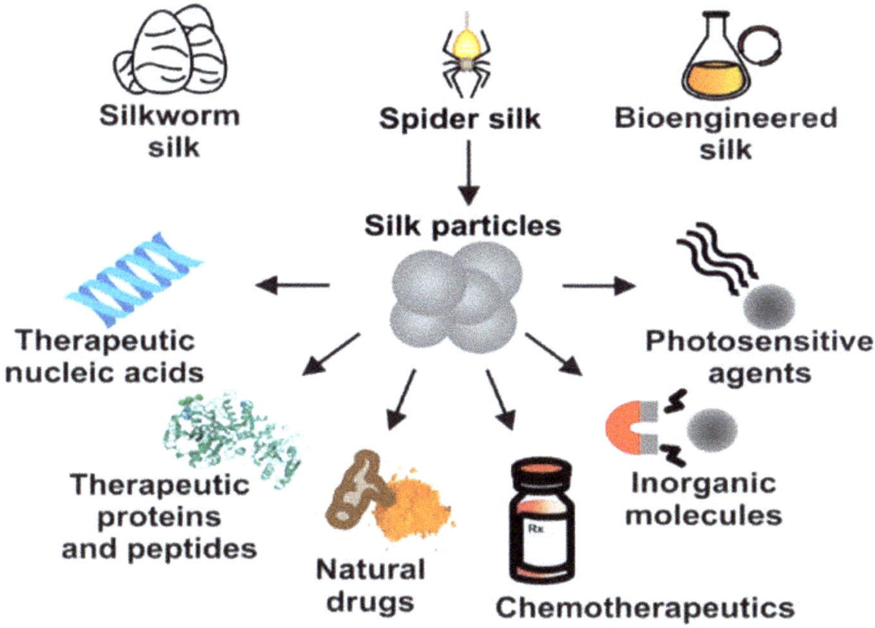

FIGURE 6.7 The proteins derived from natural silkworm cocoons, spider silks, and biotechnologically produced silk-derived proteins are used in the production of DDS for the systemic delivery of a variety of molecules used in cancer treatment.

Source: Reprinted with permission from Florczak et al., 2020. Copyright © 2020 by the authors. Licensee MDPI, Basel, Switzerland. http://creativecommons.org/licenses/by/4.0/.

Vaccine protection is a major issue in many developing countries, and this is a clear benefit. When it comes to heat resistance, the artificial silk biopolymer particles shine. The microparticle size is a requirement for this approach.

6.4.2.1 CANCER TREATMENT USING SILK PARTICLES

Today, a wide range of cancer treatments are available to patients. Patients are given a number of treatment choices depending on the type of cancer and how tough it is to treat. All of these treatment methods have significant side effects. Cancer treatments that are effective, precise, and minimally invasive have been developed in recent years as a result of extensive study. Non-traditional cancer treatments are currently available. Nanomedicine, immunotherapy, and gene therapy, magnetic hyperthermia, thermal ablation, and extracellular vesicles are only few of the examples of these technologies. As a leading cause of death, tobacco is second only to smoking as a global killer. Today's cancer research shows that the tumor microenvironment, not simply cancer cells, is a significant target for cancer treatment (TME). Nanoparticle-based drug control systems are combined with TME (Florczak et al., 2020; Pérez & Fernandez, 2015; Holland et al., 2018).

Aside from being less intrusive, nanoparticle-based cancer treatments can be more effective, precise, and less costly than conventional cancer treatments. Researchers are also examining the effects of silkworm cocoon protein nanoparticles (NPs) (such as silk fibroin and sericins) and spider silk protein NPs (spidroins), in addition to cancer research. This natural-derived material will give biocompatibility, biodegradability, and ease of production. Size, stability, drug loading/release dynamics, and functionalization with certain ligands are a few of the characteristics that can be tailored for the protein-based material, for example.

6.4.3 THE USE OF SPIDERWEB-INSPIRED DOUBLE-SIDED TAPE COULD REVOLUTIONIZE THE SURGICAL FIELD

Every year, over 1 million major surgeries are performed around the world. Using sutures to close the incision might put undue strain on the tissues, leading to infection, discomfort, and scarring. In a matter of seconds, a new adhesive may be applied to wet surfaces to treat wounds or implant medical devices. In just 5 seconds, the two-sided adhesive dries tissue and forms a firm bond. The goal of the research is to keep adhesives strong even when exposed to difficult conditions. When wet, most adhesive things, whose bandage is exposed to water, doesn't work. Adhesives were needed in a wide range of industries, including automotive, aerospace, and construction. When it comes to mending wounds following abdominal surgery, long-lasting humidity is critical. Spiders make silk in order to build a web that acts as both a home for themselves and a trap for their prey. Once the prey is entangled in these naturally occurring threads, they lose their ability to move. Spider silk has been recommended as a medical tape or adhesive by various scientists or researchers in the contemporary day because they feel it is sticky enough to cover wounds or gauzes and strong enough to stay in place for lengthy periods of time (Daley, 2018). MIT researchers heeded the advice of a questionable source in the medical area. It is possible to gather insects even on rainy mornings thanks to the webs of spiders. A new tape (Trafton, 2019) was designed using spider webs as a guide, and it works well in humid environments. This is what the BBC reported about spiderwebs (Figure 6.8), which include charged polysaccharides that quickly absorb insects' moisture. An innovative polymer was created by the researchers that quickly wicks away moisture from the tissue before bonding to simulate spiderweb patterns. Polyacrylic acid was chosen as the water-absorbing material since it is often found in disposable diapers. Water is absorbed by the acid, which forms weak hydrogen bonds that temporarily bind the constituents.

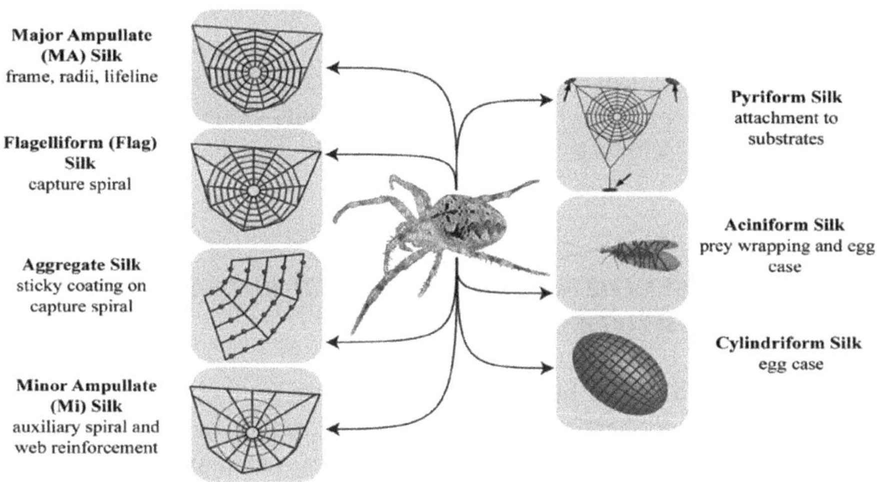

FIGURE 6.8 Spider females weave their silk: A visual guide (Araneae). Each silk variety has a unique use in mind.

Source: Reprinted with permission from Liu & Zhang (2014) https://creativecommons.org/licenses/by/3.0/

In only five seconds, N-Hydroxysuccinimide (NHS) esters sown in polyacrylic acid get over and form stronger covalent connections with tissue proteins. On the basis of necessity, the bandage is reinforced with gelatin. MIT mechanical engineer Xuanhe Zhao estimates that 230 million major surgeries are performed each year around the world. To close the wound, many of them require sutures. This tension on the tissues might cause infections, discomfort, or scarring. In trials on pigs, they effectively enhanced skin, small intestine, stomach, and liver tissue after testing the new substance first on human volunteers. Materials scientist Hyunwoo Yuk of MIT highlighted that using double-sided tape to encapsulate delicate tissues like the lung and trachea is fascinating because it can be done in about five seconds. It's possible that the tape may also be used to implant medical devices on organs such as the heart.

Researchers (Jain et al., 2014) believe that translational approaches in clinical studies will be significant. Spider silk contains 98% water, according to a new synthetic material made by scientists. Specifically, Yu Shrike Zhang of Harvard Medical School states that it uses basic engineering applications, such as suturing, in surgical operations. It is possible to employ the material in a variety of ways, including making parachutes, protective gear, and even the bodies of airplanes. Additionally, the silk proteins might be coated with antibiotics or other drugs, which would reduce the risk of infection, making them biocompatible and less likely to produce an inflammatory reaction. Implanting medical devices into organs such as the liver or heart with double-sided tape prevents tissue damage and injury. Animal testing will be conducted by the scientists in order to perfect the tape before human trials can be conducted.

6.4.4 ON THE SYNTHETIC SPIDER SILK TECHNOLOGY, AIRBUS COLLABORATES WITH AMSILK

Am Synthetic Spider Silk being a light and pliable synthetic fiber. It is now ready for commercialization. With the help of Airbus, a premier biotech business, AMSilk, has developed a revolutionary new composite material production process using artificial spider silk. As far as natural fibers go, spider silk has to be up there at the top. If it were strong enough, a spider web made of fibers as thick as a pencil might hold an Airbus A350 airplane, which weighs around 200 tons when fully loaded. Scientists have been searching for spider silk's special properties for decades. All of these efforts have been in fruitless up until this moment (https://www.airbus.com/newsroom/stories/strong-as-a-spider-web.html).

Artificial spider silk is being sold for the first time by AMSilk, a biotech startup in Munich. Aerospace is poised to take use of AMSilk, a biotech startup with Airbus. The business has already completed the development of this high-performance, completely biodegradable material for medical equipment. For implants, the human body is more readily accepted as a covering, although silk compounds in cosmetic treatments are more commonly used. In fact, a shoe prototype featuring a substantial sports clothing design was developed in 2016 for the brand. Together, Airbus, and AMSilk will create an entirely new composite material field. AMSilk currently produces silk on an annual metric tonnage scale. However, this isn't yet capable of taking it into space. Steel could be compared to it. In the battleship steel form of "spider silk," people appear to prefer.

To create silk, spider DNA is decoded and then incorporated into the bacteria that make up the silk-producing bacterium AMSilk. It is possible to make an exact copy of this product using these methods. Now, the company makes 60,000-liter storage tanks. These are filled with water heated to 37°C in order to grow bacteria. A powder is transformed into a fiber, film, or gel as a result of the process (Sadowy, 2018). The utilization of carbon fiber composites has improved to lower the weight of airplanes. Fuel consumption is expected to be minimal. AMSilk's Biosteel® fibers, on the other hand, are more flexible and shock resistant.

6.4.5 SCIENCE OF SUPERHEROES: E. COLI BACTERIA MODIFIED BY SWEDISH SCIENTISTS PRODUCE AMAZING SPIDER SILK

If you live in a house, you're more than likely to have a stash of one of the world's finest materials in your basement. Arachnid silk strands are used to make spider webs and their draglines, which are extremely elastic. Greek and Roman warriors used spider silk's antibacterial capabilities as a wound treatment.

The concept of raising spiders only for the purpose of harvesting their silk is a nightmare come true. Spiders are attempting to work, as the BBC reported. In the absence of other prey, they are likely to turn to cannibalism. A brilliant new concept is hatched by Canadian scientists. Spider genes that encode for dragline silk proteins were removed from spider and inserted into goat cells in order to produce silk in the milk. According to the researchers, the method of using goat milk to extract spider silk proteins proved successful. In contrast, a new crop of companies, such as Sweden's Spiber Technologies AB, are attempting to sell synthetic spider silk.

Spiber (https://spiber.se/) is a biomaterial with a special set of characteristics. Spiber Inc., a Japanese biotech business founded in September 2007, uses cutting-edge synthetic biology, polymer, and material science to create Brewed Protein, new protein materials. Spiber, a biomaterial with a remarkable combination of properties, is made from recombinant spider silk protein. Bioactivity spider silk can only be produced by Spiber (Figure 6.9), the only firm in the world capable of doing so. Multiple spider silk proteins can be produced using Spiber's two- and three-dimensional processing capabilities. A biomaterials business based in Stockholm is developing genetically modified microbes. The spidroin proteins found in dragline silk can be purified using GE protein purification technology, which produces huge quantities of these proteins. There are components of synthetic spider silk that can connect to cells and promote wound healing. It's been mentioned by Kristina Martinell, the director of production at Spiber, that they utilize it for bioengineering and cell cultivation. In other words, it's a custom-made biomaterial. In the beginning, Spiber uses the gene sequence that spiders employ to send spidroin. Genes are transferred into *E. coli* bacteria rather than goats for bioreactor development. Purification systems are used to remove proteins from the samples before they may be used for other life sciences applications. Pezhman Mohammadi etal (2021) extensively characterized the physio-chemical nature of the liquid–liquid phase separation (LLPS) for some proteins (Figure 6.9).

Spiber's mass is directly tied to the way pharmaceutical companies are currently employing. They had to adapt to the "stickiness" of spider silk protein in order to get the job done. This protein has a tendency to be a little heavy by nature. To get the best results, it must be used in a precise way. For the first time in 2011, By using Spiber's unique protein sample to test GE's technology, Spiber was able to test and analyze the technique. "It's amazing to see GE's equipment and potential know-how offered to small firms," Martinell says. When it comes to different types of applications, the company's method has changed through time to maintain the material liquid. When it comes to making film and foam, Spiber can also use spider silk fibers and even mesh. As powerful as mammalian muscle, the substance is. Even at boiling temperatures, it doesn't change. For example, the company uses spider silk in the field of cardiology, regenerating bone, and skin tissue as well as developing vaccinations.

6.4.6 SPIDER SILK-BASED BIOMATERIALS ARE PROMOTING INNOVATIONS IN INDUSTRIES

The large-scale biotechnological development of spider silk proteins has allowed for novel biomaterials uses to be achieved. In spider silk, tensile strength and extensibility are linked in a unique way. Artificial fibers like polyaramids and other performance fibers and polyamides aren't the only materials used in natural fiber production. This green polymer, which comprises nearly all proteins, is completely biodegradable, unlike plastics. From medicine to textiles, spider silks have a wide range of applications.

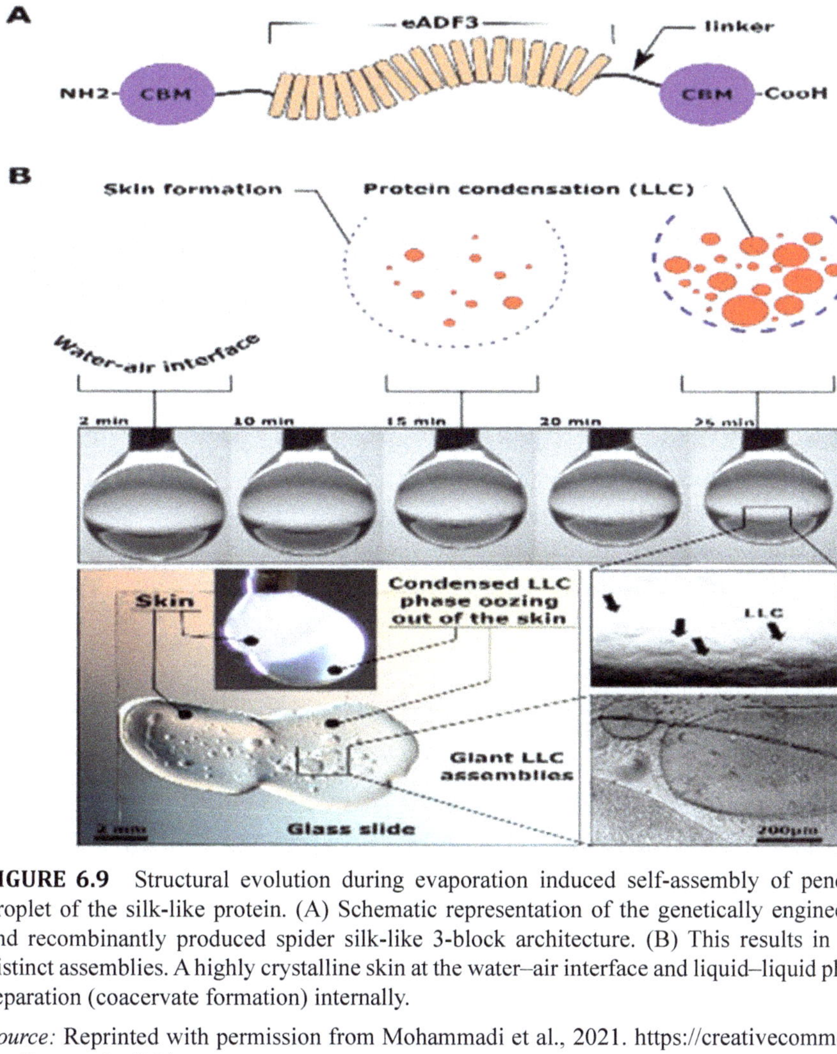

FIGURE 6.9 Structural evolution during evaporation induced self-assembly of pendant droplet of the silk-like protein. (A) Schematic representation of the genetically engineered and recombinantly produced spider silk-like 3-block architecture. (B) This results in two distinct assemblies. A highly crystalline skin at the water–air interface and liquid–liquid phase separation (coacervate formation) internally.

Source: Reprinted with permission from Mohammadi et al., 2021. https://creativecommons.org/licenses/by/4.0/.

6.4.7 SPIDER WEBS USED TO STOP WOUNDS

Scientists recognized the experiments with spider silk fibers as joints from the 18th century onwards. A significant problem in the past has been low producibility. The development of a biotechnological technology that allows their team to produce recombinant proteins produced from silk was a breakthrough. AMSilkGmbH (Martinized, Germany) launched products on an industrial scale at attractive prices. By combining this knowledge with the material's unique qualities, AMSilk has created entirely new products. There is a growing demand for the group's creative technology. Sustainable and biodegradable single biogenic materials generate one of the most extensive tests of time and reduce the burden of projected crude-oil obtained outcomes.

6.4.8 SILK PROTEINS PROCESS: BIOMIMETIC SPIDER SILK FIBERS

The University of Bayreuth biomaterials group prepared recombinant types of spider silk proteins. Morphologies such as non-woven mesh and non-woven films can be found in a wide range of forms. One of the most critical processing was producing an artificial spider silk fiber with a toughness blueprint. The commercial fiber Biosteel® developed by AMSilk GmbH is based on spider silk technology. Spider silk-based products will be the primary focus of the researcher's work. Different production hosts (bacteria, yeast, eukaryotic and insect cells) are shown, as well as various spider silk morphologies (foam, fiber, film, hydrogel, and non-woven mesh) and finally potential applications in tissue engineering (Figure 6.10).

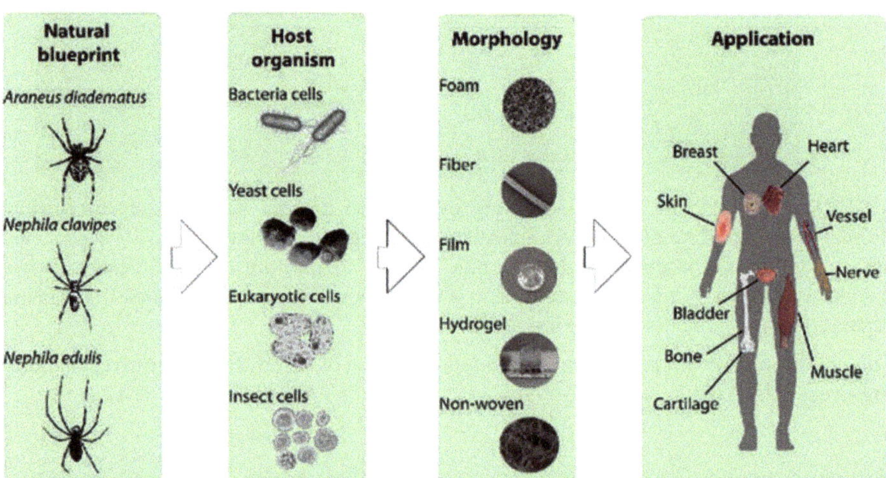

FIGURE 6.10 Schematic demonstration showing the development of the recombinant spider silk-based biomaterials. Natural spider silk serves as a blue print for recombinant spider silk production. Different production hosts (bacteria, yeast, eukaryotic and insect cells) are shown, as well as various spider silk morphologies (foam, fiber, film, hydrogel, and non-woven mesh) and finally potential applications in tissue engineering.

Source: Reprinted with permission from Sahar Salehi et al (2020). https://creativecommons.org/licenses/by/4.0/

6.4.9 SILK'S BIOMEDICAL APPLICATIONS AND DRUG DELIVERY SYSTEMS

The biological applications of materials are being studied in depth by scientists as a means of improving drug delivery. Different morphologies, such as particles and thin-walled capsules, can be used as both fixed and drug control systems. Spider silk proteins can be genetically altered to disclose cell-targeting sequences on the control system's surface, allowing for a cell/tissue-specific medication delivery.

Because of its safety, biocompatibility, controllability, and ability to be combined with other materials, silk fibroin hydrogels have garnered a lot of attention in the scientific community. Wound healing, drug administration, antiadhesive membranes, soft tissue rebuilding, and cell culture are just a few of the biomedical uses for hydrogel films. They offer a wide range of potential applications in a variety of sectors, including biomedicine.

6.4.10 BIOMEDICAL USES OF SILK IMPLANT LAYERS

Different bodily responses can be produced by the adsorption of proteins by monocytes and macrophages on the surface of a common biomaterial implant. To our knowledge, there has been no or only a minimal free reaction to spider silk in humans, reducing the danger of infection. Some people may develop an allergy to spider silk-based substances in the future. Spider silk can be used to cover up silicone breast implants that have been inserted. Breast reconstruction is achieved using silicone implants, which are currently associated with major complications. Recently, surface modification has proved to be an effective clinical approach to prevent bacterial adhesion, reduce capsular thickness, prevent foreign body reactions, and reduce other implant-associated problems. Schematic illustration of Phase 1 after implantation-initiated foreign body reaction and synthesis of a matrix around silicone breast implant shown in Figure 6.11.

6.4.11 BIOMEDICAL IMPORTANCE OF SILK FABRICATION

As a result of biofabrication, cells, and biomaterials can be combined to create tissue-like structures. In this study, hydrogels derived from spider silk appear to be biomimetic materials. Cell compatibility, elastic composition, and shear-thinning behavior that allows dispense blotting are among the potential applications for this substance, according to some. They immediately improve their fitness through biophysical cross linking.

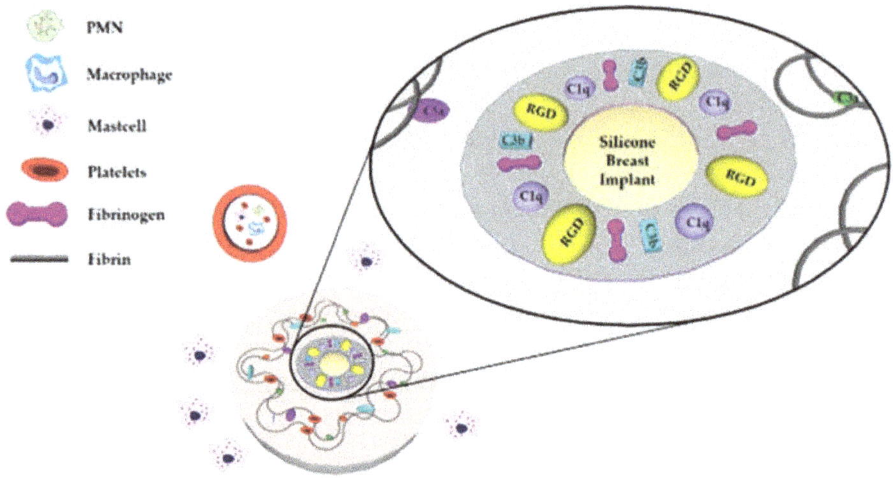

FIGURE 6.11 Schematic illustration of Phase 1 after implantation-initiated foreign body reaction and synthesis of a matrix around silicone breast implant.

Source: Reprinted with permission from Omar Faruq et al., 2021. © 2021 by the authors. Licensee MDPI, Basel, Switzerland. https://creativecommons.org/licenses/by/4.0/

6.5 SPIDER SILK: THE INCREDIBLE BIOMATERIALS

6.5.1 SPIDER SILK'S BIOMEDICAL APPLICATIONS

An implant or graft that can reliably, economically, and safely restore function to damaged bodily tissue is considered a biomaterial by definition. Attractive biomaterial that can be used in many different applications due to its bioactivity, immunocompatibility, and biocompatibility, among other attributes. In many biomedical applications, biomaterial mechanical robustness restricts its usefulness. Finally, the Biomaterial must be able to be treated in a variety of ways. Past biomaterial development advancement has fueled an innovative idea for creating an optimum biomaterial for therapeutic entities through the production of unique biomaterials. Polymers have been leading the way in biomaterial innovation.

One of nature's wonder materials is spider silk (Figure 6.12). When it comes to weight, it's a lot stronger than steel, but it's also more flexible than nylon. This natural protein is used in the construction of websites and nests.

Silk from spiders has been used medicinally for generations because of its ability to glide through the air and evade predators. Honey and vinegar were traditional ingredients in the dressings used to treat wounds in ancient times. Silk's unique characteristics aid in the delivery and healing of drugs (Cuffari, 2020; Chouhan & Mandal, 2020).

FIGURE 6.12 Appearance of natural spider silk.

Source: Reprinted with permission from Gu et al., 2020. https://creativecommons.org/licenses/by/4.0/.

6.5.2 BIOMEDICAL PURPOSES

The silk proteins sericin and fibroin are thought to be wound healing agents because of their anti-inflammatory properties. They are human body antioxidants and bio-adhesives. In the last few years, numerous research has proven that it is an excellent material for biomedical purposes. The molecular structure, moderate degradation rate, and high processability all have a role in the material's mechanical toughness. Silk sericin paved the way for the use of silk in various tissue regeneration techniques. Sericin protein can also be cross-linked, co-polymerized, and integrated into other macromolecular complexes.

6.5.3 WOUND REPAIRING

Skin gives an external emergency barrier to our internal organs. When a wound occurs, a cavity compromises the skin tissue, usually healthy structure, and function. Bioactive dressing materials, burn injuries, surgical procedures, and diabetic ulcers might cause problems if the skin is unable to heal itself. Wounded people have benefited from the wound-healing properties of spider silk. Burns treated *in vivo* with spider silk show remarkable wound healing characteristics. Certain spider silk dressing activities could result in pain due to their natural deterioration. Using spider silk fibers, which are well-known for their wound-healing properties, has become a common practice in the treatment of ophthalmic, neurological, and cardiovascular diseases. Antibacterial properties of this material are limited by the strength of spider silk. Therefore, antimicrobial chemicals that are not compatible with the silk fiber's tenacity, knot strength, and surface friction are typically used in Spider silk stitching.

6.5.4 TISSUE ENGINEERING

Developed by the biomaterial's community, tissue engineering works to create useful tissues by fusing scaffolds, cells, and chemicals with biological activity. With the use of tissue engineering, scientists hope to create functional constructions that can be used for tissue repair and regeneration. An example of FDA-approved manufactured films is synthetic skin and cartilage, but they are currently used in only a small number of human patients. There is currently little use of tissue engineering in inpatient care (Kearns et al., 2008). Heart, lung, and liver are other vital organ tissues that are reproducible in the laboratory. In order for tissue engineering to succeed, it is necessary to have bladders, tiny arteries, skin grafts, cartilage, and even implants for the trachea. These tissues are extremely useful in drug testing and research. For individualized therapy, and for minimizing the number of animals used in research, it provides essential tools.

Cell adhesion is a major mechanism in the construction and maintenance of tissues. In order to produce functional tissues, it is necessary to choose and cover the necessary physical, chemical, and biological stimuli for cell migration, adhesion, and differentiation. The biomaterials utilized in this design must degrade at the same pace as the new ECM substitute. Silk-like polymers (SLPs) can be genetically modified, and bioactive areas like fibronectin can change their physical properties.

There are numerous ways in which glycoprotein fibronectin contributes to the body. It's a key component in a number of cell biology procedures. One of the three types of fibronectins is detected in matrix metalloproteinase-2 (MMP-2) and MMP-9 proteins, and FNII is present. The FNII coding was shown to be genetically connected to the spider silk protein (6 mer). Scientists are currently working to improve the efficiency of biomaterials. Other closely comparable materials include polylactic acid (PLA), fibroin mix films, and recombinant spider silk films. When tested *in vitro*, this biomaterial did not produce any skin fibroblasts and was found to promote cell adhesion.

6.5.5 SPIDER SILK PARTICLE SURFACE ALTERATION–BIMOLECULAR CORONA FORMATION

Biomaterials are important for human health and quality of life by, e.g., supporting, replacing or restoring injured or destroyed tissues. Apart from synthetic coating materials, natural proteins such as silk or the extracellular matrix proteins collagen, laminin or fibronectin show bioactivity and trigger biological responses and cell interactions. Coatings made of native as well as recombinant spider silk proteins are promising candidates due to their intrinsic properties (Figure 6.13). They are biocompatible, biodegradable, non-toxic, processable and do not activate immune cells of the inflammatory cascade.

Spider silk-based products are known for their biodegradability, processability, and biocompatibility. These interactions between silk biomaterial surfaces and biological connections are crucial for drug delivery and implant coatings that promote tissue repair. To create a biomolecular corona, proteins, and blood can be added to the silk of spiders (Weiss et al., 2020). The biomolecular corona's composition was evaluated in relation to surface net charge, according to the research. In the biomolecular corona investigations, proteins derived from fibrinogen were found to surface-attach to positively charged spider silk materials. The enrichment of fibrinogen has a connection to blood coagulation. Blood clots can be prevented by applying negative surface charges.

Numerous spider silk proteins can be genetically engineered to alter the biomolecular corona's composition, and many of these can be utilized to carry drugs or dock cells. Biomolecular corona composition is influenced by the surface net charge of recombinant spider silk surfaces, which in turn affects visible outcomes including fibrin synthesis and the formation of blood clots (Weiss et al., 2020; Liu et al., 2021).

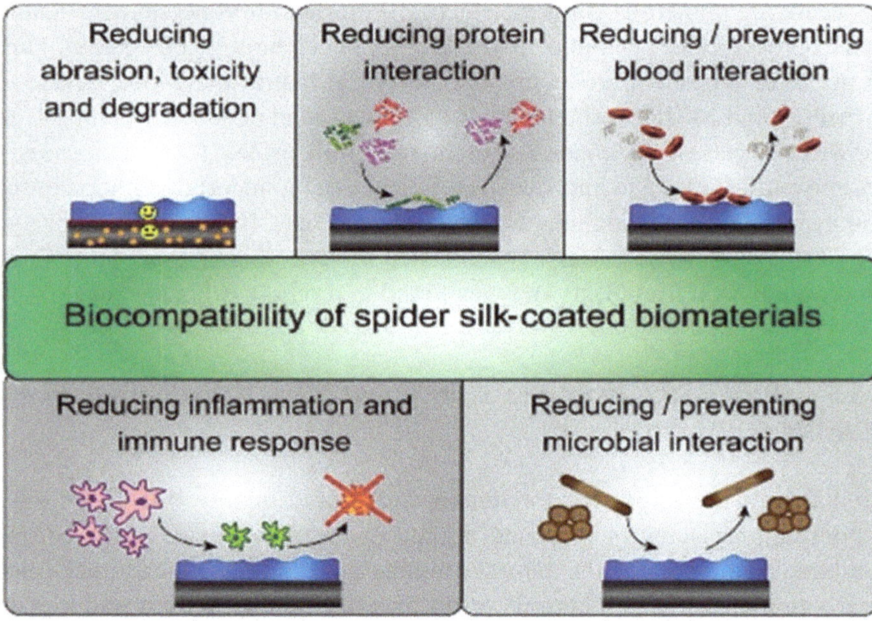

FIGURE 6.13 Spider silk coatings enhance the biocompatibility of biomaterials by reducing or preventing the adsorption of proteins, blood coagulation and microbial interaction. Another benefit is the reduction in abrasion, degradation and release of toxic compounds of the underlying material.

Source: Reprinted with permission from Trossmann et al., 2023. Copyright © 2023 by the authors. Licensee MDPI, Basel, Switzerland. http://creativecommons.org/licenses/by/4.0/.

6.5.6 APPLICATION IN MEDICINE

Using spider silk in the medical field has huge potential (Kuhier et al., 2011; Hennecke et al., 2013; Kim et al., 2012). Inflammation and cell adhesion are both inhibited by native silk, which is very biocompatible. Kerstin Reimers-Fadhlaoui and her colleagues discovered that it works well as a suture or replacement for tissues. In a recent study, Prof. Dr. Thomas Scheibel and colleagues found that coating a silicone implant with spider silk proteins reduces the risk of inflammatory and fibrotic problems after surgery. Biosensors (films) and medication delivery systems can both use silkworm silk as a matrix to stabilize enzymes and antigens for use as biosensors (films) (particles). It has been shown that bio-integrated devices can be created using conformal electronics, and silkworm silk implants can be used to generate cartilage or bones (foams).

6.6 FUTURE PERSPECTIVES

Biopolymers have been around for hundreds of years, and spider silk is one of them. Recombinant spider silk proteins can be used for a wide range of biological and industrial purposes (Kaur & Kaur, 2019). In the future, recombinant spider silk biomaterials could be used in anything from protective vests to parachutes to the building of skyscrapers. Shampoos, soaps, creams, and nail varnishes can all be made with silk fibers. Biomaterials such as insect silk, which is largely composed of large proteins, are among the strongest and most long-lasting currently known to mankind. Biodegradable and flexible with a high tensile strength, it is also biodegradable (Kaur & Kaur, 2019). According to study, spider silk contains antibacterial and hypoallergenic molecules. The biodegradability of spider silk makes it superior to manmade fibers.

6.6.1 FUNCTIONAL SPIDER SILK-BASED HYBRID MATERIALS

Silk from silkworms is often regarded as a high-quality textile with a wide range of useful characteristics. There has been a recent review of current textile invention methods by Kiseleva et al. (2019). Because of its mechanical qualities, spider silk has been studied in detail in terms of the molecular structure. This information is essential to gaining a better understanding of the inherent properties of spider silk. Fusions between natural and synthetic spider silk proteins and inorganic NPs have been described here. A better understanding of how silk proteins are formed into diverse polymers can be gained from this research, which could be used to produce new products. Additionally, the diversity of spider silk, its composition and architecture, the contrasts between the two, and the biosynthesis of natural silk are all explored.

6.6.2 ENGINEERING SPIDER SILK (BASED ON 2D AND 3D MATERIALS) IN THE PREVENTION OF MICROBIAL INVASION

It is critical to understand biofilm development in the study of antibiotic-resistant microbial strains. Developing innovative materials has considerable clout. In nature, some spider silks have microbe-resistant characteristics. Biomaterials could benefit from this. They prevent microbial adhesion on surfaces while boosting cell growth in mammals. Scientists tested the ability of recombinant SpiderSilk proteins to resist microbes, produced from wire silk proteins from *Araneus diadematus*, to be incorporated into two-dimensional films and three-dimensional hydrogels (Figure 6.14) (Kumari et al., 2020).

Microbe-repelling capabilities of proteins at the nanoscale have been discovered. Inhibition of microbial infestations by spider silk-based synthetic materials can be seen in Figure 6.14 (spider silk-based synthetic materials) (Kumari et al., 2020).

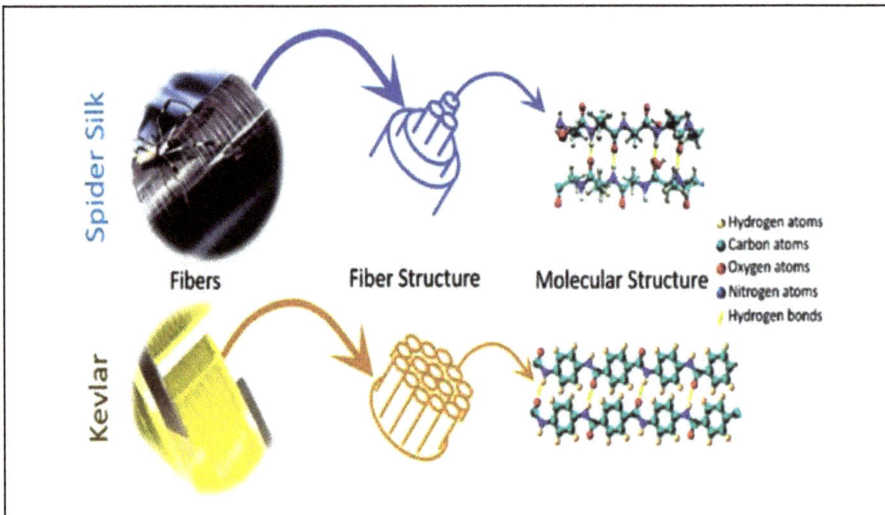

FIGURE 6.14 Schematic illustration of spider silk-based hybrid material production via biomineralization approach.

Source: Reprinted with permission from Kiseleva et al., 2020. Copyright © 2020 Kiseleva, Krivoshapkin and Krivoshapkina. https://creativecommons.org/licenses/by/4.0/.

6.6.3 SILK AS A BIOMATERIAL FOR CANCER THERAPY IS A NOVEL APPROACH

Textiles and surgical sutures are two applications for silk that have been studied by scientists (Jastrzebska et al., 2005). These include biocompatibility, biodegradability, self-assembly, and self-assembly control. It possesses a variety of mechanical qualities that make it an ideal biomedical material. Only synthetic methods allow Silk to be transformed into a variety of different morphologies, which it does in nature. Silk proteins may now be produced biotechnologically, which expands the range of possible uses for synthetic silk genes. Structure and a polypeptide sequence that allows a glimpse of silk biomaterial are present in bioengineered hybrids. For example, functional areas may include receptors, enzymes, and medicines, metals, and carbohydrates, as well as other types of molecules. Many biomedical professions, though, may benefit from the same rules. In the field of oncology, silk biomaterials have garnered a lot of attention. For cancer research, special attention is likely to be paid to developing implantable, injectable, and preferred drug delivery systems, as well as 3-D cancer models using silk scaffolds.

KEYWORDS

- **bimolecular corona**
- **biomaterials**
- **bulletproof jacket**
- **cancer therapy**
- **microcapsules**
- **spider silk**

REFERENCES

Aigner, T. B., DeSimone, E., & Scheibel, T., (2018). Biomedical applications of recombinant silk-based materials. *Adv. Mater., 30*(19), e1704636. doi: 10.1002/adma.201704636.

AM Silk Announces Partnership with Airbus, (2018). Planegg, https://www.bloomberg.com/press-releases/2018-09-12/dgap-news-amsilk-announces-partnership-with-airbus-to-develop-the-next-generation-of-composite-fibers-for-lightweight-high-p (accessed on 30 September 2023).

Asakura, T. (2020) Structure and dynamics of spider silk studied with solid-state nuclear magnetic resonance and molecular dynamics simulation. *Molecules 25*(11), 2634; https://doi.org/10.3390/molecules25112634

Basu, A., (2015). *Advances in Silk Science and Technology.* Cambridge: Woodhead Publishing.

Carlson, S., (2018). *Army to Test Body Armor Made from Spider Silk.* Defense News. https://www.upi.com/Defense-News/2018/08/20/Army-to-test-body-armor-made-from-spider-silk/3531534778738/ (accessed on 17 August 2023).

Chouhan, D., & Mandal, B. B., (2020). Silk biomaterials in wound healing and skin regeneration therapeutics: From bench to bedside. *Acta Biomater., 103*, 24–51. doi: 10.1016/j.actbio.2019.11.050.

Cuffari, B., (2020). *Role of Spider Silk in Biomedicine*. News Medical Life Sciences. https://www.news-medical.net/life-sciences/Role-of-Spider-Silk-in-Biomedicine.aspx (accessed on 17 August 2023).

Daley, J., (2019). *Double-Sided Tape Inspired by Spiderwebs Could Revolutionize Surgery.* smithsonianmag.com. https://www.smithsonianmag.com/smart-news/tape-inspired-spider-webs-could-revolutionize-surgery-180973450/ (accessed on 17 August 2023).

De Souza, E. S. J. M., Wehrspohn, R., Liao, X., Agarwal, S., Greiner, A., Khoshkhoo, M. S., & Zhou, F., (2020). *Microscopic Characterization of Polymer Fibers with Mechanical Properties Similar to Dragline Spider Silk.* Carl Zeiss Technical Report. https://www.researchgate.net/publication/343530042 (accessed on 17 August 2023).

Faruq, O., Chien, P.N., Dönmez, N., Nam, S.-N., & Heo, C.-Y. (2021). Functionalization of Silicone Surface with Drugs and Polymers for Regulation of Capsular Contracture. *Polymers 13*(16), 2731; https://doi.org/10.3390/polym13162731

Filippov, S. (2018). New Technological Revolution and Energy Requirements. *Foresight and STI Governance, 12*(4), 20–33. DOI: 10.17323/2500-2597.2018.4.20.33.

Florczak, A., Grzechowiak, I., Deptuch, T., Kucharczyk, K., Kaminska, A., & Dams-Kozlowska, H., (2020). Silk particles as carriers of therapeutic molecules for cancer treatment. *Materials, 13*(21), 4946. https://doi.org/10.3390/ma13214946.

Gerritsen, V. B., (2002). An Airbus Could Tiptoe on Spider Silk. Protein Spotlight, Issue 24, July 2002 (Monthly), https://www.proteinspotlight.org/back_issues/024/ (accessed on 17 August 2023).

Gu, Y., Yu, L., Mou, J., Wu, D., Zhou, P., & Xu, M. (2020), Mechanical properties and application analysis of spider silk bionic material. *e-POlymers 20*, 443–445.

Hennecke, K., Redeker, J., Kuhbier, J. W., Strauss, S., Allmeling, C., Reimers, K., & Vogt, P. M., (2013). Bundles of Spider Silk, Braided into Sutures, Resist Basic Cyclic Tests: Potential use for flexor tendon repair. *PLos One*, *8*, 1–10. doi: 10.1371/journal.pone.0061100.

Hinman, M. B., Jones, J. A., & Lewis, R. V., (2000). Synthetic spider silk: A modular fiber. *Trends Biotechnol., 18*, 374–379. doi: 10.1016/S0167-7799(00)01481-5.

Holland, C., Numata, K., Rnjak-Kovacina, J., & Seib, F. P., (2018). The biomedical use of silk: Past, present, future. *Advanced Health Care Materials*, *8*(1), https://doi.org/10.1002/adhm.201800465.

Hybrid Spider Silk with Inorganic Nanomaterials Aleksandra P. Kiseleva et al., (2020); *Nanomaterials* **2020,** *10*(9), 1853; https://doi.org/10.3390/nano10091853.

https://backyardgoats.iamcountryside.com/kids-corner/the-spectacular-spider-goats (accessed on 17 August 2023).

https://onlinelibrary.wiley.com/doi/10.1002/biot.201700753 (accessed on 17 August 2023).

https://www.compositesworld.com/news/airbus-partners-with-amsilk-on-synthetic-spider-silk-technology-for-lightweight-aircraft (accessed on 17 August 2023).

https://www.ctahr.hawaii.edu/biotechinfocus/images/bulletinpdf/2015_10June_I33Print2.pdf (accessed on 17 August 2023).

https://www.techquila.co.in/scientists-breed-super-goats-that-produce-spider-silk/ (accessed on 17 August 2023).

Jain, D., Sahni, V., & Dhinojwala, A., (2014). Synthetic adhesive attachment discs inspired by spider's pyriform silk architecture, *J. Polymer Sci. Part B Polymer Physics, 52*(8). doi: 10.1002/polb.23453. https://onlinelibrary.wiley.com/doi/10.1002/polb.23453 (accessed on 17 August 2023).

Jastrzebska, K., Kucharczyk, K., Florczak, A., Dondajewska, E., Mackiewicz, A., & Dams-Kozlowska, H., (2014). Silk as an innovative biomaterial for cancer therapy. *Reports of Practical Oncology and Radiotherapy, 20*(2), 87–98. https://doi.org/10.1016/j.rpor.2014.11.010.

Kaur, A., & Kamaljit, K., (2019). Spider silk: An excellent candidate as toughest biomaterial. *Research & Reviews in Biotechnology & Biosciences*, *6*(1), 31–43. http://biotechjournal.in/images/paper_pdffiles/Spi-5d5e9f44c3f8b.pdf (accessed on 17 August 2023).

Kearns, V., MacIntosh, A. C., Crawford, A., & Hatton, P. V., (2008). Silk-based biomaterials for tissue engineering. In: Ashammakhi, N., Reis, R., & Chiellini, F., (eds.) *Topics in Tissue Engineering* (Vol. 4). https://www.oulu.fi/spareparts/ebook_topics_in_t_e_vol4/abstracts/kearns.pdf (accessed on 17 August 2023).

Kennedy, H. T., (2018). *U.S. Army Gets Super Strong Spider Silk Bulletproof Panels*. https://www.biofuelsdigest.com/bdigest/2018/08/26/u-s-army-gets-super-strong-spider-silk-bulletproof-panels (accessed on 17 August 2023).

Kim, S., Miotropoulos, A. N., Spitzberg, A. D., Tao, H., Kaplan, D. L., & Omenetto, F. G., (2012). Silk inverse opals. *Nat. Photon*, *6*, 818–823. doi: 10.1038/nphoton.2012.264.

Kiseleva, A., Kiselev, G., Kessler, V., Seisenbaeva, G., Gets, D., & Rumyantseva, V., (2019). Optically active hybrid materials based on natural spider silk. *ACS Appl. Mater. Interfaces*, *11*, 22962–22972. doi: 10.1021/acsami.9b05131.

Kiseleva, A.P., Kiselev, G.O., Nikolaeva, V.O., Seisenbaeva, G., Kessler, V., Krivoshapkin, P.V., & Krivoshpkina, E.F. (2020a). Hybrid spider silk with inorganic nanomaterials. *Nanomaterials 10*(9), 1853; https://doi.org/10.3390/nano10091853

Kiseleva A. P., Krivoshapkin P. V. and Krivoshapkina E. F. (2020) Recent Advances in Development of Functional Spider Silk-Based Hybrid Materials. Front. Chem. 8:554. doi: 10.3389/fchem.2020.00554.

Kuhbier, W., Reimers, K., Kasper, C., Allmeling, C., Hillmer, A., Menger, B., Vogt, P. M., & Radke, C., (2011). First investigation of spider silk as a braided microsurgical suture. *J. Biomed. Mater. Res. B., 97B*, 381–387. doi: 10.1002/jbm.b.31825.

Kumari, S., Lang, G., DeSimone, E., Spengler, C., Trossmann, V. T., Lücker, S., Hudel, M., Jacobs, K., Krämer, N., Scheibel, T., et al., (2020). Engineered spider silk-based 2D and 3D materials prevent microbial infestation. *Materials Today, 41*, 21–33.

Lefèvre, T., & Auger, M., (2016). Spider silk inspired materials and sustainability: Perspective, *Materials Technology*, *31*(7), 384–399. doi: 10.1179/1753555715Y.0000000065.

Liu, X., & Zhang, K. Q., (2014). Silk fiber—Molecular formation mechanism, structure-property relationship and advanced applications. In: Lesieur, C., (ed.), *Oligomerization of Chemical and Biological Compounds.* IntechOpen. doi: 10.5772/57611. https://www.intechopen.com/chapters/46460.

Liu, Y., Huang, W., Meng, M., Chen, M., & Cao, C., (2021). Progress in the application of spider silk protein in medicine. *J. Biomaterials Appl.*, *4*, 08853282211038. https://doi.org/10.1177/08853282211003850.

Lucke, M., Mottas, I., Herbst, T., Hotz, C., Römer, L., Schierling, M., Herold, H. M., et al., (2018). Engineered hybrid spider silk particles as delivery system for peptide vaccines. *Biomaterials.*, *172*, 105–115. doi: 10.1016/j.biomaterials.2018.04.008.

Mohammadi, P., Zemke, F., Wagermaier, W., & Linder, M.B. (2021). Interfacial crystallization and supramolecular self-assembly of spider silk inspired protein at the water-air interface. *Materials 14*(15), 4239; https://doi.org/10.3390/ma14154239

Nguyen, C., (2016). *The Army's Future Body Armor Could be Made of 'Spider Silk.'* Tech Insider. https://www.businessinsider.com/the-us-army-wants-to-make-spider-silk-body-armor-2016-7?IR=T (accessed on 17 August 2023).

Pérez-Herrero, E., & Fernández-Medarde, A., (2015). Advanced targeted therapies in cancer: Drug nanocarriers, the future of chemotherapy. *Eur. J. Pharm. Biopharm.*, 52–79. doi: 10.1016/j.ejpb.2015.03.018.

Rosen, K. R., (2020). *The Long, Fraught History of the Bulletproof Vest*. Smithsonian Magazine. https://www.smithsonianmag.com/innovation/long-fraught-history-bulletproof-vest-180974564/(accessed on 17 August 2023).

Sadowy, B., (2018). *Airbus Partners with AMsilk on Synthetic Spider Silk Technology*. Available at: Sadowy B. (2018). Airbus Partners with Amsilk on Synthetic Spider Silk Technology. Available at: http://compositesmanufacturingmagazine.com/2018/09/airbus-partners-with-amsilk-on-synthetic-spider-silktechnology/, accessed 30 September 2023).

Salehi, S., Koeck, K., & Scheibel, T. (2020). Spider silk for tissue engineering applications, *Molecules 25*(3), 737; https://doi.org/10.3390/molecules25030737

Schierling, M., Herold, H. M., Slotta, U., Spinetti, T., Scheibel, T., Winter, G., Bourquin, C., & Engert, J., (2018). Engineered hybrid spider silk particles as delivery system for peptide vaccines. *Biomaterials., 172*, 105–115. doi: 10.1016/j.biomaterials.2018.04.008.

Silk Particles as Carriers of Therapeutic Molecules for Cancer Treatment. Anna Florczak etal (2020); Source; *Materials* **2020,** *13*(21), 4946; https://doi.org/10.3390/ma13214946.

Sofia, S., McCarthy, M. B., Gronowicz, G., & Kaplan, D. L., (2001). Functionalized silk-based biomaterials for bone formation. *J. Biomedical Materials Res., 54*(1), 139–148.

Stilwell, B., (2020). *This Amazing Body Armor is Made from Spider Silk.* wearethemighty.com/mighty-tactical/body-armor-from-spider-silk/ (accessed on 17 August 2023).

The Conversation, (2019). *Why is Spider Silk So Easy to Break When It's Supposedly Stronger Than Steel?* Academic rigor, journalistic flair? https://theconversation.com/curious-kids-why-is-spider-silk-so-easy-to-break-when-its-supposedly-stronger-than-steel-116243 (accessed on 17 August 2023).

Trafton, A., (2019). *Double-Sided Tape for Tissues Could Replace Surgical Sutures.* MIT News Office. https://news.mit.edu/2019/double-sided-tape-tissues-could-replace-surgical-sutures-1030 (accessed on 17 August 2023).

Trossmann, V.T., Lentz, S., & Scheibel. T. (2023). Factors influencing properties of spider silk coatings and their interactions within a biological environment. *J. Funct. Biomater. 14*(8), 434; https://doi.org/10.3390/jfb14080434.

Tucker, P., (2017). *US Army Boosts Spending on Genetically Engineered Spider Silk for Body Armor, Underwear*. Technology Editor. Science & Tech. https://www.defenseone.com/technology/2017/08/us-army-boosts-spending-genetically-engineered-spider-silk-possible-use-body-armor-soldier-underwear/139923 (accessed on 17 August 2023).

Vollrath, F., (2000). Strength and structure of spiders' silks. *J. Biotechnol., 74*(2), 67–83. doi: 10.1016/s1389-0352(00)00006-4.

Weiss, A. C. G., Herold, H. M., Lentz, S., Faria, M., Besford, Q. A., Ang, C. S., Caruso, F., & Scheibel, T., (2020). Surface modification of spider silk particles to direct biomolecular corona formation. *ACS Applied Materials & Interfaces, 12*(22), 24635–24643. doi: 10.1021/acsami.0c06344.

Zheng, K., & Ling, S., (2018). De novo design of recombinant spider silk proteins for material applications. *Biotechnol. J., 14*(1), 1700753. doi: 10.1002/biot.201700753.

CHAPTER 7

Butterfly Wings Inspire New Technologies

ABSTRACT

Many types of butterflies, including Morpho butterflies, are identified in Mexico and South America. The lower wing surface of Blue Morphos brown has iridescent, black, gray, or red colors, and hard blues in the upper wing surface. In most cases, animals' colors are derived from the pigments found in the food they eat. No pigments can generate iridescent blue, violet, and green colors in some butterfly species. Other animals create blue and green using microscopic structures interacting with light. In opposition, pigmentary coloration forms by the consumption of specific colors (wavelengths) of light. Morpho butterfly is usually used to produce yellow, orange, and brown colors. Scientists identified that the Blue Morpho butterfly's beautiful bright wings could extend the range of innovative technologies. As an illustration, consider the delicate, intricate, and dynamic anatomy of a butterfly wing. Materials inspired by butterflies may be useful in heat-resistant aircraft. Scientists were intrigued by the pattern on the wings of butterflies, and this led them to create a light-activated hydrogen sensor. Ultra-precise results can be achieved even at ambient temperature. When it comes to diagnosing digestive illnesses, the technology has the ability to see hydrogen abnormalities, model safety dangers, and even small volumes of gas breath. The careful study of a butterfly's wings can lead to the development of a solar cell that can produce light with two times the intensity. This concept helps the functioning of solar cells and panels. Using bio-inspired surfaces to boost light harvesting on artificial solar cells is a goal that scientists are confident they can achieve. Data analytics created by their Morpho light and temperature sensors are used to identify harmful situations. When a specific sort of light shines on the sensor, the reflected signal changes depending on

Bio-Inspired Technologies for the Modern World: Emerging Trends and Applications.
R. Ramakrishna Reddy & T. Pullaiah (Authors)

the pressure in the eyes. Chemists have long been attracted by the vibrant blue wings of morpho butterflies. Depending on the amount of pressure in the eyes, the reflected signal from the sensor fluctuates. There is something about the bright blue morpho butterflies' wings that has always drawn chemists. A stunning cerulean color reflects off their prismatic forms, which come from the stratified nanoscale protein compositions within the wings. The study of the iridescent wings led to the development of self-cleaning surfaces and explosives detectors. Researchers believe that these structures' resources can be used to produce things.

7.1 INTRODUCTION

Biological materials with different functional properties exist in Nature. After millions of years of natural selection, they've managed to hold on. The wings of butterflies are being studied by scientists primarily because of their vibrant hues. Periodic photonic homes on the scales are the cause of the iridescent colors. Shells, bones, and honeycombs present more of a challenge in terms of natural architecture. Interdisciplinary research has recently shifted its attention to these materials. Structured hues, improved sensors, photonic crystals, and solar cells are just a few of the exciting potential uses for these materials.

For millennia, living creatures have been able to recognize changes in the environment, retain and interpret information, and respond externally. Particularly noteworthy are the intelligent systems' physicochemical properties and stimuli responsiveness, such as the structural color of butterfly wings. For the Morpho butterfly's bright blue iridescence (Figure 7.1) some study groups looked at the structure of the wings (Kinoshita et al., 2008). Synthetic functional materials are synthesized based on biological structures and their intelligence systems. Known for its deep, rich blue hue, the Blue Morpho butterfly has become a cultural icon (Giraldo et al., 2016). Scientists have discovered a unique tiny pattern in the butterfly's wings that might be used for a variety of purposes, from self-cleaning tools to better piping.

Biomimicry takes its cues from a wide variety of natural phenomena. An opal-like substance that could serve as the basis for future intelligent sensors is inspired by them. There are numerous sensor and energy materials based on the micro/nanostructures and light-trapping mechanisms found in butterfly architectures. These bio-inspired sensors and energy materials have enhanced their ability to harness renewable energy, clean up the environment, and keep tabs on people's well-being in general.

FIGURE 7.1 Menelaus blue morpho (Morpho menelaus) dorsal forewings and hindwings are a bright and iridescent blue edged with black.

Photo credit: Didier Descouens. https://creativecommons.org/licenses/by-sa/4.0/.

Insects come in a variety of vibrant hues. Structural colors, which are the result of optical interference, are the most common name for them. Photonic nanostructures on the scales of butterflies' wings give them their vibrant iridescent colors. In addition to their hydrophobic properties, Morpho butterfly wing scales are extremely lightweight, resistant to mechanical stress, and thermally stable. The wavelength, incidence angle, and seeing angle are the primary determinants of optical characteristics. For the Morpho butterfly's bright blue iridescence, some study groups looked at the structure of the wings (Kinoshita et al., 2008).

The source of the blue color of the Morpho butterflies arises from the multilayer interference (Giraldo & Stavenga, 2016; Penz et al., 2012). The regular periodic ridges of the mass lamellae on Morpho butterflies are the origin of blue iridescence. By altering the surrounding media, it can change the brightness and color. Some researchers suggested that this particular property could be implemented into the engineering designs for sensitive vapor/gas sensors or healthcare visual indicators. Optical computers, less toxic pigments, and photonic papers all offer potential applications for tunable photonic circuits.

7.2 SENSORS POWERED BY LIGHT ARE NOW AVAILABLE FOR THE FIRST TIME

Unlike other hydrogen sensors on the market, a new bioinspired device has a full range of characteristics. Industrial hydrogen sensors, on the other hand, can only function at temperatures of 150°C or above. Instead of producing heat, the new sensor emits light. For medical diagnostics, the sensor can detect hydrogen at quantities up to 40,000 ppm (the level where the gas becomes possibly explosive).

Light-activated hydrogen sensors with improved room-temperature performance have been developed by scientists using inspiration from the surface of butterfly wings. Hydrogen leaks can be detected by the technique before they represent a security issue (Kaszubska, 2020; Alenezy et al., 2020). Gas in the intestines can be detected by measuring the amount of gas in the air that people exhale. Light, not heat, powers the prototype sensor developed by RMIT University researchers in Melbourne, Australia. It is based on research published in the journal ACS Sensors that show rough microstructures that mirror the surface of butterfly wings. To cite one of the researchers' views, "the new butterfly-inspired hydrogen sensor is driven by light." A lot of heat is required to get them going. It's all in their hydrogen sensor's power. In addition to operating at room temperature and being able to distinguish between a wide range of limits, they have the unique ability to selectively sense. Even the tiniest amounts can be detected by some sensors, while others are capable of detecting much larger amounts.

Medical applications and enhancing safety in the expanding hydrogen economy will benefit from this sensor's wide exposure range, according to Doctor Ahmad Kandjani, one of this group's experts. "As a flammable, hydrogen has the potential to destroy public trust in this renewable energy expert," he added. "To help advance the hydrogen market, we intend to provide reliable and stable sensor technologies that can detect even the tiniest breaches far before they become serious."

7.3 BUTTERFLY PROJECTIONS: HOW DOES THE SENSOR WORK?

Photonic crystals, minuscule spheres, are used to create novel sensors. The microscopic fissures on the outside of butterfly wings appear to be related to these hollow formations. As a result, they're able to absorb a lot of light. Photonic crystals are microscopic spheres that can be made from light rather than heat, thanks to the new sensor's power. Commercial hydrogen sensors normally work between 150°C and 400°C, according to Ebtsam Alenezy, a scientist at the University of California, Berkeley. Light can start the sensor, and photonic crystals provide structural uniformity that can be dangerous for gas detection.

Structure, consistency of output, and the fabrication stage are all vital. Our photonic crystal synthesis approach also enables our technology to be easily scaled to industrial levels, as hundreds of sensors may be manufactured at the same time. To create an electrical device, a thin layer of photonic crystals is layered on top of a titanium palladium composite.

Adding hydrogen to the chip causes the gas to become water. An electronic current can be formed as a result of this process. The sensor is able to accurately estimate the amount of hydrogen present by evaluating the amount of current. Nitrogen oxide is a concern for many industrial sensors. You can genuinely isolate hydrogen gas with this new technology, which is extremely precise.

7.4 MEDICAL TREATMENTS

Hydrogen concentrations that are associated with digestive problems should be investigated. Medical examination and estimation can benefit greatly from the use of this technology. Breath samples are typically collected and sent to laboratories for analysis as part of the usual diagnostic procedure. Investigator Sabri claims that the new chip might be a handheld device that provides instant findings. For gastrointestinal problems, the difference between healthy hydrogen levels and unhealthy ones is insignificant. As small as 10 parts per million, our sensor is capable of accurately measuring even the slightest changes. The technique has been documented, and a patent has been requested. A hydrogen sensor could be used in fuel cells, batteries, or medical diagnostics, among other applications (https://www.eurekalert.org/news-releases/615073).

7.5 SOLAR PANEL LIGHT ABSORPTION METHOD INSPIRED BY BUTTERFLY WINGS

In nature, a well-known tiny bug demonstrates coloration. Butterflies have long been a source of fascination for both the casual observer and the well-informed expert alike, thanks to their vibrant colors. As they fly, the hue of their wings (Figure 7.2) emits a sparkling sheen. The colors we see in butterflies' wings are the result of a complex chemical and structural system in their wings. The colors you perceive are based on the wavelengths that are reflected. There are two sources for these hues: the natural world and the artificial world. The first type of color is pigmented, which is a chemical pigment that absorbs and reflects light. Examples include the brown hues found in most butterflies, which may be traced back to the same pigment that causes sunburns. Only tones of brown and yellow are absorbed by the object when it is illuminated.

FIGURE 7.2 Butterflies with translucent wings illuminated from above: (A) *Cethosia penthesilea* Cramer, 1777; (B) *Polyura athamas* Drury, 1773; (C) *Vanessa atalanta* Linnaeus, 1758; (D) *Argyreus hyperbius* Linnaeus, 1763; (E) *Lampides boethius* Linnaeus, 1767; (F) *Troides aeacus* Felder and Felder, 1860; (G) *Colias palaeno orientalis* Staudinger, 1892; (H) *Hebomoia glaucippe* Linnaeus, 1758; (I) *Delias nigrina* Fabricius, 1775. Stavenga et al. (2023) Open Access.

Source: Reprinted with permission from Stavenga et al. (2023). https://creativecommons.org/licenses/by/4.0/.

The structural color of a butterfly's wings is caused by variances in its wing anatomy (Giraldo et al., 2016; Pennz et al., 2012). The butterfly's wings are covered in thousands of tiny scales that are separated from one another by air spaces. Illumination effects known as iridescence can be seen because of the structure of the object. Iridescence occurs when light is reflected many times through a transparent, multilayered surface. All of the reflected light intensifies and multiplies each other's colors, creating more vivid visual presentations. Spectrophotometers were used to determine the luminosity of butterflies and other animals. Only researching how these structural consequences influence animal behavior or development can reveal how these adaptations have helped birds survive.

7.6 SOLAR PANELS CAN BENEFIT FROM A TECHNIQUE INSPIRED BY THE WINGS OF A BUTTERFLY

The study of butterfly wing pigmentation and light dispersion showed an array of vibrant colors. Microscopic scales on the wings are covered in nano-scale structures, making the wings extremely durable. Reflections and absorption of light in these structures create stunning color combinations. In order to create new materials, such as fabrics, cosmetics, paints, and glazing, engineers employ biomimicry to adapt nature's concepts, which have evolved over millions of years. There may be a low-cost photovoltaic device that could benefit from nanostructures found in the black butterfly.

A new sort of solar cell has been created as a result of the movements of butterfly wings. It is twice as efficient in blocking light as before and could one day replace solar panels. With thick solar cells and a slant to the sun, these panels tend to be more efficient. Very thin solar cells, down to the nanoscale scale, have enormous potential. They're less efficient than solar panels, therefore, they're commonly used in watches and calculators rather than on the roofs of houses. The rose butterfly's black wings were researched by scientists (Siddique et al., 2017). They used the structure as a guide to construct thinner, more efficient solar cells. It's easy to build these cells, and they can absorb a lot of light no matter what the angle is (Figure 7.3).

Southeast Asia is the only place in the world where the rose butterfly may be found. Cold-blooded and needing sunlight to fly, its wings have grown to be extremely efficient at absorbing energy. According to Yale NUS College biology professor Vinod Saranathan: "The butterflies have been reading for millions of years and still outperform our engineering," he said. Researchers from the California Institute of Technology, led by bioengineer Radwanul Siddique, used an electron microscope to examine the nanostructures of the wings and created a three-dimensional model of it (Siddique et al., 2017). The wings are made of small scales with holes organized in a random pattern.

Those tiny holes, which are less than a millionth of a millimeter wide, enable the butterfly to sense heat and scatter light.

FIGURE 7.3 Butterfly wings inspires to generate Solar Panels.

Source: https://www.indiatimes.com/technology/science-and-future/butterfly-wings-may-have-just-inspired-us-to-build-more-efficient-solar-panels-than-ever-before-332037.html.

Butterflies have captivated humankind for millennia. The gentle beauty of bright butterfly wings belies their complex structure and functionality as the ideal foil for predators. Optical phenomena (such as interference, refraction, or diffraction) continued to produce structural colors sooner than pigment. Physical formations that interact with light to produce a specific color are the source of these colors (Figure 7.4).

7.7 WHAT CAUSES THE MORPHO TO BE BLUE IN COLOR?

Butterflies have two sources of color: pigmented color and structural color (Horton, https://animals.howstuffworks.com/insects/butterfly-colors.htm). Its wavelength is between 400 and 480 nanometers (Blue light) or nm. The slit's distance is about 200 nm. The wavelength at which constructive interference is most likely to occur is half the wavelength of blue light, which is where the distance between slits falls. Melanin-based substance absorbs light, furthering the blue coloration, which is reinforced by slits (Kinoshita et al., 2008).

In an electron microscope examination of the wings, researchers uncovered a 'disordered' pattern of tiny holes on the surface of the scales.

The incident sunlight is caught by these nanoholes, allowing a wide spectrum of wavelengths to be absorbed to their fullest. The term "photonic" refers to structures that produce light at a microscopic scale. The wing is significantly lighter because of the hollow scales.

FIGURE 7.4 Blue Morpho butterfly (*Morpho menelaus*): When the wings of a Blue Morpho reflect light, they appear to be blue.

Photo credit: Derkarts. https://creativecommons.org/licenses/by-sa/3.0/.

With a large number of nanoholes on its wings, this black butterfly is able to receive light from a wide range of angles and wavelengths. It's like a cage for the light, if you will. Imitated by using sheets of hydrogenated amorphous silicon coated with an incredibly thin layer of pores. Improved solar cells can be created using this technology. This is especially true when contrasted with an even, smooth surface. Light striking at a 50° angle might enhance the absorption rate by up to 207%. The technology's full potential would be realized in these results.

In comparison, it is a relatively straightforward process resulting from the release of binary polymer solution fragments. By the research team's estimate, it takes less than 10 minutes for the coating to work. However, the method does not allow for a doubling in the amount of sunlight that may be collected. The polymer used to make the sheet doesn't react with them. In southern regions, where sun rays rarely contact flat solar panels vertically, it could be rather useful. Thermal regulation is a primary function of the butterfly's light-absorbent wings. In order to generate more power, the researchers sought to utilize the most efficient approach for solar cells.

The researchers used an electron microscope to examine the natural structures and then applied a similar design to a thin-film solar cell's cover. Smaller-scale applications benefit from the use of a modern, adaptable material. When they shine light directly down on the surface, absorption jumps by 97%. Also, if the light is angled downward, the increase is as high as 207%. Those solar panels that aren't able to always face the sun need this feature. The little openings serve as caves, allowing light to travel deeper into the surface and be absorbed by the surfaces they encounter. This could be an important step in the improvement of this solar panel's appearance.

Consequently, it has a stronger position in the market for renewable energy. Solar cells have made it possible to move parts and generate pollution-free electricity. For more than 60 years, however, they have been plagued with inefficiency. The earliest models only absorbed roughly 1% to 2% of the sun's rays. That has risen to about 20% now, but that still means that 80% of the sun's rays are squandered.

7.8 BUTTERFLY-INSPIRED TECHNIQUE–PHOTOVOLTAIC SOLAR ENERGY MAKES CHEAPER

For example, engineers plan to find out how much solar energy is increasing, how much it costs, and how efficient solar panels can be made. Scientists are considering using the Cabbage White butterfly, *Pieris rapae*, to help with this concept. Our solar power energy can be improved by using a butterfly's wing. The tiny solar collectors on the scales of a butterfly's wing produce sunlight energy. Researchers from China and Japan are drawing inspiration from butterfly wings to develop more effective solar cells. In the future, homes, businesses, and other applications may be powered by this method.

Beautiful are the wings of a butterfly. They are mature collectors of solar energy. The black butterfly structure is being studied by Tong Xiang Fan and his colleagues. At the American Laboratories meeting in San Diego, they used an electron microscope to study the wing structure of two black butterfly species (Economic Times, 2012). They discovered that the overlapping shingles offered by the wings formed of elongated rectangular scales were quite useful. Butterfly wings have tiny holes on both sides of the scales that lead to another layer. The butterfly's second layer is illuminated by these bright spots, allowing it to absorb a lot of heat. Solar technology could benefit from their shingle-like structure, researchers said.

They've been looking into novel ways to convert light to energy in a study by the University of Exeter (Shanks et al., 2015). During their pre-flight warm-up, Cabbage White butterflies adopt a v-shaped posture. Butterflies may generate half the power generated by solar panels. The overall solar energy structure's power-to-weight ratio is increased 17-fold by reproducing this "wing-like" arrangement. A major scientific publication, Scientific Reports, published the findings of researchers from the Environment and Sustainability Institute and the University of Exeter Penryn Campus in Cornwall (Shanks et al., 2015).

Biomimicry in engineering is a well-known topic, according to Mallick & Eames (2007). Solar electricity can be developed at a reasonable cost thanks to this multidisciplinary study. When the PV panel was tested outside, the solar-to-electricity conversion efficiency was 10.2%. Using this concentrator production technology, a 40% cost reduction can be achieved in large-scale manufacture. By taking to the air first on cloudy days, the Cabbage White butterflies limit the amount of time that other butterflies may use the sun's energy to warm their flying muscles. Flight can occur on these days due to the high concentration of sun energy on their chests.

The butterfly's wings include particular sub-structures that reflect sunlight in order to ensure that the flight muscles are heated to their ideal temperature as quickly as possible. Because of this, the scientists studied how to make a lightweight reflecting material that might be used to generate solar energy by replicating the wings. At roughly 17°C, the scientists stated that the butterfly's wings should be positioned to raise its temperature. When held flat, it will elevate the temperature by 7.3°C. If solar concentrators could mimic the simple monolayer of scale cells seen on butterfly wings, they could be lighter and more efficient, according to the researchers.

According to Professor Richard French-Constant, an expert in butterfly mimicry at the University of Exeter, "This implies that the humble Cabbage White is not only a disease of your cabbages but an insect that is an authority at capturing solar energy." It is possible to view the paper, White butterflies as solar photovoltaic concentrators, authored by Katie Shanks, Dr. Senthilarasu Sundaram, Professor Richard French-Constant, and Professor Tapas Mallick from the University of Exeter, featured on the university's website (https://exeter.ac.uk). Richard French-Constant, a lecturer at the University of Exeter, has discovered the genetic switch that allows female swallowtails to appear to be distinct Monarch butterflies and, therefore, avoid being eaten.

7.9 BUTTERFLY-LIKE IMAGES CAN NOW BE DISPLAYED ON DIGITAL SCREENS THANKS TO NEW NANOTECHNOLOGY

Colorful displays of nature's nanotechnology have been inspired by butterfly wings. For ultra-high-definition screens that use less power and are more pleasant to view, researchers at the University of Central Florida have developed new technologies. National Science Foundation funds are being used to develop digital displays that use ambient light to illuminate and, hence, appear more realistic. This research was published, according to Debashis Chanda, director of UCF's NanoScience Technology Center. This display is more pleasant to use and has a more natural look than a standard computer or smartphone screen. "This is what your house looks like with a picture on the wall. The glare and brightness are absent from this model."

As opposed to LED lights hidden behind a screen, ambient light illuminates Chanda's display more quickly than strong LED lights. For the new viewing experience, Chanda insisted that she preferred prepared dishes over fresh ones. Reduce the quantity of light in computer and smartphone images in order to minimize eye strain, migraines, and other health problems.

In order to produce color, the new display makes use of a mechanism used by many species, such as butterfly wings and octopus tentacles, as well as parrots, macaws, and beetles. Butterfly, octopus, or bird coloration is derived from nanoscale features on their feathers, skin, or scale pigmentation (Daniel Boorman), As depicted in Figure 7.5. The *S. charonda* is a species of butterflies in the Apaturinae subfamily of Nymphalidae, and is called the 'great purple emperor' in English. The melanin and pterin pigments frequently found in butterflies can produce yellow, red, black and brown colors, while pure pigments cannot produce blue, violet, green and golden colors. Protein molecules do not have color, but when they are combined in a precise and controlled manner, they can produce a wide range of hues. This gorgeous color comes from the butterfly's ability to reflect light in such a way that it doesn't absorb it.

7.10 HARVESTING SOLAR POWER BY EMULATING A BUTTERFLY'S WINGS

The diffraction of natural light is responsible for the butterfly's wing hue. The dark tint, on the other hand, suggests that the butterfly's scales capture (absorb) sunlight. Butterflies can rapidly die through a light-heat transition in the black scales on their wings in extreme situations. Inspired by this natural light-harvesting system, NANOMO researchers examined the nano-structures of several butterfly black scales. When installed on Si-based solar panels, they recharged their light-trapping properties. The nanostructures of butterfly scales vary greatly among species, even though they are all the same dark hue. Gullies created between two ridges and followed by thin layers are an example of the *Ornithoptera priamus* scales. The other three species' scale structures, *Tirumala limniace, Graphium doson*, and *Papilio protenor*, all have a ridge structure with one, two, or three holes in between. Because the visible light region has minimum values varying from 1–5%, the light reflection on the black scales differs from that on the black section of *Ornithoptera priamus*. Nanostructures modeled after the black scales of butterflies can be used to make antireflective coatings for solar panels. A new method has been identified for increasing the efficiency of solar cells created by humans (https://techxplore.com/news/2020-06-solar-energy-harvesting-biomimicking-wings.html).

FIGURE 7.5 The *S. charonda* is a species of butterflies in the Apaturinae subfamily of Nymphalidae, and is called the 'great purple emperor' in English. The melanin and pterin pigments frequently found in butterflies can produce yellow, red, black and brown colors, while pure pigments cannot produce blue, violet, green and golden colors.

Source: Reprinted with permission from Mika et al. 2012. https://creativecommons.org/licenses/by/4.0/.

Photovoltaics based on silicon may benefit from the use of these structures as anti-reflective guides. Using computer simulations, it was found that the amount of surface backscattering decreased significantly after the nanostructures were applied to Si solar panels. *Ornithoptera priamus* has the best anti-reflective scale structure. The bare Si panel has a light reflection reduction of more than 35% to less than 5%. It was possible to achieve this by increasing the solar cell's maximum output by 66%. A plan like this needs to be tested in the actual world.

NANOMO's biomimicry leadership continues to be bolstered by this study. Photonics and solar energy harvesting are pioneered in the unit's utilization of fundamental physics concepts. A TUTLI subsidy for Finland's business sector was offered in order for Finland to take up NANOMO. It aids in the industrialization of bio-inspired ideas for new surface functionality requirements.

7.11 DETECTORS OF EXPLOSIVES AND SELF-CLEANING SURFACES WERE INSPIRED BY THE FLUTTERING WINGS OF A BUTTERFLY

Blue morpho butterflies' beautifully colored iridescent wings are an example of a promising field of science known as bio-inspiration, which aims to find solutions to real-world problems by studying nature. Almost no other butterfly's iridescent blue wings can rival the Morpho's in a breeze. In addition to the film's Metamorphosis cover, these species' patterns are familiar from Illustra's Metamorphosis. The nanostructure and surface chemistry seen on Morpho butterfly wings inspired the design, which incorporates both physical and chemical requirements. Because their spectacular color isn't derived from pigments but rather from optical illusions created by light-bending structures in their wing scales, scientists refer to them as structural color. The arrangement of "photonic crystals" that resemble miniature trees formed of biomolecule chitin cancels out some colors while enhancing others.

Morpho butterflies do not display their vibrant colors through pigmentation (Potyrailo et al., 2013; Bomford, 2014). Instead, tiny, transparent structures and layers on their wings' scales display the colors. These microstructures will cause specific wavelengths of light to reflect while others cancel out. The form and spacing of the layers determine which colors we see. This diffracting method means there is no fading of color over time. It also allows butterflies to adjust their coloration for camouflage, thermoregulation, and communication. Several new products have drawn inspiration from the Morpho butterflies. Manufactured by Teijin Fibers Limited, Morphotex fibers are laminated using nanotechnology. Variations in the thickness of the laminate produce different colors.

The fibers can be used in clothing and fabric. Teijin uses a similar process in Morphotone, an optically colored powder, which can provide color for cosmetics and paint. The technology requires no dyes. San Diego's Qualcomm owns a second butterfly-inspired product. Engineers at Qualcomm applied the refraction and interference principle to color display screens through a technology called mirasol. Mirasol display uses very little energy (which improves battery life) and can be seen even in bright sunlight. It is used in several mobile devices and has a future also.

7.12 THESE BUTTERFLY WING ARCHITECTURES PROVIDE INSPIRATION FOR SENSOR AND ENERGY APPLICATION DEVELOPMENT

Global population growth is wreaking havoc on the environment, harming people's health, and depleting the world's already-depleted energy supplies. As a result, environmental protection, health intervention, and technology, such as solar, wind, and wave energy, have garnered attention from around the world. There is a plethora of technologies available today for preserving the environment, managing health, and harnessing renewable energy. Most conventional technologies, on the other hand, need to be updated or enhanced. Since high-performance and efficient technologies are the subject of research studies, this is the best course of action.

Biological systems are constantly adapting to counteract the negative impacts of population growth and energy depletion. These organisms' sophisticated systems recognize inputs and detect changes in their surroundings before acting on the data. Artificial functional materials based on bio-inspired replication are guaranteed to have improved properties and performance. As a result of the butterfly's intricate micro- and nanostructures and light-trapping systems, sensors and energy materials have been developed that mimic the butterfly's selective responses to stimuli. The use of bioinspired sensors and energy materials for renewable energy harvesting, environmental cleanup, and health monitoring has never been more straightforward than it is now. Consequently, this chapter investigates numerous bio-inspired sensor and energy applications, as well as current developments in the classification of butterfly wing scale structures (Osotsi et al., 2021; Tee et al., 2017).

7.13 BUTTERFLY WING INSPIRES TO FABRICATE NANOTECH GLAUCOMA SENSOR

Second-leading cause of blindness can be tackled with eye implants. Inspiring nanostructures in the wings of a species of Butterfly have led to the production of a glaucoma implant that promises to improve the lives of patients (Masterson, 2018). The glasswing butterfly (scientifically referred to as *Greta oto*) is rarely regarded as the most beautiful bug. They resemble the monarch butterfly in that their delicate wings have a frame of orange and white stripes running across them. Glasswing butterflies lack color in their chitin, a substance found in insect exoskeletons and butterfly wings, from the time they are caterpillars. The absence of pigment renders the wings colorless and transparent, as they are so thin. The butterfly's wings do not reflect light, allowing it to pass through freely. On the theory that a butterfly's wings can extend up to 100 nanometers above its body, this device was developed. According to Andrew Masterson's research, it can be 50 to 100 times smaller than a human hair and divide 150 nanometers apart. *Siproeta stelenes* (Nymphalidae: Nymphalinae) and *Philaethria diatonica* (Nymphalidae: Heliconiinae butterflies gives an idea that the green coloration does not come from scales. By carefully checking the green areas, Scientists discovered that the costal area of the wing occasionally contained tiny, moving air bubbles and had no green color (Figure 7.6), suggesting that the green color was present in liquid form.

Longtail glasswing butterfly-inspired nanopillars were used in this study. Caltech postdoc Radwanul Hasan Siddique's research identified microscopic pillars covering see-through wing parts (https://eas.caltech.edu/news/search?keyword=Radwanul+Hasan+Siddique) of butterflies.

The pillars direct the rays of light straight through the butterfly's wings, resulting in a direct hit on the rays. If there is no reflection, the wings appear to be made of glass. It was first reported in 2015 by a team at the Karlsruhe Institute of Technology in Germany led by Radwanul Hasan Siddique (Siddique et al., 2017). Researchers say that the butterfly wings' "omnidirectional anti-reflection behavior" is unlike anything else in nature. Small height and spacing fluctuations in the surface nano-pillars were cited by Narasimhan et al. (2018) and Masgterson (2018) as the reason for the lack of reflection.

Siddique has been accepted as a resident at Caltech in the United States. With the help of an electrical engineer, Hyuck Choo, he plans to build a groundbreaking new glaucoma monitor that can be placed into a patient's eye. Those with glaucoma are permanently blinded as a result of the optic nerve being irreversibly damaged by an increase in eye pressure. On the basis of the available data, we estimate that there are approximately 11.2 million persons aged 40 years and older with glaucoma in India. Primary open-angle glaucoma is estimated to affect 6.48 million persons. The estimated number with primary angle-closure glaucoma is 2.54 million. Medications are effective in reducing pressure spikes that can lead to nerve damage, but they work best if taken just when the pressure is first rising. One or two times a year, a doctor must measure the intraocular fluid for changes that could indicate this condition. Choo and his colleagues have developed monitors that allow patients to manage their ocular pressure. Hand-held readers and micrometer-sized sensors combine to provide a system for implanting a sensor directly into the eye. Using the reader, the user may monitor their blood pressure in real time and know exactly when to take their prescription.

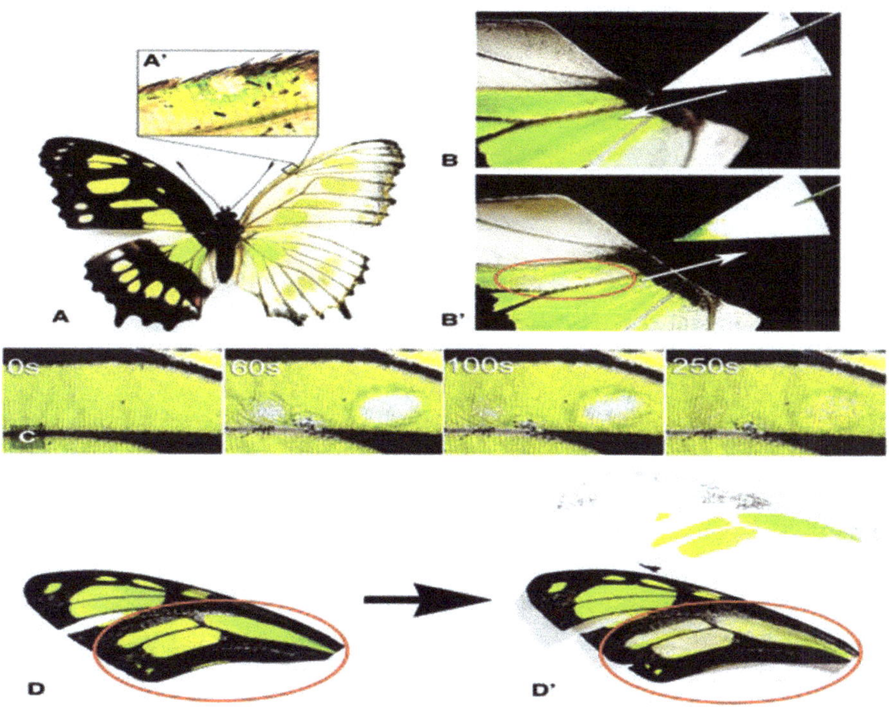

FIGURE 7.6 Wings and green liquid (A–C) *Siproeta stelenes* and (D) *Philaethria diatonica*.

Source: Reprinted with permission from Nishida et al., 2012. https://creativecommons.org/licenses/by/4.0/.

The angle is nearly perpendicular–90°, with a 5% error margin–which is a key challenge for the system in obtaining an accurate readout. As a result, the likelihood of a false reading was significant.

The problem appears to have been resolved. Siddique and Vinayak Narasimhan joined together with Choo to figure out how to adapt the glass wing butterfly's nano-pillar layout to the new glaucoma treatment. Because of the butterfly wing's symmetrical distribution of size and location, it was assumed that light would always flow through the wing without being reflected. As a result, the nanostructures generated from silicon nitride on its surface should have made the implant significantly less vulnerable to the angle between it and the scanner. There was a three-fold decrease in the number of errors in readouts after a little experimentation.

According to Choo, "nanostructures allow patients with glaucoma to check their eye pressure every day," he said in Science Daily. Second, but no less critical are the design's form and the materials it's constructed from. Many medical implants gradually lose their functionality and need to be replaced as the body tries to make sense of them. Chemical coating of implants is the conventional method of avoiding this problem. This is only a short-term solution. An alternative, more long-term approach is provided by the monitor created by Choo and his colleagues. The monitor is highly hydrophilic, which means it attracts water. Cells cannot adhere to it because it is a densely packed, slippery, inert, and harmless layer that is difficult to penetrate.

7.14 BUTTERFLIES INSPIRE ANTI-COUNTERFEIT TECHNOLOGY

The wings of a South American insect are mimicked by nanotechnology. A Canadian business is exploiting one of nature's most sophisticated structures to fight criminals: a butterfly wing. The lower lamina of the scales of related nymphalid butterflies, the Nymphalinae, plays a dominant role in the wing coloration as shown in Figure 7.7. Nanotech Security Corp in Vancouver, Canada, uses the general design of a Morpho butterfly's parts to be precise. Iridescent blue wings give it a look that is nearly impossible to replicate. It's a South American insect. The innovation was unveiled and licensed at Simon Fraser University in British Columbia (Shurkin, 2013).

Nanotechnology can be used to create iridescent hues in peacock males, for example. For the creature's wings or scales, microscopic holes that reflect different colors or wavelengths are used instead of proteins and other chemical compounds to create a hue. Morphos are known for their iridescent blues and greens because of the intricate scales on their wings. Nanotech's printed security picture can be questioned if it can be applied on any surface. Nanotech's chief technical officer, Clint Landrock, says the surfaces include plastics, metals, solar cells, fabrics, and paper. These seals could be used on medicinal pills and capsules to prevent the use of substandard products.

FIGURE 7.7 The reported *Morpho butterflies, M. epistrophus, M. helenor* and *M. cypris,* have wings with bluish colors varying in saturation and patterning (Figure 7.7 a, b, c).

Source: Reprinted with permission from Giraldo et al., https://creativecommons.org/licenses/by/4.0/.

Biomimicry can be used to adapt nature's solutions for new human inventions. Here, the cutting-edge new technology of nano-optics came to light. Nano-optics, for example, are used by the University of Michigan to print images and photos without the usage of ink (Shurkin, 2014). "Nano-hole arrays" were employed to build fake "iridescent attempts" with more surface compositions than the ones previously used. Millions of nanoholes can be mass-produced by the corporation and shaped in various ways to adjust the colors.

They employed electron beam lithography, which is analogous to manufacturing computer chips, to create unique nano-hole designs on silicon or quartz. It was in the nanometer range that they functioned. Even the tiniest bacterial cell is hundreds of times smaller than a single nanometer. From 50 nanometers to 300 nanometers in diameter, the template's holes were spaced 300–600 nanometers apart. Due to the size of the mask and its structures, this operation will take longer. The time it takes to produce a master pattern can range from a few hours to a few days. After mastering, images are developed on nickel and then transferred to a variety of media.

High-end products like artist handbags were embossed with the complete image, making it visible from a distance. Nanotech's CEO, Doug Blakeway, asserted that counterfeits could be easily distinguished from genuine items. The logo of the company would be on your bag's pin if you had one. When it was turned on, a magnificent display of color would flash on and off in front of one's eyes. The color would flare because of the spectator. Capsules and pills can be used to display the image without issue. To be certain that the medicine was genuine, you could see the brand on it. Because the image does not employ colors or pigments, the FDA does not need to approve it. Landrock marveled at the audacity of the counterfeiters. Because of the equipment required, changing the image would be difficult for an engineer. According to him, the image is brighter than any other technology, even holograms.

It has to do with how a new Ultra HD LED TV seems to an old CRT television. The most logical application of the technology is an anti-counterfeiting device on banknotes, according to Landrock. A nano-optic image was embossed on a sheet of painted paper. While some countries have employed polymer plastic banknotes, which are even more sensitive to nano-optics images, such as Australia and Canada, others have used paper banknotes. These bills have a longer life span and are more difficult to counterfeit than US paper cash. Nano-optics are unlikely to be used by the US dollar in the near future. Even though holograms and coated or polymer paper are used in other currencies, they are not used in US banknotes. In spite of this, no countries have yet agreed to participate in commercializing the technology.

Because of the scarcity of US cash abroad, the classic bills are still in use. An expert on the financial industry, Owen Linzmayer, publisher of banknote News, explained that it serves as a reserve currency for developing countries. International markets could be thrown off kilter if the United States introduces new measures. Gucci handbags, on the other hand, are exempt from these restrictions.

7.15 THE BLUE MORPHO DIDIUS BUTTERFLY'S WING PATTERNS INSPIRED THE INVENTION OF NEW SOLAR TECHNOLOGIES

New generation of solar cell technologies and other light-sensitive applications are now possible, thanks to the development of nanoscale structures inspired by butterfly wings. The blue Morpho Didius butterfly serves as the inspiration for this piece. An iridescent blue iridescence is created by small cone-shaped nanostructures on its wings. Other advancements, like as stealth and architectural applications, are spawned by this technology.

Researcher Dr. Niraj Lal from the Australian National University's (ANU) Research School of Engineering said that the team constructed similar structures at the nanoscale and employed the same principles to change the direction of light sensitively, using the butterfly wing phenomenon as an analogy. Dr. Lal is excited about the potential uses for his light-controlling technology, which might include new solar cells, architectural design, and even stealth technology.

7.16 SCIENTIFIC RESEARCH HAS SHOWN THAT PROPER LIGHT CONTROL CAN BOOST THE PRODUCTION OF SOLAR CELLS

"For the next generation of very high-efficiency solar panels, techniques are required to manage scattering, reflection finely, and absorption of distinct wavelengths of light," Dr. Lal stated. It's been difficult to get the light to travel exactly where you want it to go thus far. When it comes to solar cells, they must be able to absorb all of the sunlight's blue and green wavelengths. Other colors of light, such as red, orange, and yellow, are absorbed by the silicon layer. Such a cell set a new silicon efficiency record at the ANU in April 2017. By controlling how much light and heat enter through a window, he added, this technology could one day make opaque objects transparent to certain hues and the other way around.

There is a slew of inventive architectural possibilities for tinting windows in such a way that they are opaque to some hues but transparent to others. He claimed that the method was incredibly scalable and didn't require expensive technology. Nanostructures grow and assemble on their own, according to Dr. Lal, not under the control of a laser or electrons.

7.17 THE BUTTERFLY EFFECT

Butterflies are charming creatures. All biologists, as well as famous artists, draw attention to their beauty. Butterflies have led to manage to hypnotize culture. Researchers have made a notable effort to study the butterfly's most essential and vibrant point, its wings. An elite team of chemists has concentrated their awareness on the butterfly's wings, connected with the work of the United States Air Force. The tiny crystals of the Butterfly make up the scales on the butterfly's wings (Alexander, 2017).

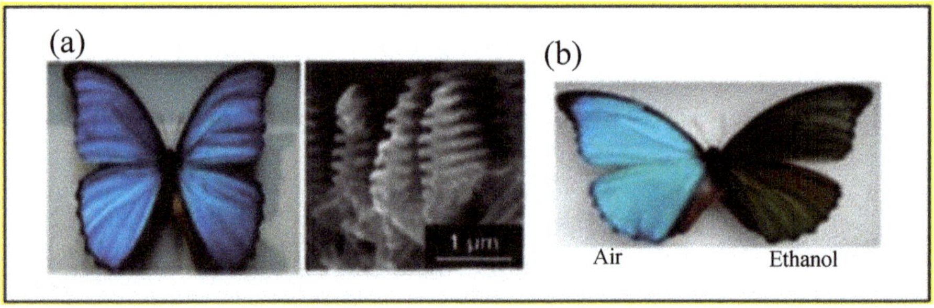

FIGURE 7.8 shows Structural colors and phenomena arising in nature are attributed to micro and nanoscale features of the surface. a) Blue wings of the morpho didius butterfly wing with a ridged structural order. Shifts to longer wavelength structural colors are displayed for the replacement of air with ethanol in b) Morpho Menelaus butterfly wings changing from blue to green.

Source: Reprinted with permission from Lonergan and O'Dwyer, 2023. © 2022 The Authors. Advanced Materials Technologies published by Wiley- VCH GmbH. https://creativecommons.org/licenses/by/4.0/.

The concept of ordered structures and associated optical properties, such as structural color, are found in materials classified as photonic crystals (PhCs). Some of the changes observed to structural coloration for both natural and artificial PhCs which can be understood through changes to the wavelength of maximum diffraction of the Bragg-Snell relation. The *Morpho didius* species, the blue beauty, is the main creature and plays a vital role. While studying this butterfly, researchers determined that the photonic crystals that give the Morpho its blue color are also able to detect chemical agents. However, they were not able to identify the chemical to the desire they wanted to. Detection limits for photonic crystals were in the parts per million range. The photonic crystals from the butterfly have a high concentration of chemicals used in the process. If it has a negative impact on human health, that is enough. But no worries with these butterfly crystals. Scientists propose to design something even better. The tools designed with this butterfly could protect the lives of soldiers and civilians under the menace of chemical warfare.

7.18 NEW TECHNOLOGIES RANGING FROM FABRICS, COSMETICS, AND SENSORS ARE INSPIRED BY THE WINGS OF THE BUTTERFLY

Beautiful iridescent wings of the tropical blue Morpho butterfly stimulate fabrics, cosmetics, and sensors, according to a recent study. If you look at the scientific literature, you'll see that they have already inspired innovative displays and many more fascinating uses. The GE Global Research Centre, the University at Albany, and the AFRL collaborated with the University of Exeter to study textiles, cosmetics, and sensors.

This experiment was made possible thanks to support from DARPA, a US government organization. Researchers have discovered that the Morpho butterfly's wings contain unique physical and chemical features. A wide range of applications could benefit from the research findings, including photonic security tags, self-cleaning surfaces, protective clothing, and industrial sensors (https://www.exeter.ac.uk/news/featurednews/title_318647_en.html).

There are numerous advantages of adding nanostructures to the Morpho wing scales. The glittering blue iridescence of the butterfly has been attributed to them, according to a study published in the journal PNAS (Potyrailo et al., 2013). Vapor molecules cling differently to the tops than to the bottoms of these structures, as revealed by these experiments. Many bio-inspired technologies rely on this ability to respond selectively to molecular vapors.

This topic was thoroughly researched by a multidisciplinary team led by DARPA Program Principal Investigator Dr. Radislav Potyrailo. This topic was explored by a wide range of scientists, including physicists, chemists, biologists, and those working in the field of materials science. When used on iridescent Morpho butterfly scales, they can indicate the presence of a surface polarity gradient. For vapor sensing, this discovery allowed them to introduce a multivariate perspective. Instead of an array of discrete sensors, a single chemically graded nanostructured sensor unit accomplishes its selectivity (https://phys.org/news/2013-09-butterfly-wings-technologies-fabrics-cosmetics.html).

"Understanding iridescence in butterflies and moths has revolutionized our knowledge of natural photonics," stated Professor Pete Vukusic from the University of Exeter's Physics Department. We can improve a wide range of applications by drawing inspiration from nature's design principles when it comes to natural photons. Professor Pete Vukusic of Exeter University's Department of Physics says, "Understanding butterfly and moth iridescence has altered our knowledge." We may use nature's design principles to improve applications in a wide range of technology. Using new photonic vapor sensing mechanisms, the researchers discovered that physical and chemical processes can coexist at the nanoscale. In this study, the team found a new mechanism in photonic vapor sensing that shows combined physical and chemical effects on the nanoscale.

7.19 LIGHT WAVES ARE USED TO CREATE THIS PAINT

Traditionally, the colors used in paint were derived from natural sources such as plants and animals. Colors like cochineal and carmine were produced by insects like these. In tempura oil painting, Renaissance artists used eggs as a binder. Instead of exploiting insects as a raw material, one company is drawing inspiration from the creatures, and it has the ability to accomplish far more than simply beauty. Cypris Materials has created a paint based on the nanostructure of the blue morpho butterfly wing, which reflects blue light waves to give the wing its blue tint (Smith, 2020; Wilcox, 2020). Light waves are reflected off the nanostructure of Cypris' paint in the same way as the wings of a butterfly reflect light. This paint does not contain any chemical pigments or dyes; instead, structural color is used.

It is probable that the paintable structural colorwork is still under investigation. When it comes to wavelengths, the size of the polymers we make has a big impact. It will "tune" itself to perfection as it is manufactured and formulated. Synthetic polymers (a chain of molecules) in the paint produce nanostructures that reflect specific wavelengths of light as it dries. This method is referred to as self-assembly. Polymers that are tailored to reflect a specific color in the visible spectrum could be utilized to make the paint. For example, ultraviolet (UV) and heat rays that are not visible to the human eye are reflected by the color red (so the paint would be precise). There are other perks beyond the obvious esthetic ones that come with this skill. It's possible for UV radiation to fade outdoor furnishings if they are not shielded from it. Near-infrared (NIR) light could be reflected back as heat in a second treatment. The coating, which minimizes heat absorption and thus A/C performance, can be applied to roofs, windows, and other surfaces. Up to 20% of the pressure can be reduced by this method, which Pearson still uses. Pearson advises using the coating on a roof, and the black hue can be preserved if desired (https://www.fastcompany.com/90557829/this-paint-inspired-by-butterflies-could-be-the-secret-to-cooler-cities).

7.20 ADOPTING THE BEAUTY OF BUTTERFLY WINGS IN THE DEVELOPMENT OF SOLAR TECHNOLOGY

Photovoltaic and solar thermal systems do not appeal to everyone. The Fraunhofer Institute for Solar Energy Systems (ISE) brightly colored modules make for a refreshing change of pace. The underlying mechanism of spectrally specific reflectance is inspired by the event that creates the shimmering colors of blue or green on the wings of the morpho butterfly (https://www.fraunhofer.de/en/press/research-news/2021/january-2021/solar-technology-with-the-beauty-of-butterfly-wings.html). The color of the solar module can be made to rise or be combined with the structure of the module. Photovoltaic devices, which harness the power of the sun's rays to create electricity, are becoming more and more prevalent. Solar photovoltaics and other renewable energy sources are now within the reach of the typical household. The notion of rooftop solar panels has changed with the growth of technology, and current solar panels have a smooth appearance to enhance curb appeal.

Solar cells are sandwiched between two sheets of glass and a second sheet of laminate polymer to make a solar panel. Because of the spaces between the solar cells, the rear sheet may be seen from the front. In most cases, the back sheet is white, which stands out against the black solar cells. There is a plethora of options available when it comes to solar panel designs. Cell type, backsheet, frame, and "bus bars" all have an impact on how panels look after they're installed. Using narrow strips joined together, solar cell busbars store the electricity generated (Fraunhofer-Gesellschaf, 2021).

7.21 IT IS POSSIBLE TO HIDE SOLAR PANELS IN THE FACADES AND ROOFS OF BUILDINGS

With this change in look, building owners and designers are reluctant to include solar panels in their designs. Facades, rather than rooftops, must be used for solar modules in order to meet Germany's "energy development requirement," which asks for an additional 2,500 square kilometers of photovoltaic systems.

Solar Energy Systems ISE in Freiburg has developed modules that are not only practical but also aesthetically pleasing as a result of this. It is possible to add color to facades or roofs very imperceptibly. With a ventilated curtain-wall facade, they can even finish off modern buildings. Solar Energy Systems ISE in Freiburg has therefore developed modules that are not only functional but also visually appealing. Facades and roofs can be enhanced gently using color-coordinated components. Thomas Kroyer, the group leader and chief scientist of the Fraunhofer-coating Gesellschaft's technologies and systems division, claims that mimicking the physical impact of butterfly wings, not the use of colors, motivated this growth. The modules' effectiveness would be greatly diminished if pigments were applied to the glass, as light would be unable to travel freely through.

Reflection of light from the Morpho butterfly's vividly colored wing Color isn't caused by pigments but rather by an optical illusion in insects native to Central and South American tropical rain forests. Its wings have an appealing surface roughness that only reflects certain wavelengths, i.e., a specific brightness, which is why this butterfly is so beautiful. The surface roughness of vacuum-based photovoltaic modules from Fraunhofer ISE is identical. The details on the cover glass may be crisp blue, green, or red, depending on the coating's tailoring. Thomas Kroyer estimates that 93% of incident light can pass through this layer, with only 7% reflecting back to provide the color effect. In Freiburg, Germany, the Fraunhofer Research Institute coined the term "MorphoColor" in honor of the bright blue morpho butterfly.

7.22 AS A RESULT OF THE IMPROVED ASSEMBLY APPROACH, THERE ARE NO MORE "INVISIBLE BREAKS"

In order to create a visually pleasing form, color alone is not enough. To make photovoltaic technologies more beautiful, Fraunhofer researchers came up with a new idea. They devised an assembly method that makes use of the reflecting qualities of roof shingles to keep the patched solar cells from glaring through the colored emergency glass. When it rains, the roof's shingles pile up and let the water pour in. Photovoltaic cells are also being made at Freiburg, Germany, in layers a few millimeters thick and then attached together to make a larger module. You don't have to worry about unsightly gaps or visible wires. Solar shingles coated with Morpho Color may be observed from a wide range of angles and still keep a consistent appearance (https://www.fraunhofer.de/content/dam/zv/en/press-media/2021/january-2021/bau-2021-ise-solar-technology-with-the-%20beauty-of-butterfly-wings_neu.pdf).

7.23 GORGEOUS AND ENERGY-EFFICIENT BUILDINGS

The vacuum-produced MorphoColour-coated protective glass can be used to make photovoltaic modules, and it can also be used as a collector for solar heat generation. This module showcases a variety of vibrant solar thermal collectors. The Morpho color-coated protective glass, which is created using vacuum technology, can be used to create photovoltaics in a solar collector. There's a good chance the color of solar thermal and photovoltaic modules may converge in the near future. To the end-user, this is a huge advantage because it provides both results from a single production line (https://www.

fraunhofer.de/content/dam/zv/en/press-media/2021/january-2021/bau-2021-ise-solar-technology-with-the-%20beauty-of-butterfly-wings_neu.pdf).

As a result of their close proximity to one another on the roof or facade, they are virtually undetectable from the ground. A facade that creates both power and heat is created when the color meets the rest of the construction and becomes a continuous wall. As a result, future homes could be both beautiful and environmentally friendly, creating more energy than they use.

7.24 THE WINGS OF A BUTTERFLY CAN BE USED AS A HYDROGEN SENSOR

The simplest atom in the universe is made up of just one proton and one electron. When it comes to atoms per cubic meter, hydrogen takes the top spot. When it provides a service, just water is produced. As a fuel that is both dependable and replenishable, hydrogen has received much attention recently.

The universe's most volatile element also happens to be its most prevalent. In order to maintain this degree of accessibility and reactivity, the cost is considerable. Anyone who has seen footage of the Hindenburg disaster can witness hydrogen's extremely reactive reactivity under the right conditions. As a result, hydrogen poses a significantly greater risk to public health and safety because of its lack of color, taste, or scent. As a result, a sensor can detect hydrogen if it is publicly accessible.

That's where a group of RMIT University, Melbourne, Australia, scientists (Lea, 2020) come in. Elias Sabri, a materials scientist, has invented a novel hydrogen sensor that is exceptionally accurate and works at ambient temperature. Even at low concentrations, it is able to detect light-activated hydrogen. An important feature is the ability to detect and inform users of probable leakage and seepages (Alenezy, 2020). Because it is powered by light rather than heat, the researchers' detector is more secure than current models. When it comes to sensors, the wings of a butterfly are a good model. Specifically, the crystals that make up their wing scales.

These crystals, which are responsible for the butterfly's amazing kaleidoscope of colors, are capable of detecting chemical compounds, according to recent discoveries. Butterfly wings, which appear to change color depending on the angle from which they are viewed, exhibit the light-refracting effect of photonic crystals. The light-scattering properties of some materials may help in the identification of substances. Even if nature is a tremendous source of inspiration, the RMIT team (Alenezy, 2020) decided to go even further in their explorations. In appearance, the RMIT prototype's photonic or colloidal crystal covering resembles that seen on a butterfly's wing. Underneath this thin layer of photonic crystals is a titanium palladium alloy. When hydrogen gas is introduced to the chip, it turns into water. This device generates electricity.

7.24.1 WHY IS THIS IMPORTANT?

The sensor is fit for use in the "hydrogen economy," agreeing to be the ultimate fuel and medical sector. There are many security companies about the widespread use of hydrogen. This new sensor can help them. The sensor is a security net for catching hydrogen leaks reliably and correctly, even at minute levels, for facing and escaping dangerous circumstances. Hydrogen can be an asphyxiant in substantial sizes and can also pop with great force if burned. This sensor alone could transform the hydrogen economy and aim for a greener future by reducing the fear around hydrogen power.

7.24.2 HOW DOES IT WORK?

Photonic (or colloidal) crystals are what make up the heart of the new sensor. These tiny spheres are hollow and are significantly related to the projections that make butterfly wings break. They are also very effective at absorbing light, enabling the sensor to take power from a light beam rather than excess heat. A more standard sensor than the standard hydrogen sensors runs at 150°C. The photonic crystals make the entire structure that just gas sensing needs. A sensor with consistent invention quality and a consistent structure will give consistent results. As a result of the use of bio-inspired shapes and modern invention methods, the sensors and technology behind them are ready for use in industry. According to this, mass production of hundreds or even thousands of these sensors might quickly and cost-effectively make them suitable for worldwide use and widespread distribution.

7.24.3 HOW ARE THEY MADE?

After a titanium palladium composite layer, a thin coating of photonic crystals was applied to electronic chips. Hydrogen is converted to water when it comes into contact with the chips. The amount of hydrogen present can be determined by measuring the size of the electric current generated as a result of this reaction. Opposite to the current sensors on the market, this technology can truly differentiate between hydrogen and other gases. It is highly particular and will not struggle in the neighborhood of nitrous oxide, unlike its competitors.

7.24.4 APPLICATIONS OF THE SENSORS

High levels of hydrogen discovered in a person's breath could signal gastro-intestinal problems. In the medical field, these sensors can also be used to control and diagnose gastrointestinal issues. Currently, a breath sample is taken and then sent to a laboratory for analysis.

7.25 DYE-FREE FABRICS WITH A BUTTERFLY MOTIF

Morphotex fiber mimics the properties of the Morpho butterfly (Figure 7.9) with the iridescent color of its wings, also similar to that of a peacock feather. Microscopic structure using polyester and nylon fibers in alternating layers so that light will bounce and scatter between the layers to reveal a rainbow of colors. As the color is purely a trick of the light, no dye is needed which cuts water usage, toxic chemicals and energy used to dye the fabric. The fabric will also never fade like dyed fabrics often do.

The wings of Morpho butterflies appear cobalt blue despite the fact that they lack pigment. This optical image shows the light refracting through the several protein layers on the scales of the butterfly wings. The microscopic structure of a butterfly's wing was the inspiration for a dress designed by Australian fashion designer Donna Sgro in 2010 using nanotechnology and the Morphotex fabric.

When it comes to photography, it's all thanks to butterfly wings. Modern gadgets allow you to view these images in virtually any light. The innovative approach uses a fraction of the water and energy of traditional dyeing.

7.26 BECAUSE THEIR WINGS ARE SO MALLEABLE, BUTTERFLIES PROPEL THEMSELVES BY CLAPPING

Butterfly migrations over a sunny meadow are awe-inspiring, but we don't know how they do it. Due to a lack of research into butterfly flying, it has been speculated that the insects employ a variety of unstable aerodynamic processes to create force (Dudley, 2000). However, fluttery flying is not the

FIGURE 7.9 Butterfly wing at different magnifications reveals micro structured chitin acting as a diffraction grating.

Source: Source: Wikimedia Commons. https://commons.wikimedia.org/wiki/File:Butterfly_magnification_series_collage.jpg. https://creativecommons.org/licenses/by-sa/3.0/.

only mode of transportation for butterflies; they can also fly with tremendous accuracy when migrating or evading predators. Butterflies have a tendency to fly away swiftly in order to avoid predators. Takeoffs for butterflies involve a lot of force creation and control, which is why their wings clap is used so frequently. For further testing, researchers created two simple mechanical clappers to capture slow-motion video of the butterflies in flight. This video shows the difference between the two when it comes to wind tunnel testing: one was rigid, while the other had wings like a butterfly. In 2021, researchers Johansson and Henningsen seized on this pattern and began to investigate the underlying aerodynamic origins of this phenomenon. Kinematics was the first step. Researchers studied the take-off flights of six silver-washed fritillaries in a wind tunnel.

Butterflies are one of nature's most amazing fliers because of their unusually short, broad, and enormous wings compared to their body size. In prior studies, it was discovered that butterflies use the upstroke wing clap to improve their force output. During the animal's upstroke, its wings clap together, creating a jet of air that pushes the animal in the opposite direction. No quantitative measurements of the clap's aerodynamics in free-flying animals exist, despite its importance in insect flight being widely recognized for the past 50 years (Ellington, 1984, 1999).

Johansson & Henningsson (2021) used mechanical clapper and flow measurements behind free-flying butterflies to estimate the wing's aerodynamic performance during takeoff. A butterfly's flexible wings support its weight during the downstroke while its upstroke and clap propel it onward. To put it another way, flexible wings increase the clap's valuable impulse by 22% and efficiency by 28% compared to stiff wings, according to Johansson & Henningsson (2021). It appears that butterflies have developed a highly effective and unique clap that explains their peculiar wing structure. They may also be able to use these findings to improve the design of artificial flapping drones, enhancing their propulsion capability.

7.26.1 FLEXIBILITY IS KEY

Just before impact, detectives spotted the wings puffing out of their pockets. They questioned whether or not the clap's jet propulsion was enhanced by the wings' compliance and this pocket. In order to find out, they made two butterfly wings out of paper and glue. Balsa wood was used for one set, which resulted in a stiff set. Latex and flexible plastic are used to produce the other sets. Links and rods were added to each pair in order to simulate the flying of a butterfly (https://www.sciencenewsforstudents.org/article/butterflies-use-jet-propulsion-for-quick-getaways).

While clapping together, Johansson & Henningsson (2021) analyzed the airflow caused by the models. There was no air pocket formed by the inflexible wings. Instead, two vortex rings were constructed. As Johansson & Henningsen (2021) describe, the flight would be "less efficient" since "more energy spent generating two rings." A stronger stream of air and just one vortex ring were produced by the flexible wing variants. It acted more like a real butterfly's wings than a rubber band. Man-made flapping drones could benefit from this research's findings by improving their propulsion performance.

KEYWORDS

- **butterfly motif**
- **butterfly-inspired technique**
- **hydrogen sensor**
- **morpho butterfly**
- **photovoltaic**
- **solar energy**
- **solar technology**

REFERENCES

Alenezy, E. K., Sabri, Y. M., Kandjani, A. E., Korcoban, D., Rashid, S. S. A. A. H., Ippolito, S. J., & Bhargava, S. K., (2020). Low-temperature hydrogen sensor: Enhanced performance enabled through photoactive Pd-decorated TiO_2 colloidal crystals. *ACS Sensors, 5*(12), 3902–3914. DOI: 10.1021/acssensors.0c01387.

Alexander, D., (2017). Butterfly wings could help us detect lethal chemical agents. *Science*, https://interestingengineering.com/butterfly-wings-could-help-us-detect-lethal-chemical-agents (accessed on 17 August 2023).

Bomford, A., (2014). *Butterfly Wings Inspire Cosmetics and Bomb Detectors.* BBC Radio 4's PM program. https://www.bbc.com/news/science-environment-28214684 (accessed on 17 August 2023).

Dudley, R., (2002). *The Biomechanics of Insect Flight: Form, Function, Evolution*. Princeton, NJ: Princeton University Press.

EconomicTimes, (2012). *Solar Energy Inspiration from Butterfly Wings*. http://www.sustainabilityoutlook.in/news/solar-energy-inspiration-butterfly-wings-14301, http://economictimes.indiatimes.com/news/news-by-industry/et-cetera/solar-energy-inspiration-from-butterfly-wings/articleshow/12605054.cms (accessed on 17 August 2023).

Ellington, C. P., (1984). The aerodynamics of hovering insect flight. III. Kinematics. *Phil. Trans. Roy. Soc. Lond. B*, *305*, 41–78. doi: 10.1098/rstb.1984.0051.

Ellington, C. P., (1999). The novel aerodynamics of insect flight: Applications to micro-air vehicles. *J. Exp. Biol.*, *202*(23), 3439–3448. https://doi.org/10.1242/jeb.202.23.3439.

Franklin, D., He, Z., Ortega, P. M., & Safaei, A., (2020). Self-assembled plasmonics for angle-independent structural color displays with actively addressed black states. *Proc. Natl. Acad. Sci.*, *117*(part 1, 3A), 202001435. doi: 10.1073/pnas.2001435117.

Fraunhofer-Gesellschaft, (2021). *Solar Technology with the Beauty of Butterfly Wings*. Tech Xplore https://techxplore.com/news/2021-01-solar-technology-beauty-butterfly-wings.html (accessed on 17 August 2023).

Giraldo, M. A., & Stavenga, D. G., (2016). The brilliant iridescence of *Morpho* butterfly wing scales is due to both a thin film lower lamina and a multilayered upper lamina. *J. Compar. Physiol.*, *202*, 381–388.

Giraldo, M. A., Yoshioka, S. C., Liu, C., & Stavenga, D. G., (2016). Coloration mechanisms and phylogeny of *Morpho* butterflies. *J. Exp. Biol.*, *219*(24), 3936–3944.

Horton, J. Where Do Butterflies Get Their Striking Colors? *Anatomy of a Butterfly Wing, How Stuff Works*. https://animals.howstuffworks.com/insects/butterfly-colors.htm *(accessed on 17 August 2023).*

https://doi.org/10.1098/rsif.2020.0854 (accessed on 17 August 2023).

https://doi.org/10.1119/1.3192768 (accessed on 17 August 2023).

https://eas.caltech.edu/news?keyword=Radwanul+Hasan+Siddique (accessed on 17 August 2023).

https://indiaeducationdiary.in/new-butterfly-inspired-hydrogen-sensor-is-powered-by-light/ (accessed on 17 August 2023).

https://interestingengineering.com/butterfly-wings-could-help-us-detect-lethal-chemical-agents (accessed on 17 August 2023).

https://sensing.konicaminolta.us/us/blog/where-do-butterflies-get-their-color/ (accessed on 17 August 2023).

https://www.exeter.ac.uk/news/featurednews/title_318647_en.html (accessed on 17 August 2023).

https://www.fraunhofer.de/content/dam/zv/en/press-media/2021/january-2021/bau-2021-ise-solar-technology-with-the-%20beauty-of-butterfly-wings_neu.pdf (accessed on 17 August 2023).

https://www.fraunhofer.de/en/press/research-news/2021/january-2021/solar-technology-with-the-beauty-of-butterfly-wings.html (accessed on 17 August 2023).

https://www.sciencedaily.com/releases/2018/04/180430131918.htm (accessed on 17 August 2023).

https://www.sciencedaily.com/releases/2020/12/201202114524.htm (accessed on 17 August 2023).

https://www.sciencenewsforstudents.org/article/butterflies-use-jet-propulsion-for-quick-getaways (accessed on 17 August 2023).

https://www.scientificamerican.com/article/a-butterflys-brilliant-blue-wings-lead-to-less-toxic-paint/ (accessed on 17 August 2023).

Johansson, L. C., & Henningsson, P., (2021). Butterflies fly using an efficient propulsive clap mechanism owing to flexible wings. *J. R. Soc. Interface, 18*, 20200854. https://doi.org/10.1098/rsif.2020.0854.

Kaszubska, G., (2020). *New Butterfly-Inspired Hydrogen Sensor is Powered by Light-Activated Sensor Delivers Precision Results at Room Temperature*. RMIT University *Science News*, https://www.sciencedaily.com/releases/2020/12/201202114524.htm (accessed on 17 August 2023).

Kinoshita, S., Yoshioka, S., & Miyazaki, J., (2008). Physics of structural colors. *Rep. Prog. Phys., 71*, 076401. doi: 10.1088/0034-4885/71/7/076401.

Lea, R., (2020). *New Hydrogen Sensor Takes its Inspiration from Nature*. https://www.azosensors.com/news.aspx?newsID=14245 (accessed on 17 August 2023).

Lonergan. A., & O'Dwyer, C. (2023). Structural colors and phenomena arising in nature are attributed to micro and nanoscale features of the surface. *Adv. Mater. Technol. 8*, 2201410.

Mallick, T., & Eames, P., (2007). Design and fabrication of low concentrating second-generation PRIDE concentrator. *Sol. Energy Mater. Sol. Cells, 91*, 597–608.

Masterson, A. (2018). *Butterfly Wing Inspires Nanotech Glaucoma Sensor.* Cosmos Newsletter. https://cosmosmagazine.com/technology/butterfly-wing-inspires-nanotech-glaucoma-sensor/ (accessed on 17 August 2023).

McDonald, B., (2017). *Science Imitates Nature to Improve Solar Cells*. CBC News. https://www.cbc.ca/radio/quirks/blog/science-imitates-nature-to-improve-solar-cells-1.4407654 (accessed on 17 August 2023).

Mika, F., Matějková-Plšková, J., Jiwajinda, S., Dechkrong, P., & Shiojiri, M. (2012) Photonic crystal structure and coloration of wing scales of butterflies exhibiting selective wavelength iridescence. *Materials 5*(5), 754-771; https://doi.org/10.3390/ma5050754.

Narasimhan, V., Siddique, R. H., Lee, J. O., Kumar, S., Ndjamen, B., Du, J., Hong, N., et al., (2018). Multifunctional biophotonic nanostructures inspired by the longtail glasswing butterfly for medical devices. *Nature Nanotechnology, 13*(6), 512–519. https://doi.org/10.1038/s41565-018-0111-5 (accessed on 17 August 2023).

Nishida, K., Adachi, H., Moriyama, M., Futahashi, R. & Hanson, P.E. (2012) Butterfly wing color made of pigmented liquid. *Cell Reports 42*(8), 112917.

Osotsi, M. I., Zhang, W., Zada, I., Gu, J., Liu, Q., & Zhang, D., (2021). Butterfly wing architectures inspire sensor and energy applications. *National Science Review, 8*(3), nwaa107, https://doi.org/10.1093/nsr/nwaa107.

Penz, C. M., Devries, P. J., & Wahlberg, N., (2012). Diversification of *Morpho* butterflies (Lepidoptera, Nymphalidae): A re-evaluation of morphological characters and new insight from DNA sequence data. *Syst. Entomol., 37*(4), 670–685. https://doi.org/10.1111/ j.1365-3113. 2012.00636.x.

Potyrailo, R. A., Starkey, T. A., Vukusic, P., Ghiradella, H., Vasudev, M., Bunning, T., Naik, R. R., et al., (2013). Discovery of the surface polarity gradient on iridescent morpho butterfly scales reveals a mechanism of their selective vapor response. *Proc. Natl. Acad. Sci. U S A., 110*(39), 15567–15572. doi: 10.1073/pnas.1311196110.

Shanks, K., Senthilarasu, S., Ffrench-Constant, R. H., & Mallick, T. K. (2015). White butterflies as solar photovoltaic concentrators. *Sci Rep., 5*, 12267. doi: 10.1038/srep12267.

Shurkin, J., (2013). *Butterflies Inspire Anti-Counterfeit Technology*. Inside Science. https://www.insidescience.org/news/butterflies-inspire-anti-counterfeit-technology (accessed on 17 August 2023).

Siddique, R. H., Donie, Y. J., Gomard, G., Yalamanchili, S., Merdzhanova, T., Lemmer, U., & Hölscher, H., (2017). Bioinspired phase-separated disordered nanostructures for thin photovoltaic absorbers. *Sci. Adv.*, *3*(10), e1700232. doi: 10.1126/sciadv.1700232.

Smith, L., (2020). *This Paint Inspired by Butterflies Could be the Secret to Cooler Cities*. The Biomimicry Institute Fast Company. https://www.fastcompany.com/90557829/this-paint-inspired-by-butterflies-could-be-the-secret-to-cooler-cities (accessed on 17 August 2023).

Stavenga, D.G., Leertouwer, H.L., & Arikawa, K. (2023). Butterfly Wing Translucence Enables Enhanced Visual Signaling. *Insects14*(3), 234. https://doi.org/10.3390/insects14030234.

Tee, S. Y., Win, K. Y., Teo, W. S., Koh, L. D., Liu, S., Teng, C. P., & Han, M. Y., (2017). Recent progress in energy-driven water splitting. *Adv. Sci.*, *4*, 1–24.

Wilcox, M., (2020). A Butterfly's Brilliant Blue Wings Lead to Less Toxic Paint. Scientific American. https://www.scientificamerican.com/article/a-butterflys-brilliant-blue-wings-lead-to-less-toxic-paint/.

Xinhua, (Xinhua/ANU), (2017). Australian National University. *Butterfly Wings to Inspire New Solar Power Technology*. Australian researchers. https://www.chinadaily.com.cn/bizchina/2017-05/20/content_29427167_5.htm; https://science.anu.edu.au/news-events/news/butterfly-wings-inspire-invention-opens-door-new-solar-technologies (accessed on 17 August 2023).

CHAPTER 8

Bio-Inspired Technologies

ABSTRACT

Engineers must be familiar with and learn a great deal from nature. Scientists and engineers alike are on the prowl for solutions to some of today's most pressing issues. Biomimicry may be able to aid and solve these problems. There are many everyday objects that have been influenced by nature. You may credit innovations for making your life a little easier. When driving, the cat's eye is essential. Percy Shaw, an English inventor, invented the cat's eye as one of many road safety measures. He was inspired by the way off-road markers are illuminated by car headlights. Instead of reflective cats' eyes, solar-powered LED versions may be seen 10 times further and run for days on a single entrance, replacing the once-ubiquitous solar-powered LED versions. As a result of brilliant cat's eyes that are being tested in accident hotspots, road networks have been given intelligent treatment for the first time in many, many decades. The tiny structure of polar bear hairs stumps experts when it comes to heat insulation. An insulator is responsible for generating the distinctive hair structure of a synthetic substance, according to researchers. Engineers can use polar bear hair as a model to create synthetic materials that retain heat just as well as the real thing. There may be many real-world applications for this new technology in the fields of architecture and aircraft after scientists devised an insulator based on individual polar bear hairs' structure. These technologies look at a wide range of subjects. To name just a few, there's the jellyfish-inspired microchip for cancer cell capture, the cockroach-inspired technology for better technology, and even the coral-inspired cement. Humanity could benefit from learning more about these qualities. Cement that self-heals is an example of an innovation. For self-healing concrete, special elements (such as fibers or capsules) and adhesive solutions are incorporated into the concrete mix. A crack will be filled with

Bio-Inspired Technologies for the Modern World: Emerging Trends and Applications.
R. Ramakrishna Reddy & T. Pullaiah (Authors)

fluids and will be instantly repaired by the fibers or capsules breaking as it occurs. The mussel-inspired grown adhesive and its use in fetal surgery were described by researchers. Robotic arms modeled after an elephant's trunk are among the other cutting-edge technologies being investigated.

8.1 INTRODUCTION

Darwin watched nature and came up with unique theories that regulate life's essentials through his study of the natural world. Nature is the best teacher, so let's take a moment to learn from her. After over 3 billion years of natural selection-based evolution, it is still the most astonishing human invention.

Human ingenuity is frequently spurred on by the inventiveness of nature. Despite its reputation as a cat phenomenon, "eyeshine" occurs in a wide range of creatures, from moths to whales. Because of the tapetum lucidum, which is a reflective layer behind the eye's photosensitive region and bounces back light particles, this phenomenon occurs.

Whatever the source of Percy Shaw's inspiration, the invention's positive influence is undeniable. Cat's eyes were erected on numerous British roadways during the Second World War as a means of guiding drivers. In order to ensure that the cat's eyes were always clearly visible, Shaw even designed a cleaning system. At the Road Studs Ltd factory, the cat's eyes continued to be installed on roads around the world. It has been voted one of the top 10 Great British design symbols, but who designed the Cats eye road stud, and where did the inspiration for the idea come from? A particularly hazardous section of road near Halifax may have given Percy the idea for the Cats eye, according to legend.

8.2 ROAD REFLECTORS: INSPIRED BY CAT EYES

Driving safety is a must in today's world. Over the last century, the reflective cat's eye has helped to prevent accidents. Cat-eye reflectors can also be used as a warning device for motorists on highways and municipal roads. For example, British inventor Percy Shaw proposed in his patent that different colors of reflectors may be used to communicate different levels of danger on the road (https://www.nkroadstud.com/news/news-information/4-functions-of.html). Motorways and intersections are marked with a variety of colors today. A British inventor, Percy Shaw, is most known for his invention of the cat's eye, a device that improves road safety. He is appreciative of his contribution to road safety innovation. Reflective cat's eyes were invented in

1934 by Percy Shaw in order for drivers to better see road directions in dark/foggy conditions. It was inspired by the way car headlights reflect off-road signage that Percy Shaw came up with the idea. A cat's eye reflector was eventually found on British highways and made necessary.

They say that cats' night vision is better than ours. In complete darkness, they have no way of knowing. Feline circling eyes are naturally equipped with muscles that allow the pupil to contract in bright light and expand fully in low light to offer the best possible illumination for the animal.

As a reflective layer underneath the cat's retina known as the tapetum lucidum, it helps maximize the cat's use of the light it does receive. There are glittering green orbs in cats' eyes when they see a modest amount of light at night. Durability expectations may have been sparked by these incredible feline stories. Many nocturnal species' night vision is improved by the tapetum lucidum, a reflecting coating beneath the retina. It permits visual pigments to absorb low-intensity light for a second time by reflecting light outward (Figure 8.1).

It is well-known that wild cats are nocturnal creatures who spend the majority of their time out in the open at night hunting for food. Yorkshireman Percy Shaw came up with the idea for the cat's eye study in the 1930s, inspired by the image of his automobile headlights reflected in the animal's eyes. For centuries, studs have relied on headlight beams to educate their students. Smart motorways and intelligent automobiles seem to be the way of the future when it comes to transportation. As of now, the road network has been updated to include cat's eyes, which have been introduced as a new method and are being tested in accident hotspots to reduce the frequency of accidents (Figure 8.2).

8.2.1 INTELLIGENT CAT'S EYES

The new reflecting cat eyes will light up in reaction to changing traffic lights, making them cleverer than ever before. According to Highways England, it makes iconic roundabouts look more like runways. Using this method for the first time, they want to increase road safety. To ease the confusion of roundabouts, these 'road studs' will be deployed. Around 170 of these were installed as part of a $3 million improvement project on Merseyside's Switch Island, where the M57 and M58 intersect with three other highways. Connecting cables and studs is automated by a controller device installed beneath the road surface. Reflective dots in the middle and on the perimeter of the road are used to indicate road markings. These are known as cats' eyes in the United Kingdom, and they glow in the dark.

FIGURE 8.1 Cats Eye vision. Public Domain.

FIGURE 8.2 Safety road Studs were inspired by a cat's eyes reflecting headlights.

Source: Reprinted with permission from Llewellyn et al. (2020). https://creativecommons.org/licenses/by/4.0/.

The influence of Percy Shaw's idea when he originally came up with it is a subject of controversy. World War II necessitated the installation of cat's eyes on many of the UK's highways. As a result, Shaw had a cleaning mechanism in place to ensure that the cat's eyes were always clear. The Road stud Ltd plant continued large-scale production and put the cat's eye on roads around the world.

The cat's eyes are still relied upon by drivers when there is limited light. However, Yorkshire native Martin Dick's revised the original idea in 1990. When he realized that the original cat's eyes didn't produce enough light in foggy conditions, he developed Solar lite road studs. Around 10 times as much light is emitted by the Solar lights as the cat's eyes do. In comparison to Shaw's original idea, these are a lot less expensive and better for the environment (Figure 8.3).

FIGURE 8.3 Rural roadway lighting alternative: Active road studs.

Source: Reprinted with permission from Llewellyn et al. (2020)© 2020 by the authors. Licensee MDPI, Basel, Switzerland. http://creativecommons.org/licenses/by/4.0/

The new purpose does not rely on the efficacy of the car headlights to function efficiently, unlike typical reflective road studs. Drivers can see up to 800 meters ahead with active studs, which use light-emitting diodes as their light source. With just four hours of charging, the lights may last for up to 10 days. It's less stressful to drive when the road studs are in operation, according to Mr. Lee, an experienced driver. The development of the cat's eyes was foreseen by drivers.

After 70 years, cat-eyes have entered a new era. They tested it in several locations across the UK and other countries. With built-in sensors and microprocessors, the intelligent road studs are able to recognize varied weather conditions and traffic speeds. Low-maintenance LED lights driven by solar cells are used to mark road borders and centerlines. Drivers may be able to react more quickly to this new technology, which is embedded in the road surface, compared to traditional cat's eyes and high pavement signs. Stumps not only passively reflect light, but they may also actively emit distinct colors of light that can be detected up to 1,000 meters away from the object. Studs can emit a white light when they detect fog. When the temperature drops significantly, it emits a blue light to indicate the presence of ice. To alert drivers of impending danger, studs can leave an orange light trail behind them after overtaking automobiles. It is also possible for the studs to talk to each other so that, for example, a car driving the wrong way can trigger red warning lights on the other side of an unknown slope or turn.

8.2.2 CAT'S EYES REPLACE WITH NEW SOLAR-POWERED LED TECHNOLOGY

The practice of placing a cat's eyes in the road will soon be history. Modified laws and new LED solar-powered lights were introduced by British Government Ministers. As a road safety test gadget, Percy Shaw discovered cat's eyes (developed in the 1930s) and was inspired by Shaw's pet cat. Instead then requiring direct illumination from headlights, the new study's cat eyes will have LEDs built right in, so they'll glow on their own without the need for external illumination. Around 10 times farther than the generally accepted model of a cat's eye, the LED studs can see up to 1,000 yards ahead. The new lights can last up to 10 days on a single charge and cost less than traditional cat's eyes over their lifetime. Reflective road studs, which are required by law, are still included in the new designs, which can be found in some of the UK's most dangerous accident hotspots. It's believed that technology can advance and provide new options if certain laws are modified.

'Solar lights,' which use LEDs, are becoming more common in places where road safety is a concern. "We hope that they will become the norm; they have demonstrable road safety advantages as well as other advantages to drivers," said Graham Muspratt, the group product manager. When road studs are used, driving becomes much less stressful since drivers can see further ahead. In addition, the road infrastructure needs to be improved.

8.3 THERMAL INSULATION MODELED AFTER POLAR BEAR FUR

A cloak made of heat-trapping porous fibers that shield a bunny from night-vision cameras is needed. Hair from a polar bear has a hollow center. They believe that the animal's own IR emissions are to blame. It is because of this structure that the bears in the Arctic are able to maintain a comfortable body temperature year-round. In several night-vision gadgets, thermal imaging cameras can be protected from bears by their hairs. IR-reflective textiles based on polar bear hair could be effective in concealing soldiers or other personnel in need of stealth protection. Synthetic thread has resulted in fibers that are too weak for this experiment. Zhejiang University is now using freeze spinning to make fibers that are strong, porous, and extremely thermally insulating. Zhejiang University Fiberin is given a polysaccharide found in shellfish called Chitosan (Peplow, 2018) (Figure 8.4).

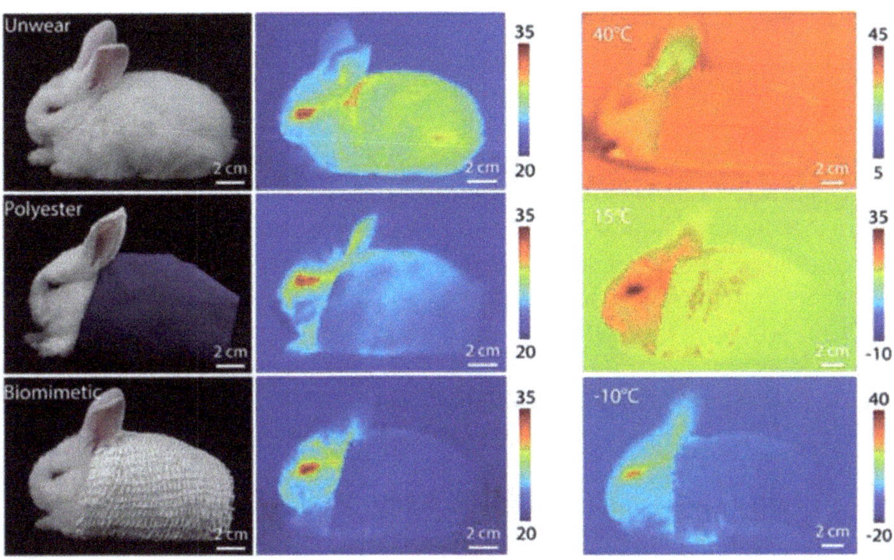

FIGURE 8.4 Thermal stealth properties biomimetic textiles and Infrared light reflectance measurement of different textile.

Source: Reprinted with permission from Ling et al., 2022. https://creativecommons.org/licenses/by/4.0/.

A cold copper ring was used to squeeze the liquid, which resulted in a frozen fiber with flat ice crystals that had been chilled to room temperature. Fibers with up to 87% porosity were created when the ice in them was sublimated during the freeze-drying process.

8.3.1 POLAR BEAR HAIR STRUCTURE

Arctic climates make it harder to survive. The fat, skin, and fur of polar bears are their only sources of heat retention. Engineers saw polar bear hair as a good example of a material that may be used as a substitute for heat-blocking materials. An infrared-absorbing insulator has been developed by researchers at the University of Science and Technology in China (USTC). In order to generate a material that can be employed in the domains of architecture and aerospace, it resembles the structure of various hairs. Hair from polar bears is hollow, unlike that of other mammals, making it unique. Each one contains a lengthy, cylindrical core that may be seen when examined under a microscope. These cavities are well-suited to their white coats because of their spacing and form. They rely on a source that is extremely heat-resistant, water-resistant, and elastic.

8.3.2 CONSTRUCTION OF NEW INSULATOR

An aerogel block consisting of carbon tubes with hollowed-out sections resembling hair was created to mimic this structure and scale it properly. Researchers claim that the novel hollow-tube design has a higher capacity than conventional aerogels while also being more heat resistant. Further enhancing its usefulness is the fact that the new material is stretcher than the hairs themselves.

8.3.3 NOVEL HEAT-INSULATING MATERIALS ARE INSPIRED BY POLAR BEARS

Scientists have created an insulator that mirrors the structure of individual polar bear hairs. In the fields of architecture and aeronautics, they are widely used. Due to luminosity, the fur of a polar bear appears white. The reason polar bears remain white all year round is because of this (Figure 8.5).

Is it possible that polar bears aren't white but rather opaque? Hollow hairs are found in the coat that covers their bodies, according to the scientific explanation. Every fiber contains air bubbles that can't be seen with the naked eye. This combination is an excellent thermal insulator and gives the appearance of white, deceiving us into thinking their fur is the same color as snow.

Scientists have created an insulator that closely mirrors the structure of individual polar bear hairs (Zhan et al., 2019). Even architecture and aircraft can benefit from its application. To keep themselves warm, polar bears use their fat, skin, and fur as insulation. Engineers use polar bear hair as a model for plastic materials that can withstand high temperatures and the natural environment, such as thermoplastic polymers.

FIGURE 8.5 Female Polar Bear near Kaktovik, Barter Island, Alaska, United States.
Image credit: Alan Wilson. https://creativecommons.org/licenses/by-sa/3.0/.

Scientists at the University of Science and Technology of China say that polar bear hair has evolved to assist in limiting heat loss in cold and humid situations, making it a viable model for a synthetic heat insulator (USTC). An elastic and lightweight material that retains heat without degrading significantly during its lifetime can be created from carbon tubes, he stated. Unlike the hair of humans or other mammals, the fur of a polar bear is hollow. The characteristic white coats of these creatures can be attributed to the forms and spacing of these cavities. Thermal insulators can benefit from the extraordinary heat-holding capacity, water resistance, and stretchability that these materials provide for themselves and their derivatives.

8.4 NATURE AND NANOSEPTIC SELF-CLEANING SURFACES

Even after billions of years, natural creatures have developed fascinating self-cleaning characteristics on their outer surfaces. Self-cleaning biomaterial, the lotus leaf (*Nelumbo nucifera*), is a symbol of cleanliness in Asian cultures. In addition to basic research, scientists are increasingly interested in the practical uses of Nature-inspired self-cleaning surfaces. Nature has produced self-cleaning surfaces. Inspiring the creation of a variety of bio-inspired artificial surfaces. In the natural world, we've found two distinct sorts of surfaces that clean themselves. The Lotus effect, superhydrophobic-induced droplet jumping, unidirectional water droplet movement, underwater-superoleophobic-based self-cleaning, slippery-based self-cleaning, and dry self-cleaning are six nature-inspired remarkable self-cleaning mechanisms (Yu et al., 2020).

Sanitizing products and solutions must be used on a regular basis to maintain the freshness of the surfaces we contact on a daily basis. Extensive cleaning imposes a financial strain while also introducing potentially harmful compounds to the ecology. When it comes to cleaning, nature excels at doing it on its own, with no help from humans. Because of its unique mechanism and exceptional versatility, scholars have been intrigued by this topic for decades. Self-cleaning was inspired by the superhydrophobic characteristic of some plant leaves. A lotus leaf is a well-known representation of this concept (Xu et al., 2016). There is a greater than 150° contact angle and a less than 2° sliding angle in the leaves of lotuses. As a result, the focus of this chapter is on nanoseptic nanotechnology-based self-cleaning surfaces.

COVID-19 infection can be prevented by keeping a safe distance, washing your hands frequently, practicing excellent respiratory hygiene (e.g., using and discarding tissues), cleaning surfaces, and keeping indoor areas sufficiently aired. Chemicals often used in cleaning and disinfection have little problem killing SARS-CoV-2, the virus that causes COVID-19. It only lasts for a brief period of time on surfaces. In order to prevent and treat viral infections, clean surfaces, especially in public settings, are necessary (Rakowska et al., 2021).

People's daily life has been profoundly affected by the global reaction to the coronavirus (COVID-19) epidemic. In several nations, regions, and towns, governments have imposed stringent measures of social distancing and shelter-in-place in an attempt to limit the spread of the virus (i.e., working from home or just staying home). Use a household cleaning product containing soap or detergent to reduce your chance of contracting an illness from a surface. Washing surfaces with soap and water is the most common method for removing most virus particles. For the most part, you don't need to disinfect your home unless you or a member of your family is unwell or has recently been in contact with COVID-19. A primary purpose of the Environmental Protection Agency is to provide Americans with COVID-19, the virus that causes the sickness. It is possible to disseminate COVID-19 by inhaling or exhaling droplets of infected secretions that settle on neighboring surfaces. SARS-CoV-2 (the virus that causes COVID-19) can survive for days on any surface, plastic or metal included. Affected individuals and the surfaces they frequently come into contact with may spread the sickness to those around them, according to the research.

Portable mats for worktops, tray tables, and bathroom vanities are among the NanoSeptic® self-cleaning goods, as are specialized items like grocery cart and public transportation grab-handle coverings (https://facilityexecutive.com/2021/07/creating-cleaner-healthier-travel-experiences-with-nanoseptic/). The virus has a diameter of 65–125 nm, which is considered nanoscale. Medical research, materials science, artificial intelligence (AI), the environment, virology, and the physical sciences are just a few of the many fields that nanotechnology has the potential to impact. Engineering the many physicochemical features of naturally abundant materials can help with a plethora of issues. "Nano-touch" is a self-cleaning surface company that serves facilities in healthcare and education as well as hospitality establishments and commercial offices.

Using NanoSeptic Self-Cleaning Surfaces in airports, hotels, and airplanes will help alleviate travelers' concerns about hygiene and sanitation. There have been unexpected viral outbreaks in recent decades. An outbreak of the newly identified coronavirus, COVID-19, has already occurred (SARS-CoV-2). Employees and visitors must be kept secure from the COVID-19 pandemic by using NanoSeptic Self Cleaning Surfaces. In the absence of cleaning materials, how can companies protect the safety of their staff and customers? We need self-cleaning NanoSeptic surfaces (Vélez-Pea et al., 2020).

NanoSeptic skins and mats from nano-touch materials (https://facility-executive.com/2017/12/nanoseptic-self-cleaning-surfaces/, https://www.clean-roomtechnology.com/news/article_page/Research_centers_worldwide_validate_effectiveness_of_NanoSeptic_surface/96878) can make dirty, high-traffic touch points like door handles and countertops continuously self-cleaning. Toxins, heavy metals, and diluted poisonous substances are not present on its surfaces. Engineered with an exclusive combination of material science, nanotechnology, and green chemistry, the materials are developed.

8.4.1 WHAT IS NANOSEPTIC?

Surface bacterial fouling has become an urgent global challenge that calls for resilient solutions. A bacteria-repellent surface (Figure 8.6) is usually achieved by introducing superhydrophobicity to remarkably lower bacterial adhesion. No harmful chemicals or poisons are required when using NanoSeptic Surfaces to oxidize organic contaminants.

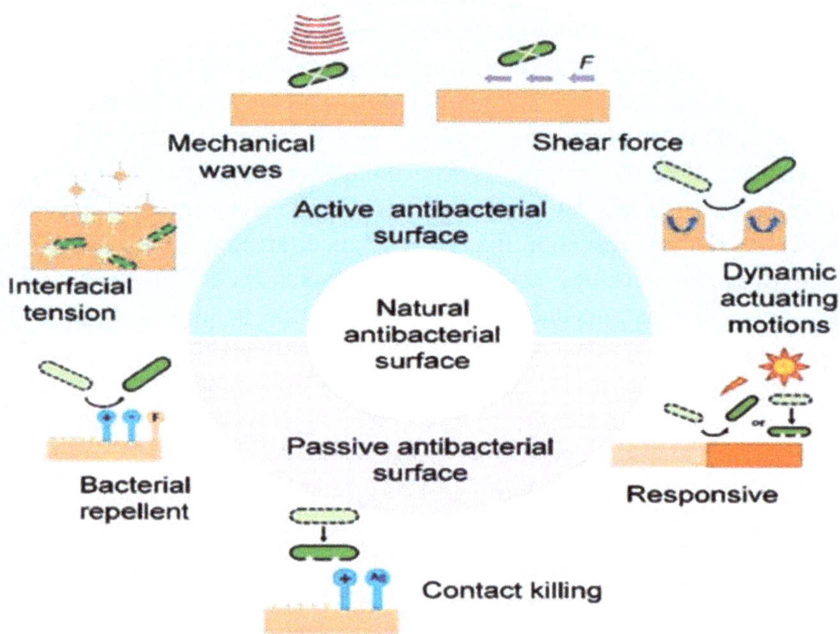

FIGURE 8.6 Scheme of nature and bioinspired antibacterial surfaces.

Source: Reprinted with permission from Yang et al., 2020. Copyright © 2022 by the authors. Licensee MDPI, Basel, Switzerland. https://creativecommons.org/licenses/by/4.0/.

Every day, employees and visitors alike will come into contact with office doors, lifts, and other materials. In order to prevent contamination, NanoSeptic surfaces must be given. High-traffic areas such as door handles, push bars, and elevator buttons should have NanoSeptic self-cleaning surfaces installed. Transparent film in various sizes can be applied to touchscreens on computers, copiers, tablets, smartphones, and breweries for individuals who work with touchscreens. NanoSeptic surfaces are free of viruses. Because of this, staff can carry out their duties without concern of contamination. Hand washing according to previous guidelines should be continued. Your office will be safe and clean if you take this precaution.

Continuous oxidation of organic materials cleans and disinfects NanoSeptic surfaces. These are the traits they exhibit:

- Daily, weekly, and even 24-hour self-cleaning surface disinfectant;
- Safe for the environment;
- Stronger than bleach in terms of cleaning power;
- Non-toxics durable.

Increasing public demand for hygienic, self-disinfecting, contaminant-free protective coverings has led to the development of these self-cleaning shielding materials. Self-cleaning material used for kitchen counters and bathroom vanities can also be utilized for doorknobs (Figure 8.7).

Door push pads, push bars, and supermarket cart handle wraps are all examples of fixed applications that benefit from self-cleaning skins. Student snack mats, travel mats, and counter mats are all options for NanoSeptic mats. A tamper-evident travel pouch holds the stories, making them convenient to take with you on the road. To the mouse pad and window-pocket counter mat from NanoSeptic are added products like a channel guide and a washable tissue box cover.

To create a non-leaching, self-cleaning surface, scientists have joined nanotechnology and material science in the development of NanoSeptic products. An oxidation process more potent than bleach decomposes dirt and contaminants when exposed to ultraviolet (UV) light. Designed to resist regular washing, these goods have a long life expectancy. Disinfectants and sanitizers will not be replaced by NanoSeptic surfaces. As a result, they work well together. Conventional methods can clean a touch point or a person's hand, but contamination occurs when the contact or airborne contaminants come into contact with the object being cleaned. NanoTouch products' self-cleaning properties are designed to limit the risk of pollutants being transferred from one contact to another.

FIGURE 8.7 Disinfection of a floor using disinfectant liquid applied using a mop.
Source: Public Domain.

8.4.2 COVID-19 COULD BENEFIT FROM NANOTECHNOLOGY

The most recent human coronavirus to surface around the world is COVID-19 (SARS-CoV-2). Nano-based protection equipment and disinfecting agents can provide much-needed protection against SARS-CoV-2. Moreover, nanoparticles can serve as a carrier for antigens or as an adjuvant, thereby making way for the development of a new generation of vaccines (https://www.ncbi.nlm.nih.gov/pmc/articles/PMC8310263/). In addition to preventing viral infection and applying spray, COVID-19 could benefit from this technology. Cotton, polyalkene, polyester, polyamide, and cellulose-based polymers, for example, can be treated with antibacterial copper and silver nanoparticles (NPs). Scientists have discovered a nanoparticle-based method that can kill influenza viruses. A variety of viruses, including human coronavirus and SARS-CoV, could benefit from this treatment.

There are two advantages of using nanomaterial for facemasks:

i. Pathogens are prevented and eliminated by the facemask's microbicidal properties, and

ii. PPE based on nanotechnology, which has a wide range of applications, works. Masks can be more easily cleaned and disinfected after they've been used because the most vital section of germs is damaged when they come into direct contact with the masks. The patents left open by the technology can be used to make a range of PPE, such as surgical aprons, medical and lab coats, foot protection, and bedsheets.

Since most viruses are destroyed when they come into direct contact with masks, it is safer to administer this material after its use because the risk of contamination is much reduced.

It is also possible to build PPE based on nanotechnology for many purposes. Other viral illnesses can benefit from nanotechnology (Vélez-Pea et al., 2020) for better diagnoses, protection, and treatment. New technologies and procedures will transform the fight against COVID-19 by providing more effective options for therapy.

8.5 CAPTURING CANCER CELLS WITH A JELLYFISH-INSPIRED MICROCHIP

Detection of cancer, bacteria, and viruses has been made possible by the tentacles of jellyfish. According to the researchers, the findings of this experiment could lead to devices that can recognize life-threatening disorders swiftly and sensitively and catch harmful cells for assessment by specialists so that they can create the best therapies to combat them. Long, sticky tentacles of jellyfish and relatives like sea anemones catch prey moving past them. Using the expertise of scientists, these marine critters developed a unique microchip that has a network of long DNA strands capable of grabbing onto proteins as they float through the water.

To learn more about how a tumor responds to treatment and which treatments work best on it, the bloodstream of a patient can be used to collect tumor cells. A blood sample must be analyzed to identify and isolate these specific cells from the rest of the cells. Microfluidic devices have been used to isolate circulating tumor cells. It has been given to a team of researchers at Brigham and Women's Hospital to study (SciTech Daily, 2012; Zhao et al., 2012). Unusual cell types, such as cancer cells, fetal cells, viruses, and microbes, may have therapeutic and diagnostic purposes in the general public. The device's long, graceful appendages take their design cues from marine organisms like jellyfish and sea cucumbers.

For his part, MIT's Karp cooperated on a microchip that uses DNA to catch cancerous cells in the circulation, among other projects. Inspired by jellyfish, whose lengthy tentacles hunt for food no matter where it landed (Figure 8.9).

Using a microchip, cancer cells can be counted and sorted to see how effectively chemotherapy or other therapies are working (Figure 8.8). After chemotherapy, doctors will be able to evaluate how resistant the tumor is or if it will spread to other parts of the body. Additionally, the microprocessor can count cells 10 times faster than current technology. It is now being tested on malignancies in the laboratory.

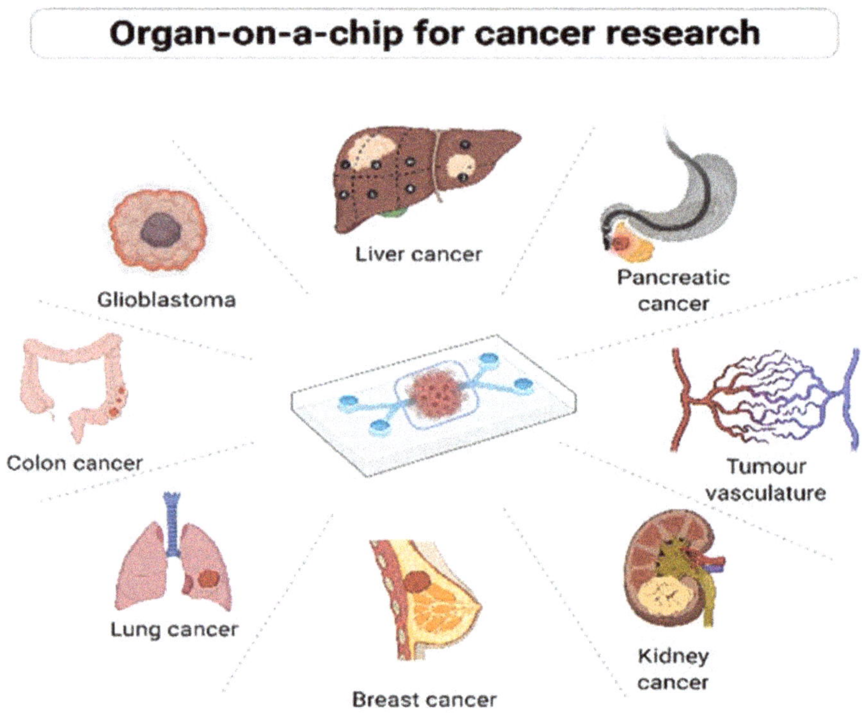

FIGURE 8.8 New diagnostic and treating tools, organs-on-chip are quite promising regarding the achievement of more personalized medicine.

Source: Reprinted with permission from Gonçalves et al 2022. Copyright © 2022 by the authors. Licensee MDPI, Basel, Switzerland. https://creativecommons.org/licenses/by/4.0/.

This is a brief summary of what it says (https://prezi.com/4gxovjaasjjy/jellyfish-inspired-microchip/?frame=7c54f20ae2235c2782244072406a9b74327333ed).

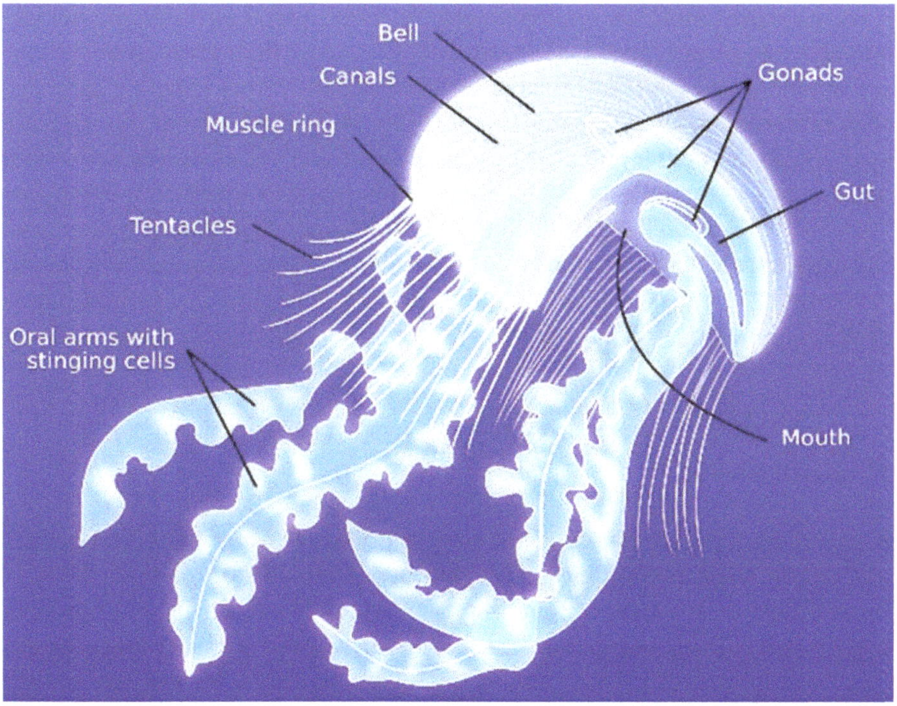

FIGURE 8.9 Jellyfish are mainly free-swimming marine animals with umbrella-shaped bells and trailing tenticles (under the Creative Commons Attribution-ShareAlike License 4.0).

Source: Reprinted with permission from Gonçalves et al. 2022. Copyright © 2022 by the authors. Licensee MDPI, Basel, Switzerland. https://creativecommons.org/licenses/by/4.0/.

8.5.1 HOW IT WORKS?

Chip coated with long repeating DNA strands for use in microfluidic systems. It attaches to certain proteins on cancer cells as they float through the bloodstream and stops them from spreading:

- Even while antibodies or synthetic nucleic acids were used in other microfluidic devices, they could not capture big entities like complete blood cells.
- These cells can be grabbed by the jellyfish-like apparatus.
- Tentacles of lengthy, three-dimensional DNA were the key.
- It's giving them a herringbone pattern that's modeled after the sticky formations made by jellyfish.

Long repeating DNA sequences can be covered by a microfluidic device. It attaches to certain proteins on cancer cells as they float through the bloodstream and stops them from spreading. Jeffrey Karp (principal researcher), MIT researchers, and researchers at Brigham and Women's Hospital are involved in its development. They have created a jellyfish-like robot that follows and moves like a jellyfish by the Max Planck Institute for Intelligent Systems (MPI-IS). The "Jellyfishbot" was the name of the team's innovation. Untethered robots often have bells and tentacles that resemble their natural counterparts (https://tectales.com/bionics-robotics/jellyfish-inspired-robot-to-deliver-drugs-inside-body.html).

8.6 COCKROACHES CAN TEACH US HOW TO CREATE BETTER TECHNOLOGIES

Research shows that cockroaches have the ability to get back up on their feet before their neurological systems have time to inform them how. A valuable lesson for those attempting to create better multi-legged robots is that their legs may recuperate on their own without the assistance of the command center (Shipman, 2014; Shi et al., 2019). Cleaning up a dirty kitchen or bathroom is a cockroach's favorite pastime. In a split second, they're scrambling out of the room and out of your sight. These insects also drive research into antibiotics, robotics, and mechanical limbs, so pest controllers aren't the only ones involved. The sole of your shoe isn't going to stop a running cockroach from recovering its footing (Boyle, 2013) (https://www.popsci.com/technology/article/2013-02/watch-how-cockroaches-are-helping-scientists-design-better-robots).

A cyborg cockroach was created by scientists at North Carolina State University by combining insects with a machine. Electrical pulses can be used to mimic obstacles in front of an insect or the sensation of something crawling up behind it to control its actions by placing electrodes into the insect's antennae and rear-end sensory members known as cerci. Californians in Irvine put a cockroach on top of a ping-pong ball that moved freely on top of a miniature, wheeled buggy. The researchers used electrical signals to compel the cockroach to return and move its legs in specific directions, which drove the buggy with the ping-pong ball's movements. In the event of an earthquake or an explosion, a remote-controlled cockroach may be used to carry cameras or sensors into the affected areas.

8.7 CEMENT-LIKE CORALS

A new kind of environmentally friendly concrete was produced using coral as a catalyst. One of the most extensively used building materials in the world is concrete. Construction crews use twice as much of it as any other building material, including wood, steel, plastic, and aluminum. Coral reefs play an important role in biodiversity protection because they provide a sanctuary for marine life. Many animals can find safe haven in these structures, which can resist even the most violent storms. Coral reefs are held together by a unique form of 'cement,' according to a new study. In coral reef-building organisms vast amounts of Calcium carbonate (CaCO3) is present, which can be placed in some of the cements in concrete. Concrete cylindrical specimens are made by filling the cylindrical mold with mortar in three layers (Figure 8.10). The processing of CA into granular is described by Bunyamin et al. (2022), and ash is depicted in Figure 8.11. Red coralline algae have been stabilizing reefs for at least 150 million years by producing a firm calcareous skeleton. It's sturdy, weather and fire-resistant, and reasonably priced. The snag: Cement production, one of the primary components of concrete, accounts for only 6% of all man-made carbon dioxide emissions. Cement substitutes using carbon dioxide may be possible in the future. Such a remarkable fact. Inspired by the corals, Calera is trying to do this.

According to Stanford University biomineralization expert Brent Constantz, corals developing reefs inspired him to make new cement for structures (see: https://earthsky.org/human-world/brent-constantz-builds-cement-like-corals-do/). Cement production reduces atmospheric concentrations of carbon dioxide, a potent greenhouse gas. Saltwater reacts with atmospheric CO_2 to form carbonate, which is then coupled with calcium from the installation's carbonation process to form a stable solid. Coral skeletons and cement have a similar manufacturing process (Figures 8.10 and 8.11).

In order to manufacture cement, Costantz devised a method that closely resembled this movement. When a power company releases CO_2 into the atmosphere, the CO_2 gets dissolved in saltwater, resulting in the formation of carbonate. The seawater's carbonate and calcium combine and solidify to form an environmentally beneficial cement. Calera is the name of a sophisticated process that could be used in new or existing power stations to absorb and store CO_2 emissions. Although the solids created by this approach cannot completely replace the Portland cement component of concrete, the technology is a wonderful breakthrough in the realm of greener building materials.

FIGURE 8.10 Concrete cylindrical specimens are made by filling the cylindrical mold with mortar in three layers.

Source: Reprinted with permission from Bunyamin et al. (2022). https://creativecommons.org/licenses/by/4.0/.

Biodiversity hotspots include coral reefs (University of Erlangen-Nuremberg, 2020). As they are able to withstand violent storms, they provide a safe haven for a wide variety of animals while also protecting densely populated coastal areas from storm-driven waves. However, how are these reefs, which are made up of coral that is sometimes fragile, so stable? A specific form of 'cement' is to blame, according to researchers from the Friedrich-Alexander Universität Erlangen-Nürnberg (FAU) and the University of Bayreuth (Chen et al., 2015; Science Daily, 2020). This is the first time that researchers have been able to identify the culprit. A calcareous skeleton made of red coralline algae has stabilized reefs for at least 150 million years.

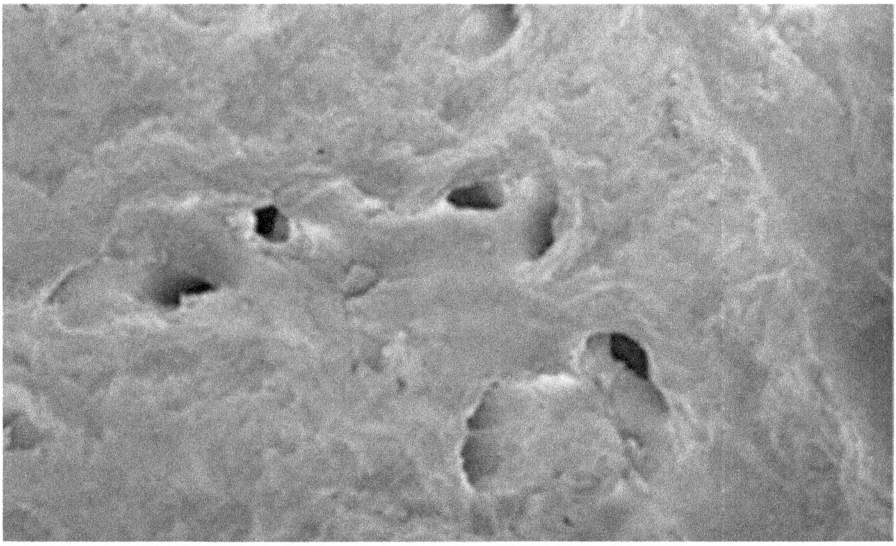

FIGURE 8.11 CA is utilized in place of cement and fine sand in the concrete.

Source: Reprinted with permission from Bunyamin et al. (2022). https://creativecommons.org/licenses/by/4.0/.

8.8 BACTERIA CAN BE USED TO FIX CRACKS IN SELF-HEALING CONCRETE

This material is commonly employed in construction because of its ability to resist compression loads while requiring steel to resist tension stresses; it is also subject to cracking because of its brittle nature. Sulfate, chlorides, and carbonates can enter a building through cracks, which can be harmful. Corrosion is caused by these dynamic fluids penetrating into the concrete, which weakens its structural integrity. However, the entry of dangerous elements causes the need to close these fissures through structural improvement, which may not be considered a failure of the concrete itself. The increased costs of repairs have prompted academics to look at alternatives to crack sealing, such as increasing spending on crack healing. In studies on self-healing, organic and inorganic compounds have been committed to seal fissures. Bacteria are one example of an organic approach for filling in concrete fractures. At the time of its construction, Roman civilization relied primarily on concrete, which was one of the most durable building materials. Over time, cracks in concrete might appear due to drying shrinkage, thermal movement, or other factors, all of which can cause serious problems. An innovative solution to this problem was devised by Dutch scientists Henk Jonkers and Eric Schlangen. Bacteria found near active volcanoes became familiar to scientists. As our bodies restore bone through mineralization, they have created bio-concrete that heals itself utilizing these bacteria (Li & Yang, 2007; Stott, 2009).

Despite our best efforts, concrete will eventually develop cracks. It's not surprising, given the well-known brittleness and variability of the concrete matrix. Because of these, even if there is no external load from the temperature gradient in the early-age temperature gradient, bleeding and shrinkage can lead to fissures in the material itself. Microcracks in concrete, particularly in the epidermis, remain a major concern and challenge for civil engineers worldwide (Alagaifi et al., 2021). If the concrete structure has microcracks, harmful materials could enter and deteriorate it. In order to stop the growth of cracks, they're developing a new sort of concrete that's both durable and effective. Scientists are becoming interested in self-healing concrete, an environmentally friendly concrete that can repair itself (Figure 8.12).

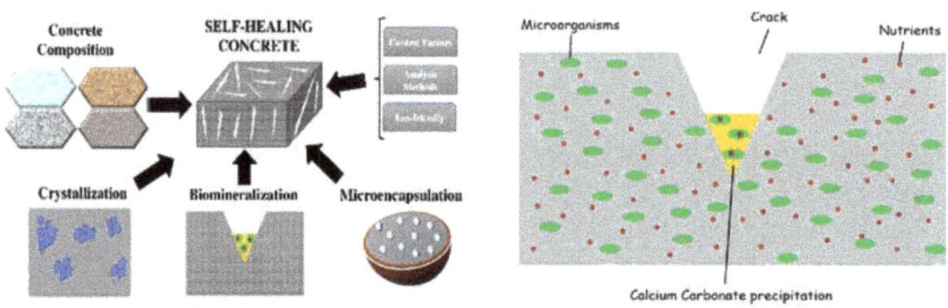

FIGURE 8.12 The self-healing concrete fills up the cracks and Healing of cracks through calcium carbonate precipitation by microorganisms.

Source: Reprinted with permission from Roque et al. (2023). Copyright © 2023 by the authors. Licensee MDPI, Basel, Switzerland. https://creativecommons.org/licenses/by/4.0/.

8.8.1 A LOOK INTO THE FUTURE OF CONSTRUCTION WITH THREE SELF-HEALING BUILDING MATERIALS

Bacteria and calcium lactate will coagulate as a result of the combination of water and concrete. Like conventional concrete, it does not crack or absorb water during the mixing process and only comes into play if the concrete cracks or absorbs water. Calcium lactate is digested by bacteria, which produces limestone to cover fissures. As little as three weeks is all it takes for bacteria to mend 0.8-millimeter-wide cracks. Most advantageously, a special liquid produced by the bacteria can be sprayed into the crevices of existing structures and streets.

Bridges and motorways, for example, require significant and inconvenient maintenance that has only lately been developed to extend the life of a structure. The entire structure could also be updated on a regular basis, but this is not a long-term solution because of the volume of CO_2-releasing concrete used in the present construction.

Construction's lifespan could be extended by using three new self-healing materials that are revealed. It also includes Erik Schlangen's asphalt, which uses a small amount of reason heating to re-calibrate itself. Researchers at TU Delft and MIT recently discovered that some metals have self-healing properties, which they believe can be used to patch up cracks in concrete (Vermeer, 2021). Concrete that is self-healing can return to its former state even if it has been damaged by cracks. Bio-concrete is a substance that, with the help of microorganisms existing in it, biologically provides minerals like limestone and heals concrete surface fissures.

This concrete's primary ingredients are bacillus, calcium lactate, nitrogen, and phosphorus. These bacteria can persist in concrete for up to a century or more (Figure 8.13).

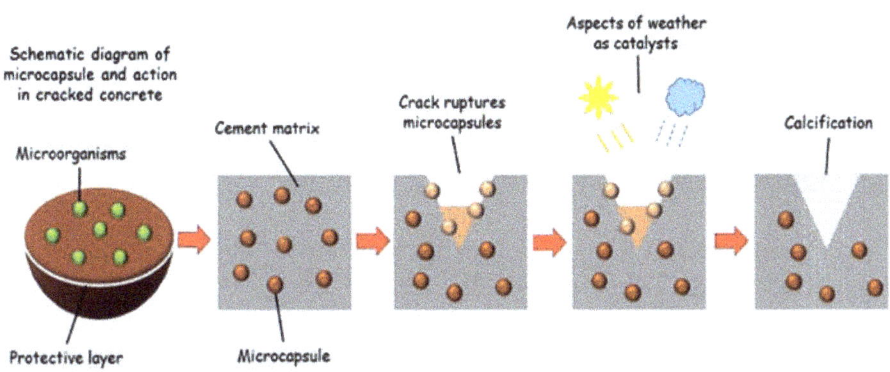

FIGURE 8.13 A schematic model of a microcapsule and how it contributes to the healing of cracks in concrete.

Source: Rique et al. (2023). Open Access.

If the concrete foundation is breached and water seeps in, bacteria will flourish. The bacteria transform calcium lactate into insoluble limestone as soon as they awaken. The resulting limestone seals and compresses the surface.

Because of the bacterial conversion, steel-reinforced concrete constructions last longer.

8.8.2 INNOVATION

A new method of micro-crack mending, proposed by Jonkers et al. (2010), makes use of this technology. It becomes a building material when it mixes with microorganisms that produce limestone. Water pours into concrete through fractures, causing it to deteriorate over time. Bacteria are dispersed throughout the concrete, where they can remain dormant for up to a century. This reaction is aided by the presence of water and moisture in the air. Limestone that is generated by bacteria responds when it is exposed to water and heat. Capillary motion will carry the agent into the crack. The crack between the two faces has been repaired. These results are referred to as bio-mineralization, and they show promising outcomes for self-healing concrete.

In a variety of ways, the application is available. To fix small cracks (0.5–0.8 mm wide), a spray can be used, or a mixture can be poured in while the concrete is being mixed prior to construction. A concrete canal and irrigation system produced using self-healing concrete was built in Ecuador by Jonkers despite the fact that this approach is currently unavailable on the market. Additionally, research has begun on another method that focuses on self-healing concrete's resistance to carbonation and chloride intrusion.

8.9 ORGANIC WASTE PRODUCTS ARE USED TO MAKE BIO-BRICKS

Waste management for agricultural crop residues is becoming increasingly evident to be a significant issue. Producing countries benefit from byproducts like sugarcane bagasse, wheat straw, and rice husks. India alone reports more than 600 metric tons (MT) of agricultural trash. The most frequent methods of disposing of this agro-waste are landfilling, incineration, and composting, all of which have serious negative consequences for the environment. Waste from agriculture may be used to make building materials, as new research has demonstrated, as well as real-world solutions to some of agriculture's most pressing problems.

KIIT School of Architecture in Bhubaneshwar and the Indian Institute of Technology in Hyderabad have developed bio-bricks created from agricultural waste (Special Correspondent, 2021) (https://www.thehindu.

com/news/national/telangana/indias-first-bio-brick-based-building-inaugurated-at-iit-h/article36258196.ece). Waste management and the development of environmentally friendly, long-lasting building materials are driving forces in their development. Priyabrata Rautray of IIT Hyderabad and Ar. Avik Roy of KIIT School of Architecture, Bhubaneshwar, presented their findings at the International Conference on Engineering Design (ICED-2019) in Delft, Netherlands (Figure 8.14).

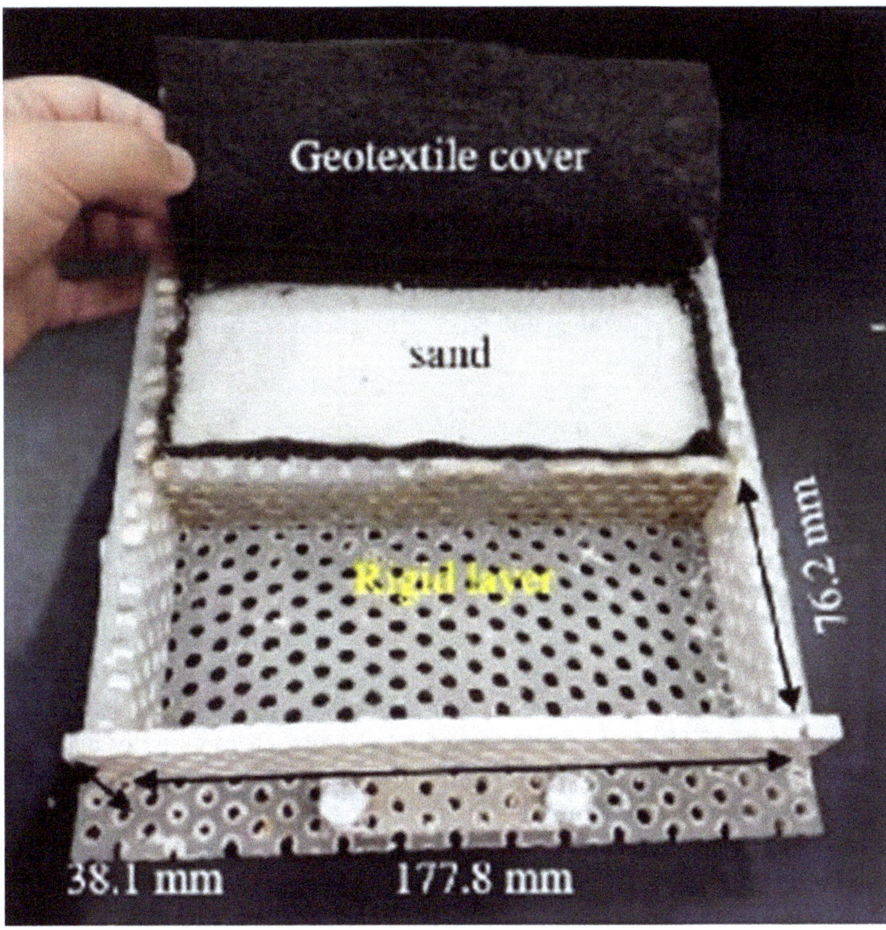

FIGURE 8.14 Preparation of the bio-brick sample in a rigid mold in full contact.

Avik Roy, KIIT School of Architecture, Bhubaneshwar, says it's not just better for the environment to use bio-bricks instead of clay bricks. These bio-bricks are useless when compared to traditional bricks made from kiln-fired clay. The carbon dioxide that these bricks absorb throughout the course of their lifespan makes them carbon sinks. The research team worked 900 grams of sugarcane bagasse to construct a single block. Rather than repurposing this much bagasse, burning it releases 639 grams of carbon dioxide. To make it even more environmentally friendly, the lime used in the production of each bio-brick absorbs 322.2 grams of carbon dioxide from the air during the curing process. Timber and metal frames can be combined to produce low-cost dwellings that are not load-bearing. Additional benefits include excellent heat and sound insulation, which is appropriate for nations like India with a humid climate, as well as the ability to keep a constant humidity level in the buildings. Additionally, bio-bricks can serve as insulation or decorative boards in addition to their original purpose of making bio-bricks. According to Ar. Rattray, "this renewable resource might be put to many other uses." For their "Bio-Bricks," which are created from agricultural waste products, scientists from IIT Hyderabad and KIIT Bhubaneswar are aiming to improve the load-bearing capacity.

8.10 MUSSEL-INSPIRED USE OF GLUE DURING FETAL SURGERY MAY IMPROVE PATIENT SAFETY

It is possible to treat congenital defects in the fetus by fetal surgery and fetal therapy while the fetus is still in its mother's womb. In the beginning, they invented the majority of the biomaterials used in fetal surgery today. There are numerous opportunities for biomaterials engineers to manufacture materials tailored for the specific problems and purposes of maternal-fetal surgery as the field progresses from open operations to minimally invasive approaches.

Toxic-free glue based on the sticky proteins found in mussels and other species outperformed commercially manufactured goods, pointing to the possibility of surgical glues for the repair of staples and suture staples. Every year, more than 230 million major surgeries are performed around the world, and the United States alone treats more than 12 million severe wounds. Sutures and staples are used to close about 60% of these wounds.

Among the drawbacks of sutures and staples compared to adhesives, says Purdue University chemical and biomedical engineer Julie Liu, "inherent loss of surrounding healthy tissue" is one of the most notable. Water impedes the bonding process of most adhesives, making them ineffective under damp conditions. Adhesives for medical use need to meet additional criteria in addition to eliminating this issue: they must be non-toxic and biocompatible.

Motivated by the animal world, scientists have felt strong glue to make doing surgery on fetuses safer. Researchers synthesized the adhesive after studying how mussels kept a tenacious grip on slippery rocks. An adhesive that can hold the amniotic sac from tearing after surgery could help the fetus live in the womb longer. The infant could have a better future if this were to happen. The use of a strong glue derived from a mussel adhesive has made surgery on developing fetuses far less risky. Choosing whether or not to perform surgery on a fetus is a difficult one. If the amniotic sac is penetrated, the fetus is at greater risk of disease. The substance secreted by mussels, also known as nature's glue, is a type of liquid protein that hardens rapidly into a solid water-resistant adhesive material. Many features of this process have inspired the development of synthetic materials for medical applications (Figure 8.15).

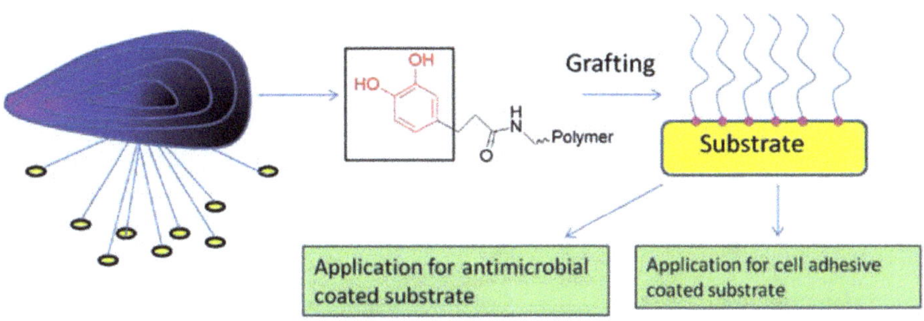

FIGURE 8.15 Application of mussel-inspired substrate-coated materials.

Source: Reprinted with permission from Kaushik et al., 2015. Copyright © 2015 by the authors. Licensee MDPI, Basel, Switzerland. https://creativecommons.org/licenses/by/4.0/

Using a glue generated by studying the tenacious hold of mussels on slick rocks, researchers hope to save the lives of pediatric patients in the future. The amniotic sac, which is very weak, is one of the most crucial dangers in doing surgery on fetuses, according to Balkenende (Scientist, the University of California, Berkeley in the US). After the surgery, the little hole where the scope penetrates the membrane can begin to break. Early labor is expected if it is cut simply, according to Balkenende. Because these procedures take place in the second trimester of pregnancy, early delivery raises the risk of fetal morbidity because the fetus is not yet completely developed. To help the fetus remain in the womb longer, researchers have suggested that an adhesive that might prevent amniotic sac ripping may be beneficial.

Spina bifida, a disease in which the neural tube (the section that becomes the baby's head and spinal cord) does not develop or shut correctly, is a clear-cut example of a major congenital deficiency that makes a fetus a candidate for surgery. It used to be necessary to cut and open a woman's belly and womb to correct these issues. Nowadays, this procedure is no longer necessary.

Doctors are now able to make progress through a small opening using thin endoscopic tools. Yet the instruments must enter the amniotic sac in order to deliver the fetus, which is so delicate that it cannot be fixed or effectively stitched shut. One of the most challenging aspects of the recovery process is applying surgical adhesive. Because the membrane is moist, it's a challenge to seal it with an adhesive. The glue relies on a mussel-created adhesive that can cling to rocks and other underwater structures for long periods of time without degrading. Even in stormy seas, mussels may cling to rocks and ship hulls thanks to the sticky compounds they produce (Figure 8.16).

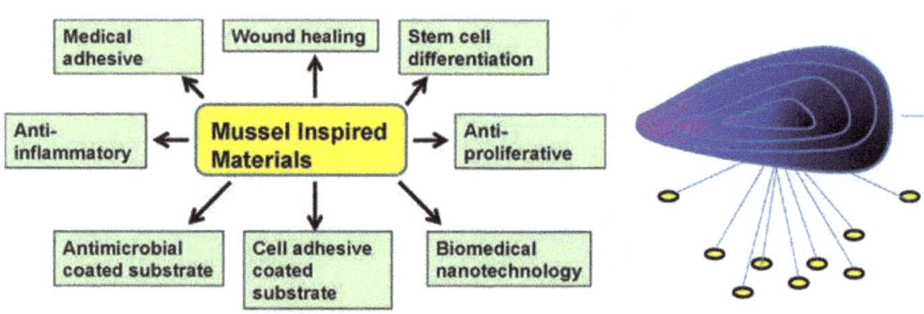

FIGURE 8.16 Mussel glue that saves lives.

Source: Reprinted with permission from Kaushik et al., 2015. Copyright © 2015 by the authors. Licensee MDPI, Basel, Switzerland. https://creativecommons.org/licenses/by/4.0/.

Syringes are used to administer the solution. They used fragments of a cow's heart as a model for the amniotic sac to conduct their research on this topic. They overlapped sections of the moist, filmy membranes with the solution in the syringe. The substance became stretchy as soon as it came into contact with water. An hour after the glue had hardened, the pieces were held together. In spite of this, the researchers still needed to find out how the approach may work in practice.

8.11 UNDERWATER GLUE INSPIRED BY MUSSELS

With inspiration from the natural adhesive secreted by shellfish such as mussels and barnacles, MIT engineers have developed an even stronger waterproof glue that can repair ships and even cure wounds and surgical incisions. Sticky proteins in these mussels allow them to adhere to rocks or ship hulls, allowing them to survive underwater. A common term for a mussel's byssus structure is "beard," which is a scientific term. The secreted adhesive cement is used by the mussel to adhere itself to surfaces. In general, there are two kinds of mussel proteins: three and five from the foot (Figure 8.17).

FIGURE 8.17 A bed of blue mussels, Mytilus edulis in the intertidal zone in Cornwall.

Source: Public Domain.

Scientists at the Massachusetts Institute of Technology (MIT) managed bacteria to develop a hybrid substance that combines mussel protein and bacterial protein, which are found in biofilms, to build their novel waterproof adhesives. Bacterial growth on a surface forms these slime layers, which are visible here as well. Compared to the sole mussel, this combination has a stronger underwater adhesion. One study claims that this is the largest known protein-based underwater adhesive to be derived from genetic coding, and the other claims that it is the most biodegradable.

The researchers' goal is to build a new foundation. The University of Washington's Timothy Lu, an associate professor of biological engineering ele,ctrical engineering, and computer science, is developing materials that include many functional domains to see if this improves material performance.

Curli fibers are a structure created by combining bacterial proteins with two distinct foot proteins. Proteins with a fibrous structure that can combine to build larger and more complex mashes. Researchers allowed these proteins to incubate to form denser structures after they had been isolated from the bacteria. In both dry and wet situations, the resulting structure designs can bind. The adhesives were studied using atomic force microscopy (a technique that examines the surface of a sample using a needle tip).

Wet glue with independent absorptive and cohesive moieties, according to a UCLA expert, will be the ultimate result. In addition to being completely original, this work of art is also quite thorough and accurate.

8.12 ROBOTIC ARMS ARE INSPIRED BY ELEPHANT TRUNKS

A new robotic arm was inspired by the elephant trunk's adaptability and wide range of movements. Around 40,000 muscles make the elephant trunk exceptionally robust despite the absence of any bones. On the basis of these findings, scientists developed a new prosthetic arm that can assist the disabled and assist with heavy lifting in agriculture, among other uses. The invention of the robot arm was one of the most significant technological advances of the 21st century. Manufacturing processes were streamlined, ushering in the age of automation. There are, however, some drawbacks to being based on the human arm. It's made up of a lot of little bits, so it's very stiff. The large arm is also a danger to people; thus, collisions are more likely to be avoided.

The elephant trunk has been widely imitated by bionic robots (Figure 8.18) because of its strong dexterity and stiffness adjustability. Elephant's trunk inspired Festo, a German developer of bionic arms. In addition to solving this problem, an elephant trunk-inspired bionic Handling Assistant opens up a wide range of new options. A "third-hand system" known as the Bionic Handling Assistant by Festo can be employed in a variety of ways. This financing is available to both for-profit and non-profit organizations, including hospitals, clinics, and universities.

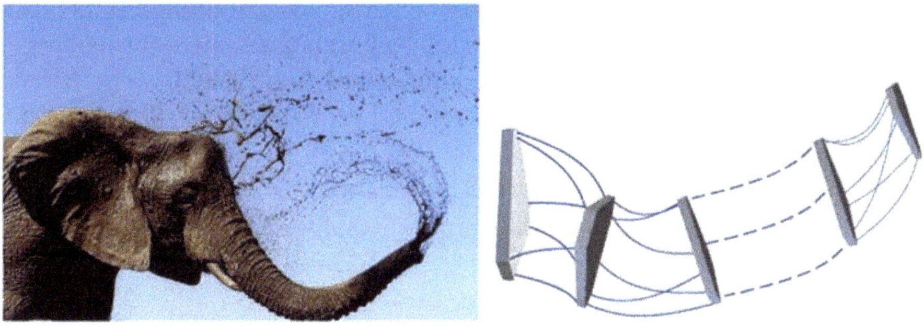

FIGURE 8.18 Elephant trunk robot based on flexible series-parallel structure.

Source: Reprinted with permission from Huang et al., 2022. Copyright © 2022 by the authors. Licensee MDPI, Basel, Switzerland. https://creativecommons.org/licenses/by/4.0/.

There is a strange explanation for the bionic "trunkfour" 's claws; they are used to pick things up. This invention, like an elephant's, has the ability to stand out and grab something with its trunk in any way it sees fit. Using three fin-shaped fingers that can be compressed to a smaller size, Festo created a new form of gripper. An object can be imprisoned by these digits, which need less force to grab and less risk of harm when they are wrapped around it (Trafton, 2014; Cui et al., 2017).

Metal and matching motors aren't present in the bionic treatment associate, which means it's believed to be safer for delicate objects. The direct connection between humans and machines is no longer in question, which is a major benefit over heavy industrial robots. The system is now stable even in the event of a collision. In order for it to function properly, it must have these three components. The hand axis with a ball joint and the adaptable Fin Gripper attachment allow for a smooth transition. It allows for greater flexibility and an unprecedented mass/payload ratio, allowing for efficient use of resources in both manufacturing and service.

Hollow plastic chambers that expand and contract in response to changes in air pressure allow the Bionic Treatment Associate to move through a remarkable range of motion in all three dimensions of the human body. Working with humans in a workplace is made easier with this tool's moderate forces and yielding when pressed. "Hazelnuts to grapefruits are the three gripping item sizes now being developed by scientists. FinGripper is already being used by several clients in their product lines," according to the website of Festo (Hoang et al., 2020).

Inventive robotic arms were inspired by the trunks of elephants (Lavars, 2020). Elephants' trunks 40,000-plus muscle bundles make it a strong, flexible limb that can handle heavy loads while being nimble and precise. Festo has made good use of the inspiration by building a system that connects flexible materials, acceptable control, and efficient movement. The Bionic Learning Network, a collaboration between Festo and some of the world's most renowned universities, institutes, and development organizations, is behind this project. The elephant trunk application, for example, exemplifies the intrinsic biomimetic focus of the cross-disciplinary teams.

8.12.1 LIKE THE TRUNK OF AN ELEPHANT, THE COILING ROBOTIC GRIPPER HOLDS ONTO ITS PREY

In nature, helical manipulators that can handle a wide range of objects are commonplace. A helical soft-fabric robotic gripper with adjustable stiffness and sensory input, inspired by nature, which is flat, scalable, thin, and lightweight, is presented in this research.

This gripper can handle a wide range of objects of various sizes and weights thanks to computerized clothes engineering technology and a modest hydraulic source. For high load capacity, a thermally induced variable stiffness structure (VSS) is used to complete a softening/stiffening cycle in 24 s, which is one of the fastest discoveries documented. The gripper uses a very flexible liquid soft metal sensor with a 15-fold increase in sensitivity over prior designs to provide real-time touch detection (Figure 8.19).

The gripper was able to firmly grasp and recover a wide variety of objects, including ones that weighed up to 220 times the gripper's mass. As the elephant trunk gripper—which is a convergent type that grabs and adsorbs—is soft, there is no risk of injury in collaboration with humans. It can also handle objects in different sizes ranging from deliberate parts or boxes without complex machinery or sensors.

8.13 MOSQUITO BITE HELPS CREATE THE IDEAL INJECTION NEEDLE

In some seasons, mosquitoes will bite us, and we won't feel it. It's based on biomimicry, which develops technological solutions influenced by nature to make a painless injection needle. Mosquitoes' ability to penetrate and drain blood from their prey was researched by engineers at Ohio State University's biomimicry laboratory. Bharat Bhushan, an Ohio State University scientist who worked on the painless microneedle project, explains that they used their engineering expertise to find out the method mosquitoes employ to puncture humans without hurting them. In addition, the engineers used micro-indentation to measure the hardness of various portions of a typical mosquito's lengthy beak (Figure 8.20).

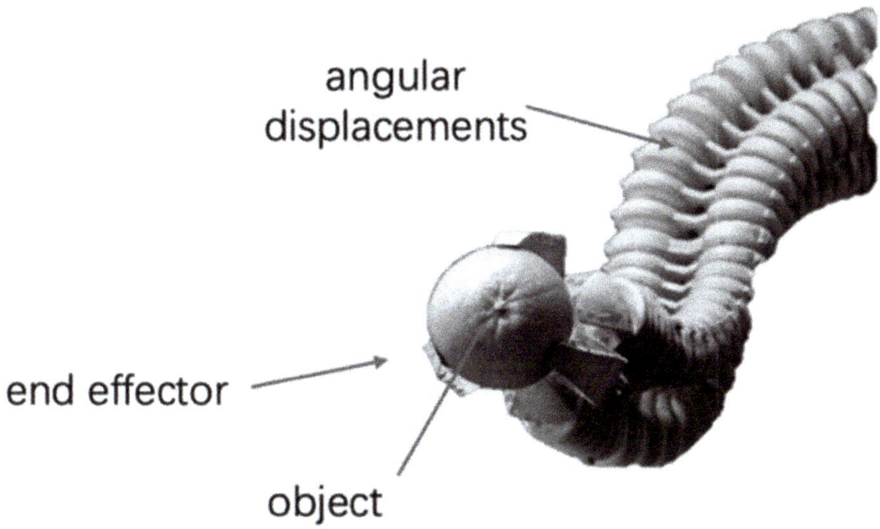

FIGURE 8.19 Bionic gripper, which can handle a wide range of objects of various sizes.

Source: Reprinted with permission from Su et al., 2022.

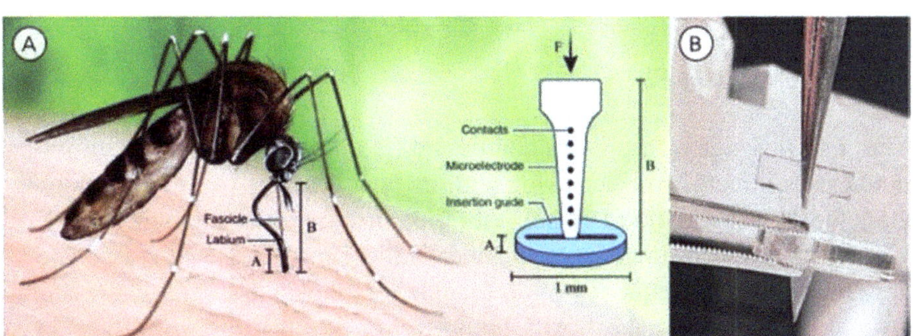

FIGURE 8.20 Overview of mosquito-based biomaterials and medical devices. (B) Photograph of a dummy polyethylene microelectrode probe inserted through a guide. Courtesy of Dr. Capadona Lab.

Source: Reprinted with permission from Dixon and Vondra, 2022.

The proboscis of a mosquito aims to be an intelligent machine. There are some regions where it is trained closer to the mosquito's head. Bhushan explains, "The skin is less irritated by the gentle machine." In order to puncture the skin, mosquitoes first use their tips to select a suitable location and then insert a serrated tube. This tube is vibrated at a rate of 15 Hz by a mosquito with a movement of less than 0.1 mm. As unpleasant as sawing through the skin may sound, it's actually less painful because of the wave. A mosquito can puncture the skin with only a third of the pressure required by an injection needle to do the same job. Anxiety-inducing skin reactions are reduced when the skin is subjected to less stress.

8.13.1 UNDERSTANDING WHY WE DO NOT KNOW THE MOSQUITO BITE

The itch is caused by the saliva that enters the bloodstream as a result of the bite. There are seven segments to a mosquito's proboscis, including two that hold the skin in place, two that cut, and a central tube that suctions blood. This device eliminates the sensation of discomfort that is often felt when a foreign object enters the body through the skin. Three of these elements have been replicated to make a motorized needle by scientists. Biomimicry at its finest, this motorized needle includes two serrated tubes and a center straw. The serrated edges and vibrating tubes decrease the amount of contact with the skin. This method has a significant impact on reducing pain.

8.13.2 PAINKILLER

The numbing agent, a protein compound, is applied by the mosquito after the beak is injected into the skin. With its beak's sensitive tip, the mosquito aims for the nearest blood vessel. Vibrating motion is used to puncture blood vessels after mosquitoes have located them (Figure 8.21). It's a considerably more precise frequency of up to 5 Hz that mosquitoes employ this time around. Once within the host's body, the mosquito must overcome even more formidable defenses (Holloway, 2019). As a result, rapid vibration is unlikely, but it is also not required. There's a noticeable difference in the elasticity of blood vessel walls compared to skin. The mosquito begins drawing blood as soon as the tip of its snout enters a blood vessel.

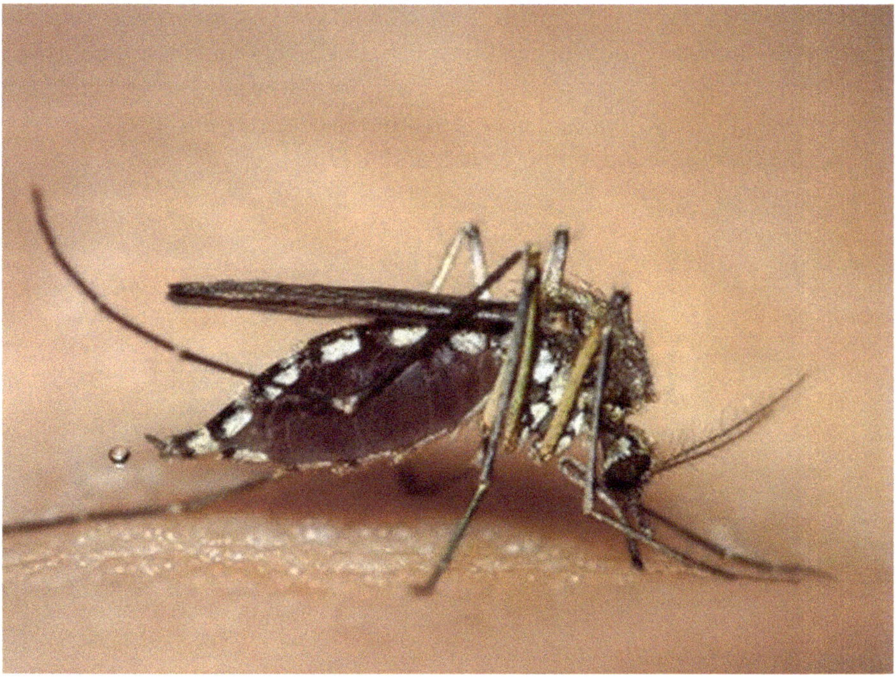

FIGURE 8.21 Female *Ochlerotatus notoscriptus* feeding on a human arm, Tasmania, Australia. Mosquitos apply vibrating motion to puncture blood vessels after mosquitoes have located the blood Vessel.

Source: Wikimedia Commons. https://creativecommons.org/licenses/by-sa/4.0/.

8.13.3 DOUBLE-NEEDLE AND THE DESIGN OF A MICRONEEDLE?

A double needle has the support of the scientific community. Drugs and/or blood are administered via one needle, while an anesthetic is administered via a second. The first needle, like the mosquito's proboscis, has a softer tip, while the surrounding area is serrated. Vibrations cause the skin to be punctured (Figure 8.22). The needle must now be obtained in accordance with this plan.

8.14 THE GRID IS MANAGED BY THE INNOVATION-HIVE MIND

To the extent that an owl may fly at night, inspiration can be found in all kinds of unlikely places. A seemingly limitless supply of creative ideas can be found among the world's 30 million different species. In order to improve the lives of humanity, people need to retain an open mind when it comes to listening and observing nature. Be aware of how nature has evolved and how it might lead to inventions and new ideas. In other words, pay attention, take notes, watch, understand, think, create, and invent.

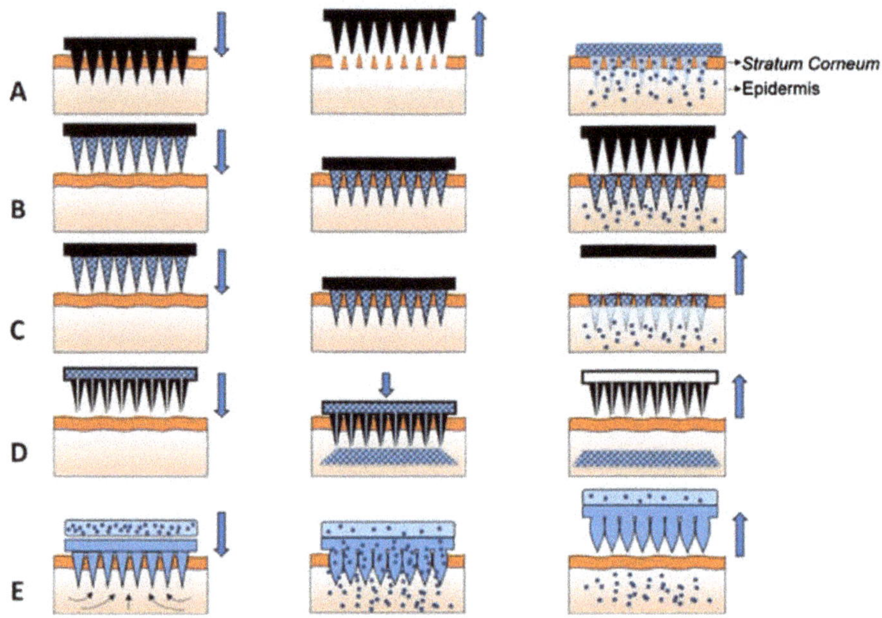

FIGURE 8.22 A schematic representation of five different Microneedles (MNs) types used to facilitate transdermal drug delivery.

Source: Reprinted with permission from Gadziński et al., 2022. https://creativecommons.org/licenses/by/4.0/.

Hives inherently have bees that know what their job is. Honeybees create a typical organization that works efficiently to build hives and support their colonies by communicating and coordinating their methods. Swarm Logic is the collective intellect of all animals that operate together in groups. Typically, different types of construction equipment operate independently of one another. They rely solely on a single thermostat and have no idea what else is going on in the plant or how it is currently operating. Because they are out of sync, they frequently wind up operating at the same time, wasting power. Current controllers, which take their cues from bees, use a wireless mesh network, mostly among the devices that consume the most power, to communicate and lay out the energy demand, resulting in intelligent, networked, and energy-responsive assets.

The electromagnetic fields around power lines are interfere with bee behavior. Bees that are more aggressive and safe guard the Power thefts. Honeybees interact and coordinate behaviors to form a typical organization that efficiently builds hives and feeds their colonies. During this procedure, additional energy is wasted. Inspired by the bees, present controllers have a wireless mesh mainly among the power-consuming devices that let them communicate among themselves and layout the energy demand–resulting in intelligent, networked, and energy-responsive assets (O'Connell, 2018) (Figure 8.23).

The hive mind governs the grid (Anonymous, 2016). Bees are occupied to the hilt. Individuals are able to detect what the colony needs to do and start doing it instinctively despite their low brainpower. For example, the electrical grid has several elements that don't communicate with each other, which creates a challenge with multiple human infrastructures. The entire grid is not monitored by grid elements. Using Regen Energy, a company's reserved power-sucking gadgets and machines are transformed into a network capable of supporting demands during expensive peak-power hours. In order to enhance productivity, the company provides wireless controllers that interact with each other to keep the bees in the hive in sync.

FIGURE 8.23 A wild colony of Honey bees in Chandigarh Photo credit: Sarbjit Bhaga; Overhead power line.

Photo credit: Sarbjit Bahga. https://creativecommons.org/licenses/by-sa/4.0/.

Photo credit: Varistor60. https://creativecommons.org/licenses/by-sa/4.0/.

KEYWORDS

- **injection needle**
- **new insulator**
- **road reflectors**
- **robotic arms**
- **self-healing concrete**
- **waterproof glue**

REFERENCES

Algaifi, H. A., Bakar, S. A., Alyousef, R., Mohd, S. A. R., Wan, I. M. H., Shahidan, S., Ibrahim, M., & Salami, B. A., (2021). Bio-inspired self-healing of concrete cracks using new B. pseudomycoides species, *J. Materials Res. Technol.*, *12*, 967–981. https://doi.org/10.1016/j.jmrt.2021.03.037.

Anonymous, (2016). *Hive Mind Manages the Grid.* http://bio-inspiredd.blogspot.com/2016/01/hive-mind-manages-grid.html (accessed on 17 August 2023).

Boyle, R., (2013). *How Cockroaches Are Helping Scientists Better Robots.* Popular science. https://www.popsci.com/technology/article/2013-02/watch-how-cockroaches-are-helping-scientists-design-better-robots/ (accessed on 17 August 2023).

Bunyamin, B., Kurniasari, F.D., Munirwan, R.P., & Jaya, R.P. (2022). Effect of coral aggregates of blended cement concrete subjected to different water immersion condition. *Advances in Civil Engineering 2022*, Article ID 2919167, 10 pages. https://doi.org/10.1155/2022/2919167

Chen, D. A., Ross, B. E., & Klotz, L. E., (2015). Lessons from a Coral Reef: Biomimicry for structural engineers. *J. Struct. Eng.*, *141*(4), 02514002. doi: 10.1061/(ASCE)ST.1943-541X.0001216.

Cui, M., Ren, S., Wei, S., Sun, C., & Zhong, C. (2017). Natural and bio-inspired underwater adhesives: Current progress and new perspectives. *APL Materials*, *5*, 116102. https://aip.scitation.org/doi/full/10.1063/1.4985756 (accessed on 17 August 2023).

Dixon, A.R., & Vondra, I., (2022) Biting Innovations of Mosquito-Based Biomaterials and Medical Devices: *Materials* **2022,** *15*(13), 4587; https://doi.org/10.3390/ ma15134587

Faheem, H., (2011). *Effectiveness of Using Cat-Eye Reflectors as Warning and Traffic Calming Devices in Local Roads and Highways*. Conference Paper, 3rd International Conference on Road Safety and Simulation. Indianapolis, USA.

Festo, (2012). *Robotic Arm Inspired by the Elephant Trunk.* https://asknature.org/innovation/flexible-gripper-inspired-by-the-elephant-trunk/ (accessed on 17 August 2023).

Gadziński, P., Froelich, A., Wojtylko, M., Bialek, A., Krysztofiak, J., & Osmalek, T. (2022). Microneedle-based ocular drug delivery systems–recent advances and challenges. *Beilstein J. Nanotechnol. 13*, 1167–1184. https://doi.org/10.3762/bjnano.13.98

Giannaros, P., Kanellopoulos, A., & Al-Tabbaa, A., (2016). Sealing of cracks in cement using microencapsulated sodium silicate. *Smart Mater. Struct., 25*, 084005. https://iopscience.iop.org/article/10.1088/0964-1726/25/8/084005 (accessed on 17 August 2023).

Gonçalves, I.M., Carvalho, V., Rodrigues, R.O., Pinho, D., Teixeira, S.F.C.F., Moita, A., Hori, T., Kaji, H., Lima, R., & Minas, G. (2022). Organ-on-a-chip platforms for drug screening and delivery in tumor cells: A systematic review. *Cancers 14*(4), 935; https://doi.org/10.3390/cancers14040935

Grabinski, S., (2019). Improved visibility–solar-powered road markers on polish roads. *IOP Conference Series: Materials Science and Engineering, 661*, 012150. doi: 10.1088/1757-899X/661/1/012150.

Hoang, T. T., Phan, P. T., Thai, M. T., Lovell, N. H., & Do, T. N. (2020). Bio-inspired conformable and helical soft fabric gripper with variable stiffness and touch sensing. *Advanced Materials Technologies, 12*, 2000724, https://doi.org/10.1002/admt.202000724.

Holloway, J., (2019). *Mosquitos Inspire Pain-Free Microneedle*. https://newatlas.com/mosquito-micro-needle/55207/ (accessed on 17 August 2023).

Huang, Q., Wang, P., Wang, Y., Xia, X., & Li, S. (2022), Kinematic analysis of bionic elephant trunk robot based on flexible series-parallel structure. *Biomimetics 7*(4), 228; https://doi.org/10.3390/biomimetics7040228

https://bbbprograms.org/media-center/dd/nad-nanoseptic-self-cleaning-claims (accessed on 27 September 2023).

https://facilityexecutive.com/2021/07/creating-cleaner-healthier-travel-experiences-with-nanoseptic/ (accessed on 17 August 2023).

https://www.linkedin.com/company/nanotouchclean (accessed on 27 September 2023).

https://www.nk-roadstud.com/news/news-information/function-of-the-reflective-cat-eyes.html (accessed on 17 August 2023).

Iqbal, D.M., Wong, L.S., & Kong, S.Y. (2021) Bio-cementation in construction materials: A review. *Materials 14*(9), 2175; https://doi.org/10.3390/ma14092175

Jonkers, H. M., Thijssen, A., Muyzer, G., Copuroglu, O., & Schlangen, E., (2010). Application of bacteria as self-healing agent for the development of sustainable concrete. *Ecological Engineering, 36*(2), 230–235. https://doi.org/10.1016/j.ecoleng.2008.12.036.

Kaushik, N. K., Kaushik, N., Pardeshi, S., Sharma, J. G., Lee, S. K., & Choi, E. H., (2015). Biomedical and Clinical Importance of Mussel-Inspired Polymers and Materials: *Mar. Drugs 13*(11), 6792-6817; https://doi.org/10.3390/md13116792

Lavars, N., (2020). *Coiling Robotic Gripper Grasps Objects Like an Elephant's Trunk*. New Atlas. https://newatlas.com/robotics/coiling-robotic-gripper-elephants-trunk/ (accessed on 17 August 2023).

Lee, B. P., Messersmith, P. B., Israelachvili, J. N., & Waite, J. H., (2011). Mussel-inspired adhesives and coatings. *Annual Review of Materials Research, 41*(1), 99–132.

Li, A., Mu, Y., Jiang, W., & Wan, X., (2015). A mussel-inspired adhesive with stronger bonding strength under underwater conditions than under dry conditions. *Chem Commun (Camb)., 51*(44), 9117–9120. doi: 10.1039/c5cc00101c.

Li, V. C., & Yang, E. H., (2007). Self-healing in concrete materials. In: Van, D. Z. S., (ed.), *Self Healing Materials*. Springer Series in Materials Science. Springer, Dordrecht. https://link.springer.com/chapter/10.1007/978-1-4020-6250-6_8 (accessed on 17 August 2023).

Ling, X., Osotsi, M.I., Zhang, W., Wu, Y., Jin, Q., & Zhang, D. (2023) Bioinspired materials: From distinct dimensional architecture to thermal regulation properties. *Journal of Bionic Engineering 20*, 873-899.

Llewellyn, R., Cowie, J., & Maher, M. (2020). Active road studs as an alternative to lighting on rural roads: Driver safety perception. *Sustainability, 12*, 9648.

Messersmith, P. (2016). Biological polyphenols as inspiration for biomedical adhesives and coatings. Front. Bioeng. Biotechnol. Conference Abstract: 10th World Biomaterials Congress. doi: 10.3389/conf.FBIOE.2016.01.02469

O'Connell, C. (2018). *Technologies Inspired by Nature.* https://cosmosmagazine.com/technology/technologies-inspired-by-nature/ (accessed on 17 August 2023).

Peplow, M., (2018). Polar bear hair inspires stealth fabric. *Adv. Mater., 96*(9). https://cen.acs.org/articles/96/i9/Polar-bear-hair-inspires-stealth-fabric.html (accessed on 17 August 2023).

Rakowska, P. D., Tiddia, M., Faruqui, N., *et al.,* (2021). Antiviral surfaces and coatings and their mechanisms of action. *Commun. Mater., 2*, 53. https://doi.org/10.1038/s43246-021-00153-y.

Roque, B.A.C., Brasileiro, P.P.F., Brandao, Y.B., Casazza, A.A., Converti, A., Benachour, M., & Sarubbo, L.A. (2023). Self-healing concrete: Concepts, energy saving and sustainability. *Energies 16*(4), 1650; https://doi.org/10.3390/en16041650

SciTechDaily, (2012). *Jellyfish-Inspired Microchip Can be Used to Capture Cancer Cells.* https://www.sciencedaily.com/releases/2012/11/121112171312.htm (accessed on 17 August 2023).

Shi, H., Yu, Z., Ma, J., Ni, C., & Shen, X. (2019). Properties of Portland cement paste blended with coral sand powder. *Construction and Building Materials, 203*, 662–669.

Shipman, M., (2014). *Cockroach Cyborgs Use Microphones to Detect, Trace Sounds.* NC state News, https://news.ncsu.edu/2014/11/bozkurt-roach-biobot-2014/ (accessed on 17 August 2023).

Special correspondent, (2021). *India's First Bio-brick Based Building Inaugurated at IIT Hyderabad.* The Hindu. https://www.thehindu.com/news/national/telangana/indias-first-bio-brick-based-building-inaugurated-at-iit-h/article36258196.ece (accessed on 17 August 2023).

Stott, R., (2014). *Three Self-healing Materials That Could Change the Future of Construction.* Arch Daily, Architecture News, https://www.archdaily.com/552763/three-self-healing-materials-that-could-change-the-future-of-construction (accessed on 17 August 2023).

Su, H., Hou, X., Zhang, X., Qi, W., Cai, S., Xiong, X., & Guo, J. (2022). Pneumatic soft robots: Challenges and benefits. *Actuators 11*(3), 92. https://doi.org/10.3390/act11030092

Trafton, A., (2012). *On the Hunt for Rare Cancer Cells.* MIT News Office. https://news.mit.edu/2012/capturing-rare-cancer-cells-1112 (accessed on 17 August 2023).

Trafton, A., (2014). *Engineered Proteins Stick Like Glue—Even in Water.* MIT News Office. https://news.mit.edu/2014/new-adhesives-stick-in-water-0921 (accessed on 27 September 2023).

University of Erlangen-Nuremberg. (2020). *The Cement for Coral Reefs: Vital Role of Coralline Red Algae for Coral Reefs.* ScienceDaily. www.sciencedaily.com/releases/2020/11/201103115652.htm (accessed on 17 August 2023).

Van, D. H. D., (2018). *Self-healing Concrete in Full Development.* https://www.biobasedpress.eu/2018/02/self-healing-concrete/ (accessed on 17 August 2023).

Vélez-Peña, E., Pérez-Obando, J., Pais-Ospina, D., Marín-Silva, D. A., Pinotti, A., Cánneva, A., Donadelli, J. A., et al., **(**2020**)**. Self-cleaning and antimicrobial photo-induced properties under indoor lighting irradiation of chitosan films containing Melon/TiO_2 composites. *Applied Surface Science, 508*, 144895.

Vermeer, C. M., Rossi, E., Tamis, J., Jonkers, H. M., & Kleerebezem, R., (2021). From waste to self-healing concrete: A proof-of-concept of a new application for polyhydroxyalkanoate. *Resources, Conservation and Recycling, 164*, 105206. https://doi.org/10.1016/j.resconrec.2020.105206.

Xu, Q., Zhang, W., Dong, C., Sreeprasad, T. S., & Xia, Z. (2016). Biomimetic self-cleaning surfaces: Synthesis, mechanism, and applications. *J. R. Soc Interface., 13*(122), 20160300. doi: 10.1098/rsif.2016.0300.

Yang, X., Zhang, W., Qin, X., Cui, M., Guo, Y., Wang, T., Shi, Z., Zhang, C., Li, W., & Wang, Z. (2020). Recent progress on bioinspired antibacterial surfaces for biomedical application. *Biomimetics 7*(3), 88. https://doi.org/10.3390/biomimetics7030088

Yu, C., Sasic, S., Liu, K., Salameh, S., Ras, R. H. A., & Van, O. J. R., (2020). Nature–inspired self–cleaning surfaces: Mechanisms, modeling, and manufacturing. *Chemical Engineering Research and Design, 155*, 48–65. https://doi.org/10.1016/j.cherd.2019.11.038.

Zhan, H. J., Wu, K. J., Hu, Y. L., Chen, J. F., Ni, Y., & Yu, S. H., (2019). Biomimetic carbon tube aerogel enables super-elasticity and thermal insulation. *Chem., 5*, 1871–1882.

Zhao, W., Cui, C. H., Bose, S., Guo, D., Shen, C., Wong, W. P., & Ha, K., (2012). Bioinspired multivalent DNA network for capture and release of cells. *Proc. Natl. Acad. Sci., 109*(48), 19626–19631 https://www.pnas.org/doi/full/10.1073/pnas.1211234109#:~:text=https%3A//doi.org/10.1073/pnas.1211234109 (accessed on 17 August 2023).

Zhao, Y., Jin, L., Zhang, P., et al., (2018). Inverse displacement analysis of a hyper-redundant elephant's trunk robot. *J. Bionic Eng., 15*, 397–407. https://doi.org/10.1007/s42235-018-0030-z.

CHAPTER 9

Biological Interesting Investigations

ABSTRACT

Radiometric dating reveals that the Earth is approximately 4.5 billion years old. Using radiometric methods, the age of the first rocks on Earth has been determined. In Australia's Jack Hills, researchers have discovered tiny minerals called zircons dating back 4.4 billion years. These minerals have survived even as the rocks that once held them have eroded away. Minerals dating back 4.3 billion years were discovered in Australia. Bioinspiration can be applied to any problem, from the smallest to the largest scales, using scientific accuracy and creativity. Research in fields other than biology and technology is sparked by the bioinspiration phenomena. It is a framework that encourages students to explore new areas of study. Natural selection has resulted in the emergence of a huge array of materials, structures, models, systems, and processes that are ideally suited to specific tasks. Through the approach of biomimicry, or bioinspiration, engineers can apply nature's design principles to very complex and sophisticated engineering models at various length scales. MIT engineers have created a biocompatible and strong glue that can be used to seal damaged tissues and put an end to pain. The sticky component promotes barnacles to exploit the picture as a means of clinging to rocks. They both used hydrogel-based materials that mimicked the lobster's underside structure, the most difficult hydrogel to learn in nature. Bio-inspired innovation is now an ingrained part of architecture, water purification, and medical technology. Textiles and surfaces derived from bioinspiration for use in green technology and many other fields of human endeavor. There is a slew of cutting-edge technologies based on natural phenomena. It is noteworthy that Bioinspiration is a new scientific endeavor. Scientific and engineering principles are used to create an impact.

Bio-Inspired Technologies for the Modern World: Emerging Trends and Applications.
R. Ramakrishna Reddy & T. Pullaiah (Authors)

9.1 INTRODUCTION

Ideas inspired by means or laws operating in biological organisms are called Bioinspiration (Whitesides, 2015). Bioinspiration is utilized when designers, inventors, and developers turn to biology and Nature in the hopes of observing patterns, forms, features, and mechanisms that help in solving problems. One of the more prominent examples of bioinspiration is the use of this philosophy in the progress of new ways to organize computation.

Research in technology and fields outside of biology is sparked by the bioinspiration phenomena. It provides fresh avenues for investigation. Bioinspiration offers a wider range of applications than simply generating new ideas. It has the ability to select subjects that are reasonably easy to do research on. Additionally, this inquiry can lead you to places that can benefit you. Research-based on bioinspiration (Whitesides, 2015) has provided numerous applications. It can be used in labs with limited resources because of its low cost. For example, it connects the cultural synergies and technical disparities of different regions.

Relevant ideas frequently emerge when scientific and technological data gleaned from nature is analyzed. Newton and Kepler began to grasp the first empirical theories of gravity. These academics studied the movement of the Sun and the planets. Electromagnetism got its start with experiments by Michael Faraday and James Clerk Maxwell, who showed how electrical currents interact with magnetic fields using magnets. The study of heat transmission and mechanical work led to the development of thermodynamics. Quantum mechanics was born out of research into the spectrum qualities of light. The molecular foundation of inheritance, cell features, higher animal outcomes, and illness have all emerged as key areas of research in biology as a result of the discipline's discoveries.

9.2 ONE OF NATURE'S GREATEST ENGINEERING FEATS IS THE HUMBLE EGG

Related terminology such as "bioinspiration," which is a creative approach to design based on biological notions, has emerged in the last several years. There are numerous challenges that living organisms face that can be readily solved by using bioinspiration as a solution. In the future, it will serve as a bridge between biology and a variety of other professions. It's an intriguing way to rebalance the benefits of diverse sectors in scientific and technical discovery through bioinspired research. When it comes to coming up with fresh ideas, the possibilities for using bioinspiration are virtually endless. It looks like a good idea and could be used to replicate it. Researchers are working to produce human antibodies by altering the immune system of the ostrich (Koyama & Tennyson, 2018). An ancient species, the ostrich, has developed an immune system capable of responding to many antigens and producing large amounts of immunoglobulin Y antibodies over the course of 23 million years. Here's an illustration of how bioinspiration might be used.

Calcium carbonate, which makes up around 97% of an eggshell, is held together by proteins. Depending on its mineral content and protein content, the eggshell's rigidity varies (see https://journals.co.za/doi/pdf/ 10.05.10520A JA1019578554). Because of the eggshell's three-layer structure, it is both tough and permeable, protecting the fragile interior while still enabling oxygen to enter. Hunter-gatherer groups primarily use ostrich eggshells as water containers in the ethnography. The hunters may take a drink of water from the ostrich eggs before continuing their pursuit of an animal over long distances. For the San people, the hard, thick shell of the ostrich egg was a boon. The egg's proteins break down naturally, and calcium carbonate returns to the earth when it splits. As a zero-waste container, the egg is ideal. An Ostrich egg contains roughly 2,000 calories, 45% fat, and 47% protein. Ostrich eggs, which contain 16.29 IU/g of vitamin A, are a great source of these vitamins and thiamine. Ostrich eggs (https://medium.com/@amnaath-araliabbasi/benefits-of-ostrich-egg) contain minerals, including zinc, calcium, iron, magnesium, and manganese, making them a better choice for human growth. Anyone on a low-sodium diet should choose ostrich eggs, which are completely free of sodium.

9.2.1 BIOINSPIRED MATERIALS CAN BE USED TO ADDRESS WATER PURIFICATION

Only 2.5% of the planet's total water is fit for human consumption, or 0.007% of the planet's total water. Water extraction accounts for 70% of the world's freshwater supply. Industry accounts for 20% (including energy), while households, institutions, towns, and small and medium-sized businesses account for 10% (Whitesides, 2015; Gonzalez-Perez & Persson, 2016). When water is held in a reservoir, it is possible to use it directly. Freshwater has numerous applications. As a result of the greater difficulty in accessing these sources, humans have relied less on them for their freshwater needs. There is a shortage of access to safe drinking water and sanitary facilities that contributes to water scarcity. Technology and environmental issues will be critical in addressing future challenges.

Figure 9.1 shows different methods of wastewater treatment. Biological treatment methods use biological agents like plants and microorganisms in this way to remove the harmful pollutants. Biomimetic membranes are the only ones to make use of amphiphilic self-assembly. As with reverse osmosis (RO), the membrane separation is also maintained. About 0.007% of freshwater is usable for human consumption. It's just 3% of the world's water resources. According to the 2016 UN World Water Development Report, 1.8 billion people will be affected by water scarcity by 2025. In the coming years, freshwater supplies are expected to be depleted, while salt-water resources cannot currently be tapped. Adding specific water filtration methods helps meet the new issues it presents for water purification. There are clear reasons why bioinspired materials are being used in medicinal applications. Despite the fact that they aren't as prevalent, these materials are still excellent options for water filtration (Whitesides, 2015; Gonzalez-Perez, 2016; Bhushan, 2019; Brown & Bhushan, 2016).

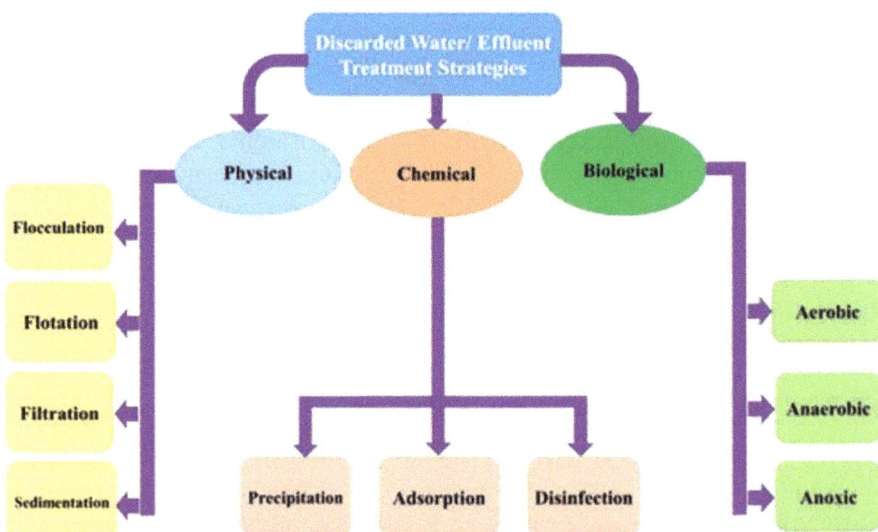

FIGURE 9.1 Different methods of wastewater treatment: Biological treatment methods use biological agents like plants and microorganisms in this way to remove the harmful pollutants.

Source: Reprinted with permission from Ravi Kant Bhatia et al., (2022).

9.2.2 WHAT MAKES UP NATURAL MEMBRANES?

Phospholipids are the fundamental constituents of cellular membranes. This marriage led to the creation of the Greek term amphiphilic. A substance is amphiphilic if it possesses both a hydrophilic and lipophilic component. Proteins and numerous block copolymers, which have both hydrophilic and lipophilic sections, are examples of frequent amphiphiles. When molecules are amphiphilic, their self-assembly structure is influenced by their amphiphilic nature. Spiral micelles, bilayered vesicles, and planar vesicles are the most prevalent supramolecular structures (Figure 9.2). The development of mesoporous materials utilizing these architectures can aid in water filtering.

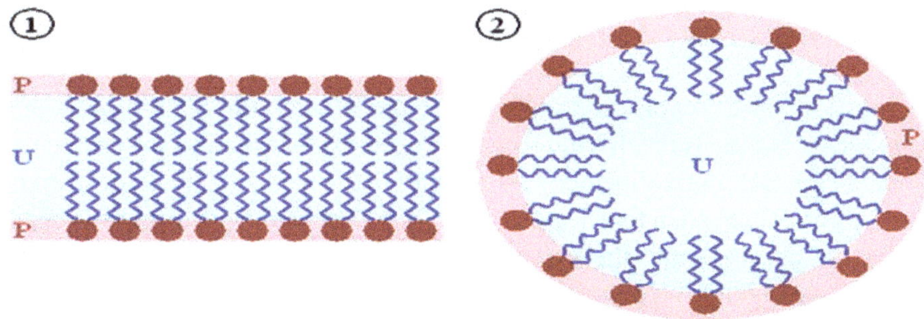

FIGURE 9.2 (1) Representation of the lipid bilayers; and (2) micelle.

Source: Stephen Gilbert–GFDL https://creativecommons.org/licenses/by-sa/3.0/deed.en

9.2.3 BIOMIMETIC MEMBRANES

For example, the aquaporin membrane emerged as one of the most promising biomimetic membranes (Abaie et al., 2021). Membrane-based water filtration technology is widely used by these companies. As a result of biomimicry, existing challenges can be overcome. The permeability, selectivity, and energy consumption of biomimetic membranes are all too high. The homogenous nature of these membranes serves as a criterion for their classification (isotropic and anisotropic). One nm RO, two nm nanofiltration, and so on are examples of various nanofiltration techniques. RO is a frequently utilized, low-cost water filtration method. To remove impurities from water, you can use RO. It may remove up to 90% to 99% of contaminants from water using a semi-permeable membrane. Anisotropic biomimetic thin-film composite membranes are used in RO water purifiers.

Aquaporin Z, a water channel found in nature, can be inserted into the polyamide composite membranes of a triblock copolymer to create biomimicry materials. Carbon nanotubes, aquaporins, and high water flow can also be employed to improve water filtration. Aquaporins and carbon nanotubes are more costly to manufacture than polyamide composite membranes. It is a significant disadvantage in this instance. The typical size of new biomimetic membranes with artificial water channels is 5 to 25 nm less than that of a conventional composite membrane. According to a recent study, this consists of one layer of polyamide, polysulfone, and polyester totaling approximately 140–150 nm (Abaie et al., 2021). If tiny, portable filters need to be created, this approach is superior to the standard method.

9.3 BIOMIMICRY IN ARCHITECTURE – DURIAN FRUIT

Biomimicry is a viable alternative to conventional building design and technology. Nature is the driving force in building, which results in spectacular works of art. Solution refinement and development by nature are at the heart of biomimicry. There are a few necessities for any long-term construction project. Structural performance, thermal environment, water efficiency, zero-waste systems, and energy supply are all necessary for these systems to function.

Biomimicry focuses on finding solutions. Over the course of a billion years of study and development, biological organisms improved and evolved through natural selection. Technology, systems, functions, and procedures that can be embodied are the answer to the challenge of coping with nature (Lee, 2011; Rajshekhar Rao, 2014). In many cases, these issues are similar to those faced by human beings. They're always looking for fresh approaches to design, and their solutions to similar problems have always been more cost-effective.

9.3.1 IT INSPIRES ARCHITECTURE

Inspiration can be an ambiguous concept and is used in many disciplines. It has a significant role in art and architecture. A broadly agreed definition of the term "inspiration" is critical to the study of the concept. It arouses motivation in others by providing direction or generating a force or influence that motivates others. A variety of structures (Celadyn, 2018; Knippers & Speck, 2012; Helms et al., 2009), works of art, and artifacts have been cited as sources of inspiration in the literature. Inspired by chestnuts and durian are two examples of tree fruits that could inspire architecture. In Figure 9.4, a durian-inspired building may be seen in Singapore (Figure 9.3). The durian is recognized as Singapore's national fruit. Figure 9.4 shows the beginnings of the Esplanade structure (near Marina Bay), which began as two glass domes. As a result of the architectural change, the structures took on the look of a durian that had been halved and covered in pointed aluminum shades.

FIGURE 9.3 The southeast Asian plant Durian: The king of fruit.

Source: Image courtesy: Jonny Clow

It is possible to find solutions to many problems using nature's design principles. Biomimicry is a term that refers to the deliberate construction or replication of systems, typologies, and aspects of flora and fauna in order to arrive at sophisticated design solutions. Biomimicry in architecture refers to mimicking the design of trees, animal abodes, or naturally created structures like stalactites, as well as stimulating processes that occur in nature. Some of the best instances of biomimicry architecture can be seen all over the world, thanks to designers who studied nature and applied it to their designs. Let's take a closer look at these amazing structures.

9.3.2 SINGAPORE'S ESPLANADE THEATRE

In the design of the facade of the Esplanade Theatre (Figure 9.4), the prickly thorns of the skin of the Durian plant were taken into consideration (Figure 9.3). The fruit within is shielded from the elements by this prickly skin. Singapore's newest architectural icon is the Esplanade. Durians are known for their unusual dome-shaped roofs topped by triangle-pointed shades. The shape of the building has elicited strong reactions from Singaporeans, critics, and foreign tourists. The outside of the theater changes during the day to enable natural or diffused light while limiting glare and thereby preventing overheating of the structure.

FIGURE 9.4 Singapore's Bay esplanade theatres.

Image courtesy: Banku. https://creativecommons.org/licenses/by-sa/3.0/deed.en

Nature has a powerful pull-on architect. Architects can now use new construction methods and materials to identify and reproduce complicated structures thanks to new technology. The biomimetic design takes its cues from the natural world.

9.4 BONE HEALING MATERIALS THAT DEGRADE NATURALLY

Bone defect repair has been significantly associated with the second generation of biomaterials for almost half a century (Wei et al., 2020; De Melo & Habibovic, 2018; El-Rashidi et al., 2017). The biodegradability of biodegradable materials makes them ideal for application in bone tissue engineering. Also destroyed and absorbed into the transplant are biomolecules produced by the degradation of bone tissue. To facilitate the creation of bone tissue, it modifies the regenerative microenvironment. A stress-shielding effect is avoided while tissue regeneration is stimulated, as the graft's intuitive characteristics diminish over time and biological stress is transferred to growing bone tissue. Thus, the degradable biomaterial saves the harm and financial burden of a second procedure. According to recent studies, biodegradable polymers, ceramics, and materials based on magnesium are all hot topics right now.

The majority of bone is composed of cells, fibers, and a matrix. The improved bone matrix contains collagen, which offers tensile strength. Composed primarily of calcium phosphate, the mineral composition of bone provides its compressive strength. Numerous calcium salts have accumulated in the intercellular material. When this tissue becomes highly active, the skeletal system of the body is produced. The substance soothes and safeguards numerous organs. Several reasons, including trauma, orthopedic surgery, osteoarthritis, osteoporosis, and primary tumor resection, can induce bone abnormalities and loss. There are several regenerating tissues and organs in the human body. They cover the skin, liver, glands, and other organs in addition to the blood and bones. In the majority of instances, small injuries or damage to bone tissue can be restored (El-Rashidy et al., 2017).

Advances in microsurgery during the past few decades have been well-recognized and appreciated. There is merit in the challenge of bone tissue. It reaps the benefits of working and repairing harm to its structure. The physical and biological qualities of the excellent tissue are mimicked by the outcome of similar materials. Bone autograft substitutes are in short supply. Thus, new tissue engineering approaches fill that gap. Biodegradable matrices made from polymers are critical to the formation of new tissue in the transplantation domain because of their fundamental adaptability (i.e., scaffolds). Biocompatibility and bioactivity are often better in natural polymers. A profusion of unique supporting materials results from their collecting and further engineering. Hydrogels and scaffolds with fibrous or porous architectures are included in these systems.

Filippi et al. (2020) highlight that a wide range of natural polymers are now employed in bone tissue engineering and explore them in detail. In this manner, biomolecules are functionalized and synthesized. It's important to list both the positive and negative aspects of each application in order to assess and contrast their true applicability potential. When the scaffolds are colored with various cell types, structural renewal happens both *in vitro* and in nature. Some early clinical data can be seen here. According to a collection of studies, advances in natural polymer research will have a substantial influence. They develop materials that can be translated into effective tissue regeneration, with the expected outcome in the near future being the repair of bony deformities.

9.5 INSPIRATION FOR NEW IDEAS: AIRBUS LEARNED A LOT ABOUT STRENGTH OPTIMIZATION FROM BONES

The Airbus Group is utilizing 3D printing to help develop a more robust, lighter-weight galley partition that replicates the structure and growth of cells and bone. By decreasing the weight of the aircraft, the approach enables each aircraft to reduce annual CO_2 emissions by 4,65,000 metric tons (MT). Airbus asserts that the bionic division is revolutionary if you wish to save money on fuel. By keeping a close eye on the physiology of bones and doing research, we can discover what else Mother Nature has to teach us about the development of materials that are both environmentally friendly and extremely powerful (Biomimicry, 2016).

9.5.1 WHY BONES SHOULD BE INDUSTRIAL DESIGNERS' BEST FRIEND?

As a result of mechanical stress and loading, bones grow. Bones must be strong yet light at the same time. Bone is a mixture of hard calcium-based mineral (hydroxyapatite) and elastic protein (hyaluronic acid) (collagen). In contrast to most synthetic materials, which are all the same, this material's structure is tailored to the direction of the greatest stress rather than being uniform like most of them. Look at these two instances, both influenced by the natural world: The Golden-fronted woodpecker's porous frontal bone absorbs shock to protect its brain from harm. Despite their small weight, woodpecker skulls must be large enough to deal with the beating of their beaks against the wood. A shock-absorbing spongy bone covers the frontal region of the skull, which protects the brain from harm. With a number of holes, this bone is less compact than a more compact bone. High-impact forces make up the beak. As a result, it's more than simply a skull; it's the entire design.

A stress-dispersing microstructure in a horse's metacarpal bone prevents structural weakening caused by a hole. Using just dense bone for the construction of horse leg bones would result in excessive weight. The hydroxyapatite ceramic microplatelets give bone its stiffness and hardness, while the collagen proteins that bind it together give it its toughness. Bone is a foam-like cellular material. Bone is often thought of as a rigid material, yet it actually has a certain degree of flexibility. To create strong materials, nature frequently combines rigid and pliable elements. The next phase in innovation might be a submissive and functionally graded material.

Because of a latticework of fibers that manage tensile and compressive pressures, bone's interior structure is both strong and lightweight. There is a wide variety of natural materials and structures that are capable of sustaining themselves. Polyurethane-filled cylinders, laminated composite plates that slide under pressure, and much more are just some of the many examples of innovative design in this field. There is no end to the inventiveness of Mother Nature.

9.6 THE USE OF BIOINSPIRED MATERIALS AND COMPONENTS IN GREEN TECHNOLOGY AND SCIENCE

In the context of biomimetics, it implies that it replicates live organisms. The term "bioinspiration" has a calming effect on engineering. It's a really interdisciplinary field. All kinds of things are engaged in this idea, including their biological activities, structures, and principles. There are biological materials of different shapes and sizes, from atoms to macroscales. A material's physical and chemical qualities, along with its surface structure and morphology, play a vital influence in determining the material's features and characteristics. There are a broad variety of materials, covers, and products that have multi-purpose qualities. Bioinspired materials and surfaces, which can be used in green science and technology, are becoming increasingly popular.

Biomimetics encompasses a vast array of academic areas. When it comes to simulating natural processes through the use of materials and procedures inspired by biology, chemistry is a crucial component of the science. Materials science and engineering are both involved in the development of novel gadgets and applications that utilize natural processes. Material innovation based on biological processes is the focus of this discipline. There is a lot of focus on nanoscience and nanotechnology. Biomimicry has piqued a lot of people's curiosity in a wide range of industries. Biomimicry goods are expected to bring in billions of dollars in revenue each year.

9.7 BIOPLOYMERS AND FUNCTIONAL POLYMERS THAT MIMIC NATURE

It is common in nature to use "joining" as a method of adhering things together. The technical adherence of proteins, cells, and organisms to surfaces is one such example of this phenomenon. The problem: How might concepts found in nature be put to use in adhesive bonding and surface engineering applications?

Standard synthetic adhesives used in medical adhesive bonding are concerned with adherence in aquatic settings. In order to create new adhesive systems, the structural components of adhesives from, for example, blue mussels can be combined with synthetic polymers. In addition, histological preparations are joined, wounds are cared for, and renewable raw materials are applied.

9.8 TACKLING COVID-19 FROM THE POINT OF VIEW OF MATERIALS SCIENCE

An important link exists between SARS-CoV-2 and materials science. This study's findings will aid in the advancement of antiviral research and treatment. Bhushan (2018) evaluated prior attempts in materials science to build imaging systems and microfluidic tools to analyze viral architecture and transmission.

Their research focuses on building simple, accurate, and inexpensive viral detection technologies. In this study, researchers investigated the role of materials in the development of antiviral medications, including the significance of artificial material platforms for organoids and organs on a chip, drug delivery, and immunization, and the production of medical equipment. Innovations based on materials science contribute to the study of SARS-CoV-2, but they also serve as platforms and tools for future study, protection against, detection of, and treatment of other viral diseases.

Rapid growth and adaptability are two characteristics of viruses. To stop the spread of viruses, we need multidisciplinary approaches that include rapid, precise detection, effective protection, and regular treatment updates. Detection, protection, containment, and therapy can only be performed in unison by a worldwide network. Emerging viral infections have historically relied heavily on materials science to build instruments for rapid response. Using point-of-care equipment and treatment and vaccine tactics, they are able to decipher the structures of viral proteins.

To understand antiviral analysis, materials science is crucial. This impacts the value of scientific collaboration. As a global epidemic, the SARS-CoV-2 virus is being tackled by experts in virology, epidemiology, and medicine. Bhushan (2018) explored viral transmission, infection, and illness progression (Figure 9.5) from a materials-science perspective and from the perspectives of viral protection, detection, and therapeutics (Figure 9.5). Finally, he spoke about the necessity of materials science in the future study of infectious viruses and viral illnesses. By using universal platform technologies, an effort will be made to link material science. You need it if you want to learn how viruses propagate and how to defend yourself from them as well as to diagnose them and treat them so that you may avoid them in the future.

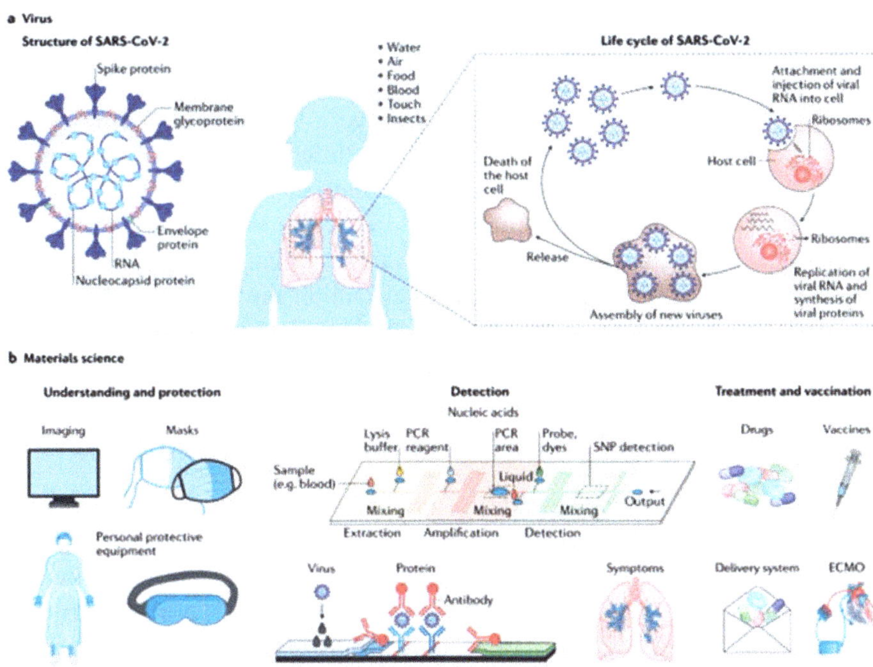

FIGURE 9.5 Transmission and replication cycles of the SARS-CoV-2 B virus are described in detail. SARS-CoV-2 investigational tools and protective gear are both made possible by advances in materials science.

Source: Reprinted with permission from Zhongmin Tang et al., 2020. Copyright © 2020, Springer Nature Limited

The study of viral structure and life, protection, detection, treatment, and vaccination relies heavily on materials science (Bhushan, 2018). For instance, materials science plays a crucial role in various facets of antiviral research, including viral biology and transmission, detection, treatment, and immunization. This is particularly true for materials science. Viruses must be isolated and purified in light of the emergence of new viral dangers.

In order to identify and purify viruses from blood and saliva samples in advance of future viral outbreaks and to speed and coordinate our reaction, we need hand-powered ultra-low-cost centrifuges. In viruses, electron microscopy and cryo-electron microscopy must be employed, along with confocal microscopy, in order to examine them. In this setting, real-time dynamics, including RNA replication, protein synthesis, and single-virus transmission from cell to cell, must be examined. The structures are examined using high-resolution imaging or time-resolved wide-angle X-ray scattering equipment.

Because of new materials-enabled technologies such as thermal-proximity coaggregation and time-resolved wide-angle X-ray scattering, studying drug-virus interactions has never been easier. These instruments are viewed as more potent since they are sensitive and have a high spatial and temporal resolution. When it comes to high-resolution microscopy, however, spatial resolution should not be traded for time resolution. The portability and low cost of the Octopi, Bite scope, and scale-free vertical tracking microscopy equipment make them perfect for this research. It is challenging to investigate viruses at the nanoscale because of the diffraction limit of portable optical imaging equipment. As a result, the geographical resolution may not be as critical in this scenario as the time resolution.

9.9 A PUFFER FISH-INSPIRED ULCER AND TUMOR MONITOR

Researchers at MIT have created an ingestible tablet that expands to the size of a ping-pong ball in the stomach, allowing it to remain in the body and collect data for weeks. Monitoring stomach pH levels and spotting the growth of tumors and ulcers are merely two of this tablet's potential uses.

Tumors and ulcers can be studied using the design, which was inspired by the pufferfish. Two layers of hydrogel are used to create the pill. The inner one acts as a sponge, while the outside one serves as a shield. Inflating and deflating the pill is surprisingly comparable to the pufferfish's capacity. It can be made to contract and pass through the patient's digestive tract without injury by giving them a sip of calcium solution. This research was inspired in part by the pufferfish's biological mechanisms. Yuk et al. (2017) came up with a tablet that may be used for both diagnostic and therapeutic reasons. Ping-pong-ball-sized worms make their way to the stomach (the pufferfish also expands on-demand, but it does so to scare its enemies).

It is estimated that the new device can remain in the stomach for around one month. A number of physiological parameters can be recorded using the monitor that's been installed. It sends alerts when certain ailments, such as ulcers or tumors, are detected (through a wireless micro-device inside the pill). "Puffer Pills" are too large to travel through the digestive tract, therefore, the micro-sensor can operate in a stomach environment for four weeks without issue. Afterward, the new pill will be released. Once the pill has been swallowed, it returns to its former size, allowing it to continue its journey into the gut.

The tablet can grow its volume by up to 100 times in 15 minutes, according to US researchers who published their findings in the scientific journal Nature Communication (Yuk et al., 2017). On the outside, they used a highly resistant nano-crystalline structure bio-gel, while on the interior, they used a sodium polyacrylate bio-gel, which absorbs stomach acids and is suitable for retaining a micro-sensor (there are many different kinds of micro-sensors commercially available). Research on the new tablet will focus on a wide range of illnesses. It was found to be extremely durable and capable of transmitting accurate data regarding the stomach's temperature and bacterial population during numerous tests. It's been shown that a micro-video camera instead of a sensor in the new pill could provide exceptional accuracy in monitoring the development of lesions like neoplastic lesions. In addition, a variety of biomedical devices are being ingested. For up to one month after use, ingestible tablets inspired by pufferfish can expand to the size of ping-pong balls.

The pill has a wide range of options. For example, a thermometer used in recent research may hold any number of sensors. pH levels can be used to determine the presence of bacteria and viruses in the stomach. It's possible, say experts, that the pill will include a small camera. As a result, ulcers, tumors, and other disorders could be halted. While gastric balloon weight reduction treatment now requires the use of an endoscope to place a rigid gastric balloon down the patient's throat, the new delivery method could make this procedure less intrusive.

9.10 WOUND HEALING ADHESIVE INSPIRED BY SLUGS

As a defense tactic, the slug leaves a muddy track. No matter how fierce a predator is, the slug's slime-like material (Figure 9.6) keeps it intact. This ability inspired a team of engineers at Harvard University to develop an incredibly strong surgical adhesive. After surgery, organs and tissues are sutured together to prevent additional injury. Two types of polymers are used to make the glue, which is a correct claim. Alginate and polyacrylamide are made from seaweed extracts. The weight-to-strength ratio of these materials is poor.

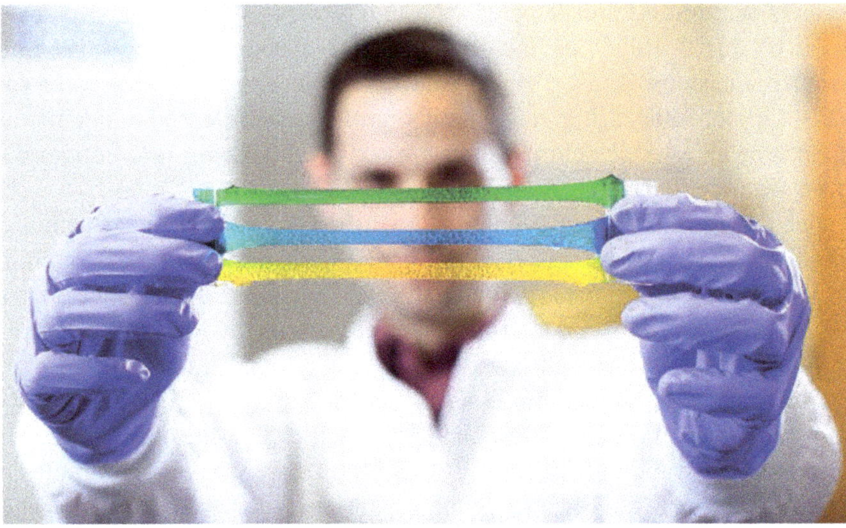

FIGURE 9.6 This viscous mucous is extremely potent. A severe predator is incapable of separating the slug.

Image courtesy: Wyss Institute at Harvard University

When they enter and develop a strong link, they are at their best. The glue can endure three times as much stress as previously assumed, according to extensive testing. Even the most modern medical adhesives get dislodged when it is used to fix a dynamic organ like the heart. There are numerous ways in which these bio-inspired inventions can be applied. We're deliberating on the finest ways to treat patients in the field of medicine. For many, it will be years before they can be used in the operating room. In the minds of engineers, their inventions will alter the way we practice medicine.

9.11 MIT RESEARCHERS FIND LOBSTERS' BELLIES AS STRICT AS INDUSTRIAL RUBBER

Lobster is a delicacy in the New England region of the United States, especially in Maine, and is renowned for its unusual appearance. In addition to being incredibly powerful and remarkably flexible, the underbelly of this crab makes it stand out. MIT researchers utilized it as a source of inspiration for a novel design for body armor. An armored crustacean appears vulnerable when it's flipped on its back because of its translucent membrane. When the animal moves through the rocky seafloor with its claws and flippers, the soft membrane protects it from cuts and scratches thanks to a microscopic structure similar to that of plywood (Christopher, 2019). Lobsters can whip their tails back and forth because the membrane is a natural hydrogel, which is 90% water. The rest is mostly made up of the fibrous substance chitin, which is found in many shells and exoskeletons. Because of this, predators have a hard time eating and dissecting the tail. Other natural hydrogels found to be tougher than lobster membranes include collagen and animal skins. Tread, hose, and belts made of industrial rubber composites are all on par in terms of tensile strength. Lobsters, with their strong yet flexible membranes, could serve as inspiration for the development of more flexible body armor, particularly at the elbows and knees. Lobster membrane-like materials can be used to improve the performance of soft robotic systems and tissue engineering. A new understanding of how one of nature's most persistent species manages to persevere has been gained as a result.

A new sort of flexible body armor for humans might be guided by a material as tough as industrial rubber, which is used to create car tyres and garden hoses. To protect the knees and elbows, these braces are made to fit snugly over the joints.

9.11.1 IT'S NOT IN THE FOOD CHAIN; RATHER, IT'S ON THE OCEAN FLOOR

With boiling lobster, you know how difficult it may be to crack the claws and shells without a unique mix of instruments and technology. To protect itself from underwater predators, lobsters boil their shells in bright red or blue colors.

The underside of a lobster tail reveals a shell that is loose and thin if you flip the tail over. Despite its appearance, it is sturdy, healthy, and flexible. The lobster's body is protected by a thin layer of sand, rocks, and shells as it crawls over the ocean floor to evade predators and those annoying lobster traps. In contrast, the lobster's tail demonstrates that its membrane is both flexible and stretchy by allowing it to move its tail in any direction to avoid unwanted intercourse. The underbelly of a lobster is as durable as industrial rubber (Figure 9.7).

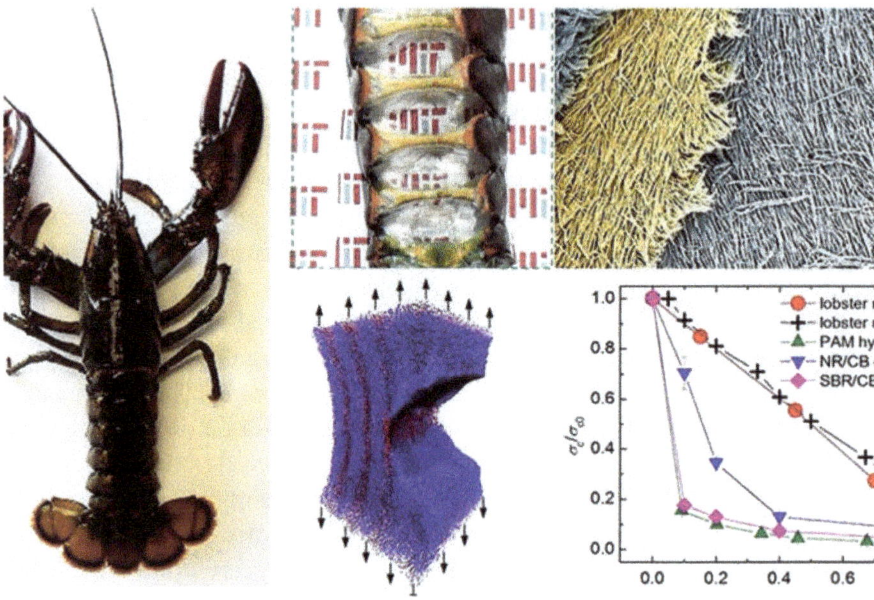

FIGURE 9.7 A graphic demonstrates the toughness of lobster underbelly tissue in an experiment at MIT's mechanical engineering department.

Source: Reprinted with permission from Wu et al., 2019. Copyright © 2019 Acta Materialia Inc. Published by Elsevier Ltd.

A membrane composed of a hydrogel that is 90% water and 10% fibrous chitin captivated MIT researchers, who teamed with scientists from Sichuan University and Harvard. The military needed a new type of body armor that was both flexible and protective (Wu et al., 2019). COMSOL Multiphysics® can be used to conduct these kinds of investigations, however, it's possible that the study provided here was conducted with a different software.

9.11.2 A STUDY OF LOBSTER SHELL BIOMECHANICS

The flexible membrane was initially put through its paces in a series of mechanical tests. There was an intriguing discovery that the membrane was resistant to fracture even when stretched and cut severely (for example, by a particle of sand or a fragment of shell). Electron microscopy was utilized to analyze the membrane at extreme levels and yielded a number of important results. The shell has the look and feel of plywood. The membrane had about 10,000 chitin layers despite its minuscule size of a quarter of a millimeter. Even though it was barely a quarter of a millimeter long, the membrane comprised almost 10,000 layers of chitin. The chitin strands in each successive layer were oriented 36° away from one another. The membrane becomes so tangled up because of the unique composition of this material. Its astonishing pliability is made possible by the abundance of water. In addition to new-and-improved body armor, the researchers who studied the lobster belly membrane also opened the door to advancements in soft robotics and bioengineering.

9.12 BIOMIMICRY-GRASSHOPPERS INSPIRED ENGINEERING-INNOVATION

In biomimicry, engineers use biological organisms as inspiration for their designs. There are numerous advantages to learning from nature. For starters, understanding evolution is an exhilarating process. Different mechanisms such as natural selection, genetic drift, and others enable organisms to adapt to their environment so that they might have diverse, desirable features to withstand various environmental circumstances around the globe. Because the features of species on the Earth have become so complicated and developed during the long history of development on Earth, it's interesting to learn from nature. A common sight in parks and gardens is grasshoppers. They are able to jump in any direction because of the open space. However, they need to go out and find food, a place to call home, and a mate with whom to lay their eggs. Engineering designs based on grasshoppers have been used to assist in solving human problems. Robotic arms based on the legs of a grasshopper, for example, have been used to build automobiles. Grasshopper-like actuator arms in robots can store energy when they contract.

Biomimicry of the grasshopper legs is the most useful invention. The leg has plenty of energy reserves. When it jumps, it contracts and then releases its power. Designing a system to replicate the grasshopper legs' ability to accelerate at a fast rate of speed as a means of transportation has been the most difficult part of this project (Chuan et al., 2017). The object's versatility has been demonstrated by its resemblance to the legs of a grasshopper. Some examples of their use include bionic legs for humans and vehicles that can move on their own power.

Muscles can conserve energy by contracting and releasing the energy they store. The jumping motion is triggered by relaxing. In spite of their reputation as crop-damaging pests, grasshoppers have been shown to be extraordinary organisms thanks to their jumping apparatus in a study of grasshopper legs. In a split second, the grasshopper's body speed increases to a high-velocity decline. The energy they've accumulated is released when they're relaxed.

9.13 MEDICAL GADGETS BASED ON BIOINSPIRATION

Improvements in medical technology (Sheng et al., 2017) have altered stents, left ventricular assist devices, and pacemakers, which have revolutionized the treatment of human diseases such as coronary artery blockage and heart failure. Despite the importance of these devices in the prevention and treatment of thrombosis, their mechanical operation and stable electronic physiology remain challenging and unresolved.

Fortunately, nature provides us with a wealth of bioinspired physico-chemical ideas that we may use to design unique synthetic materials. The tools we use to get around the problems we can't solve on our own. There has been tremendous development made in bioinspired techniques due to their ability to design and fabricate innovative gadgets that can be used in a variety of fields. Bioinspired devices (Figure 9.8) are finding increasing use in various sectors of research.

The rising tide of public health concerns has steadily shifted the focus of many countries' development strategies to precision medicine. From a medical perspective, the term "precision medicine" includes a wide range of definitions. Biocompatibility and anti-thrombotic qualities of medical devices and materials are becoming more diverse and urgent than ever before. New medical device research subjects are emerging as a result of this new trend in materials, mechanical systems, and sensors (Chen et al., 2019). Medical equipment requires biomaterials, biointerfaces, and biosystems, which are still an issue.

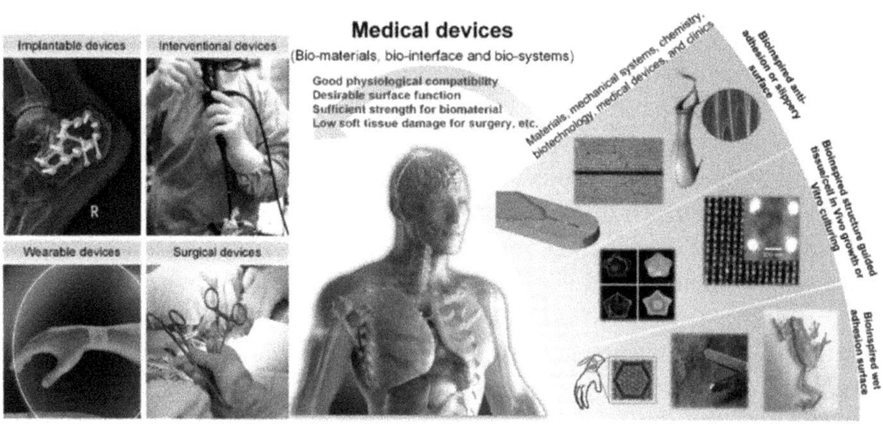

FIGURE 9.8 Diagrammatic illustration of bio-inspired medical device techniques.

Source: Reprinted with permission from Sheng et al., 2017. Copyright © 2017 Published by Elsevier B.V.

9.14 SIGNIFICANCE OF BIONICS IN MEDICAL DEVICES

The medical device typically refers to implantable medical devices, interventional medical devices, surgical equipment, etc., that interact directly or indirectly with soft tissue. All requirements for the medical device must take into account the necessity of the life system, which comprises the human body.

- Excellent physiological congruence;
- Anti-fogging and anti-adhesion are two hot surface functions;
- Energy sufficient for biomaterial;
- Low tissue trauma for surgical and other uses.

The objective of medical treatment is to restore the operation of the body's life-support system or to recover health from sickness. If you take the necessary precautions, the quality of healing will improve. Tissue and organ physiological and functional pathways are evident. Motivation can be aided by the specifics of these physiological mechanisms, hierarchical structures, and the behavior of entire living organisms. The development of medical equipment is the basis for bionics. Medical devices have recently adopted bioinspired approaches. Bioinspired materials, bioinspired functional surfaces, and bioinspired sensors play significant roles. The Scientist investigates the significance of bionics in medical devices by focusing on applications of bio-inspired materials.

Biomimicry, also known as bioinspiration, is an effective method for reinventing technology, particularly in biomedical engineering. There is a new wave of prospective medical gadgets incorporating natural elements. Here are nine bioinspired innovations. They have an effect on the evolution of medical technologies.

9.14.1 OCTOPUS-INSPIRED SUCKER FOR TISSUE GRAFTING

There is a possibility that fragile tissue grafts will be destroyed or infected before they can be transplanted into a patient. This is one of the greatest obstacles involved with transplanting such tissue. Extremely thin grafts are extremely difficult to use and can delay the patient's recovery. Hyun Joon Kong of the University of Illinois's department of chemistry and biomolecular engineering and his team have developed a sucker (Figure 9.9) that can pick up and release grafts with minimal force. Their device utilizes a temperature-responsive hydrogel layer, which is heated up using an electrical heater, causing the gel to shrink. The gel then expands and adheres to the graft, similar to how an octopus wraps its tentacles around its prey. After applying the hydrogel to the desired area, it is heated to release the suction on the graft.

9.14.2 BIOHACKING: TYPES AND SAFETY

Biological experiments (such as gene editing or the use of medications or implants) have altered the features or capacities of live organisms, especially when undertaken by individuals or groups outside of a typical medical or scientific research environment.

Dave Asprey is a biohacker, a person who uses science and technology to enhance their health and happiness (https://lewishowes.com/podcast/top-5-biohacks-for-happiness-health-with-dave-asprey/). According to Asprey, there are around 1,00,000 biohackers in the world, and he is a superstar among them.

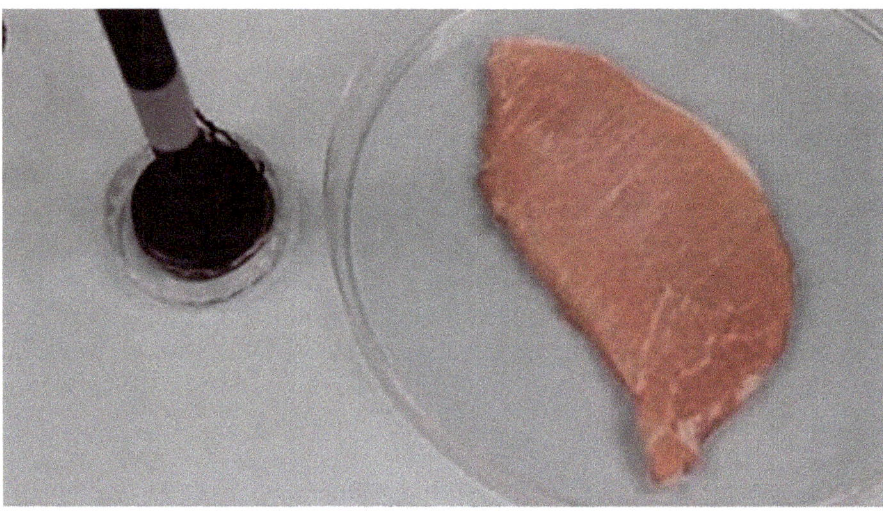

FIGURE 9.9 The octopus-inspired suction device can pick up and release a graft with minimal force.

Source: Reprinted from Touchstone (2020), TechXplore.com. https://techxplore.com/news/2020-10-octopus-inspired-sucker-thin-delicate-tissue.html

It's a form of "do-it-yourself biology" that aims to increase productivity, health, and well-being by purposeful pauses in the body's natural processes. The ancient practices of meditation and intermittent fasting (Figure 9.10), two common biohacking strategies, are still used today. A more technical approach to biohacking is used by some biohackers who are trying to improve themselves while correcting defects and becoming superhuman. Do-it-yourself health events are becoming more extreme as technology for technical Biohacking opens up on the market.

Who knows the motivations behind those who engage in self-mutilation? Some people do it to treat their health difficulties because they don't feel well and don't want to rely on traditional therapy alone. Others are looking to harness the power of engineering to test out new ideas and gain greater autonomy over their own bodies. It's also done by those who wish the aging process to stop and desire to live forever.

The first examples of wearable electronics are powered by human energy and are wristwatches. Smartwatches, activity trackers such as Fitbit and Runtastic, innovative eyewear such as Google Glass, and head-mounted displays are contemporary examples of wearable technology. The next natural step for wearable technology is the development of implantable devices. Biohacking or "biology + hacking" refers to any implant or intervention in the human body that enhances performance or health.

Despite the fact that it may sound like a scene from a science fiction film or a popular Netflix show like Black Mirror. There is a vast selection of embedded technology in development or on the market. They are comparable to electronic tattoos, electronic stamps, password pills, memory chips, magnetic implants, and guiding/GPS gadgets. They contribute to who we are. Humans use medical, dietary, physical, and electronic methods to control their biology. It's a common belief after Biohacking that what we put into our bodies can have a significant impact. For the "outputs" (such as enhanced memory, less stress and concentration, and enhanced performance), we must alter and provide our bodies with improved "inputs."

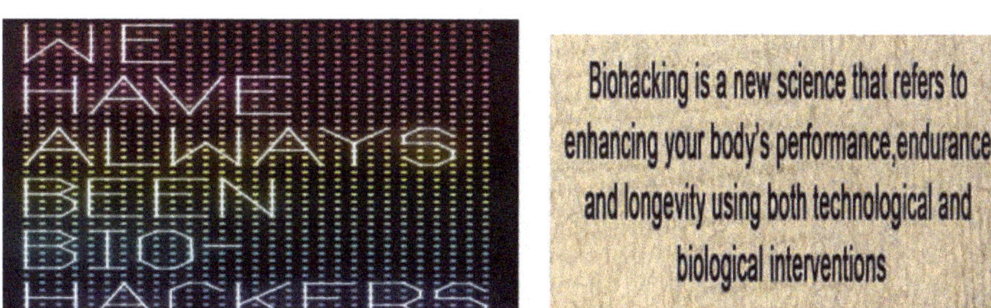

FIGURE 9.10 Biohacking.

Photo credit: Akasharc1995. https://creativecommons.org/licenses/by-sa/4.0

Biohacking is referred to as "citizen" or "do-it-yourself biology" (Figure 9.13). A common approach among "biohackers" is to gradually enhance one's health and well-being by making little modifications to one's diet or way of life. Everything from weight loss to improved cognitive function can be achieved through biohacking. However, the greatest approaches to biohacking are based on a thorough understanding of how your body functions and a cautious approach. As a broad phrase, biohacking (Gangadharabatla, 2020) is undeveloped. It encompasses everything you do to improve your overall health, from what you eat to how much sleep you get. A method for controlling the body's functions. Through biohacking, you can employ tools, data, technology, and science to become the best version of yourself.

Biohacking comes in a variety of shapes and sizes. Nutrigenomics, DIY biology, and grinder are the three most common types.

9.14.2.1 NUTRIGENOMICS

Your genes and the food you eat are the focus of Nutrigenomics. One of the most often used forms of Biohacking is based on the controversial but widely accepted concept of mapping out and optimizing your body's overall genetic expression through the use of various nutrients. Distinct nutrients have different effects on your mental, emotional, and behavioral health.

9.14.2.2 DIY BIOLOGY

DIY biology, commonly known as DIY bio, is a form of Biohacking guided by individuals with scientific training. Using the advice of these biohackers, non-experts can undertake systematic self-tests outside of a laboratory or medical facility.

9.14.2.3 GRINDER

According to the grinder, a subculture of biohacking, every part of the human body can be hacked. By injecting their bodies with chemicals, implants, and any other devices they may discover, mills strive to become "cyborgs" whose bodies work as they desire. The chemicals in your body can be altered through exercise, nutrition, medication, supplements, and even fasting to improve your health, cognition, and lifespan. A biohacker is a usually healthy person who decides to alter their physiology through various do-it-yourself approaches in order to increase the efficiency of their mind and body.

9.14.3 BIOHACKING YOUR DIET–GUIDE FOR BEGINNERS

An introduction to biohacking diets by Kayalvizhi Arivalan (https://www.femina.in/wellness/weight-loss/biohacking-diet-202545.html) is available. 'Biohacking,' on the other hand, is a term used to describe the process of altering one's lifestyle in order to get the most out of one's physical body. Biohacking can expedite weight loss and enhance cognitive and physical performance. Professionals recommend adjustments to a person's physiology, food, exercise, and way of life in order to improve their health and well-being. Nutrigenomics, the study of how food interacts with a person's genes, is central to modern biohacking of diets. Since every person's body is made up of unique combinations of cells and genes, a unique diet is required for each one. We're bringing you the easy steps to a successful hack.

9.14.4 OLD FASTING

Another strategy is traditional fasting. Any nutritional imbalance drives the body to adapt. The consequence could be beneficial or detrimental, but the body adapts as best it can. When fasting challenges the body, the body activates its survival mechanism by utilizing stored fat instead of glucose from food. This way, there is an increase in energy levels in the body and mind. It also helps on a cellular level by eating all the bad cell contents to make energy through a process called autophagy. This helps the body create new cells (https://www.newindianexpress.com/lifestyle/health/2021/sep/01/biohack-your-bod-2352506.html).

Similarly, consuming certain minerals such as zinc can improve our immune system. Fitbits and digital wearables that are used to track how we are functioning physiologically are also part of biohacking. Exercise and physical activity can help release good cortisol hormone, which helps in improving health.

9.14.5 IT'S ALL IN YOUR GENES

Everyone wants a healthy and long life. But the genes we inherit from our parents might make us suspectable to certain illnesses. What if there was a way to 're-engineer' the genes that can cause a certain disease? Scientists say that it is possible to alter your genes through a certain diet, medication, exercise, or even supplements. Scientists have found that our body is programmed by our genes. Thanks to modern research through genomics, we now not only have the ability to interpret these codes but also re-engineer them. Nutrigenomics, pharmacogenomics, epigenetics, and gene-editing technologies are silently incubating in our city to shape the future of healthcare and lifestyle and to increase productivity. The future drugs would have fewer side effects and would work on gene silencing, a process of regulation of gene expression in a cell to prevent the expression of an unhealthy or misfit gene.

9.14.6 INTERMITTENT FASTING

Biohacking can help you lose weight more quickly and improve your brain and body performance. In order to improve one's health and well-being, professionals suggest alterations to a person's physiology, exercise, diet, and way of life. Nutrigenomics, the study of how food interacts with a person's genes, is at the heart of a biohack in diets today (https://www.femina.in/wellness/weight-loss/beginners-guide-to-biohacking-diet-202545.html#:~:text=A%20biohack%20in%20a%20diet,measures%20for%20an%20effective%20hack).

9.14.7 THE POWER OF SUPERFOODS

In addition to their health benefits, researchers believe that the high antioxidant and micronutrient content of superfoods may also help people lose weight. 'Superfoods are powerhouses with vital vitamins and minerals,' adds Brindha, who believes that consuming nutritious foods that fulfill the appetite could make individuals adhere to their eating patterns.

9.14.8 TRY AN ELIMINATION DIET

After eliminating a certain food from your diet for a period of time, you can gradually reintroduce it to see the effects on your body. Brindha recommends an elimination diet for biohacking your food because it's the greatest way to discover out what's generating inflammation in your body. She goes on to say that this diet keeps you away from any meals that you're worried might trigger any unwelcome symptoms. The elimination diet improves mood, mental agility, and performance by relieving symptoms such as diarrhea, constipation, and bloating. A nutritionist or doctor should be consulted before making major modifications to your diet, says the expert.

9.14.9 IS BIOHACKING A SCAM OR IS IT A LEGITIMATE METHOD OF IMPROVING HEALTH?

The answer to a broad question like this is contingent on what the biohacker has to give. The scientific community's attention to it. For example, studies have revealed a connection between the practice of mindful meditation and decreased levels of stress and chronic pain. Meditation has been proven to be a successful biohack in clinical trials that followed the scientific method.

9.15 IS BIOHACKING SAFE?

Biohacking may be safe in some cases. Taking specific vitamins or changing your diet, for example, can be a safe option. A medical practitioner may be able to ensure the safety of a person's mods, such as RFID implants. Some biohacking techniques are dangerous, if not outright unlawful. In the world of academic research, studies that are considered unethical by the scientific community are often the focus of DIY bio- and grinder projects.

Even if you're only testing on yourself, it's still considered forbidden to conduct experiments on other people. Because of the potential for negative outcomes or injury, this is a well-respected authority in the field of biology. Biohacking, according to a 2017 Brookings Institute paper, is both a boon to the scientific community and a hazard to the public. The long-term effects of genetic modification or other experiments on humans can be difficult to predict without regular, controlled testing.

When it comes to using biology with a hacker attitude, biohacking (https://superhumantalks.com/what-is-biohacking/) may be considered. The phrase has yet to be given a standard definition. The following are some of the alternative definitions:

- You and technology are 'symbiotically intertwined.'
- 'Biotechnological enhancement of your physical well-being.'
- Achieving the best possible results by creating the right conditions.
- 'Biohacking involves tampering with your body,' says one proponent.

Biohacking (Chuah et al., 2016) is those who bio-hack through self-experimentation risk important health outcomes should anything go awry. Intermittent fasting is a common bio-hack generally used to boost wellbeing. But it is, failing to ask a nutritionist or complete fasting correctly can result in eating disorders and weight loss. It's common to perform a biohack without planning. The groundwork of it people do before the whole impression of a bio-hack confirms that the hack will be victorious. Other biohacks rely on research and technology from the scientific community and several issues. They have also demonstrated both controversial and notable hazards. It enables scientists to make precise modifications to any DNA or bacterial genome. Ashley Hamer, for example, CRISPR, although severe and potentially valuable, inappropriate use of gene editing is unethical. Editing an organism's DNA chain is not simply a challenging undertaking. It results in ongoing genetic modifications in the future. Similar to damming a river to generate hydroelectric power, unanticipated events can frequently occur downstream. Several further hacks will also make headlines in biohacking communities. There are no other hacks that may be difficult to choose a safe bio-hack. Biohacks usually come with extraordinary assertions.

Heart disease risk may be reduced, mental function may improve, and more. Many of these claims have been proven to be true. Care and patience must be taken to assure their safety and effectiveness if this is the case. Biohacking may have taken off in the ever-changing Silicon Valley, but millions of Americans have also incorporated it into their daily lives. The above-mentioned Biohackings, such as diets, meditation, and innovative technologies, are among the most prominent. Others relate nootropics, such as beverages containing caffeine like Bulletproof coffee or simple vitamins and supplements like vitamin D. Biohacking is a campaign devoted to one kind of DIY science. Getting science and the scientific method to the people is ambitious. The options of what could be found and invented without the conditions of grant money and specific hypotheses are endless. Biohackers are more engaged in body modification. It's a subset of and the potential for creating cyborgs.

9.16 SILICON VALLEY

Using a variety of methods, including nootropics (smart medications), intermittent fasting, light therapy, and blood testing, Silicon Valley techies have perfected the art of getting the most out of their bodies. Dave Asprey, the creator of Bulletproof Coffee, is one of the most well-known biohackers. "Biohacking is the practice of altering the environment around you and within your own body. It wants to know everything about you so that it can help you develop your body, mind, and life in a holistic way." What I'm saying is, "It's about improving your biology."

Most people associate Asprey with his morning cup of coffee, which includes "Brain Octane" oil and grass-fed butter. It goes well with ketogenic diets heavy in fat and is the most potent thing you can act to favorably. All aspects of life are affected. According to the investigators, if you consume a diet that keeps your energy levels up and your mind concentrated, you'll free up a lot of your valuable willpower to focus on the things that you choose to focus on. The following locations offer red light therapy: In-office or at-home use of red-light treatment devices are all options available to patients of rheumatologists or dermatologists.

9.17 BIOHACKING YOUR BONES: OSTEOSTRONG

People need to talk more about heart health. Cardiovascular disease is something that everyone should be aware of and take steps to prevent. Today, skin health is all about sunscreen use and collagen-boosting nutrients in our everyday diets. Weight loss, inflammation, cognition, gastrointestinal health, and the aging effects of a poor diet and a lack of exercise. All of these are at the forefront of our thoughts. But how often do we think about bone health?

Bone density changes throughout time, although most individuals aren't aware of this. Men and women gain more bone mass than they lose up to the age of 30. Bone density and strength are being worked on. It's a different story when we're in our 30s. Women's bone density decreases by roughly 2% per year after the mid-30s. For a few years after menopause, it continues to affect certain women, but it does not affect those who are at high risk of developing osteoporosis. Men's bone density diminishes over time, according to scientific research. They lose bone mass until they are around 65 years old. Even though men's bone mass lasts longer, they are more resistant to acquiring osteoporosis as they age.

9.18 HUMAN ENHANCEMENT, BIOHACKING, AND TRANSHUMANISM: WHAT ARE SOME COMMON USES?

There are a multitude of methods by which people might develop themselves. Humans can improve their appearance in a variety of ways, such as via wearable technology or nanotechnology implants. They can also use software like a brain-computer interface (BCI) or gene therapy programs to aid their recovery. Tracking one's sleep patterns and nutrition can be used to alter one's biological makeup in biohacking applications. Blood transfers from a young person to an elderly person have been tried by some proponents as a way to slow down the aging process. Transhumanism, on the other hand, maybe the most extreme form of human growth. When it comes to these kinds of things, you're dealing with sophisticated systems like cryonics or the preservation of the human body long after it has died.

9.19 BIOHACKING FOR HEALTH AND HAPPINESS

9.19.1 WHAT IS BIOHACKING?

Biohacking is the practice of employing methods drawn from fields like biology, genetics, neuroscience and nutrition to enhance physical or mental performance, improve overall health and well-being, or achieve a specific health outcome (like weight loss), according to Brea Lofton, a registered dietitian and nutritionist for health and wellness company Lumen. Sometimes biohacking is also called do-it-yourself (DIY) biology (https://www.forbes.com/health/wellness/biohacking/). It covers a broad range of techniques from the less invasive practice of intermittent fasting to drastic and invasive procedures, such as the implanting of microchips or injecting of gene-editing enzymes. As a broader concept, biohacking encourages experimental biotechnology outside the confines of traditional research environments. Away from the hospital and university labs, some biohackers attempt to realize their dreams of becoming superhuman, circumventing what they consider traditional 'flaws', including aging, weight gain, or chronic illnesses.

9.19.2 WHY IS BIOHACKING DANGEROUS?

Biohacking, in essence, degrades the quality of scientific study. In most cases, the medications fail to pass valid toxicity tests. Because it provides services to patients while also putting their own safety at risk, it should be avoided. It is impossible to accurately predict how a medicine will interact with the human body without extensive pre-clinical testing in the laboratory.

9.19.3 WHAT ARE THE DANGERS OF BIOHACKING, TRANSHUMANISM, AND HUMAN ENHANCEMENT?

Since there is such a high demand for these ideas of human growth and development, many people are turning to the self-administration of chemicals and methods on their bodies, putting them at risk for things as in subsections.

9.19.3.1 THREATS TO HUMAN SAFETY

There are currently no regulations or safety measures in place to govern these operations. They could be held responsible. Proponents may be putting their own or volunteer health at risk. Imagine a group of biohackers using gene editing in an unsupervised context, such as a citizen science lab, to take control of their own bodies. What if something goes wrong? Who would bear the brunt of the burden?

9.19.3.2 ETHICAL CONCERNS

No laws have been established about how much human experimenting in group laboratories should be permitted, notwithstanding the voluntary nature of the experiments. Whether or not human enhancement practitioners are breaching the human right to life is a matter of debate.

9.19.3.3 PRIVACY ISSUES

In addition, implanting a chip raises a number of privacy problems. As a recent example, a communication chip that can elude army scanners was developed a few years ago. The chip, on the other hand, is extremely easy to hack and manipulate devices that are close.

9.20 PERFORMANCE ENHANCEMENT FOR HUMANS

To be included in the Human Performance Improvement category, one must hack the body, the mind, or both. The term "body hacking" refers to a variety of methods for improving one's health and physical performance. It incorporates not only physical activity but also a healthy food and way of life. Mind hacking is a term used to describe techniques for enhancing one's cognitive abilities. Strategies and techniques to boost your brain's speed, productivity, or memory.

Similarly, Consciousness Hacking refers to the use of technology for spiritual development. New genetic material can be added to our DNA using gene treatments, essentially rewriting our biological instructions. You've made an error in editing your DNA. A tumor can be induced by dangerously messing with your body. It's unnerving to watch people inject themselves with untested gene therapies. In addition, there's the problem that the outcomes of these biohacking "experiments" aren't founded on solid scientific data. If a person decides to self-experiment, the government cannot step in. Chemicals can be sold as research compounds, despite the fact that it is illegal for a corporation to promote a medicine if it is prohibited.

9.21 BIOHACKING AT ITS MOST EXTREME

There are some biohackers that go a step further than simply changing their diet and taking cold showers. Human development is the process through which people use technology to alter their bodies.

Grinder biohackers use magnets, chips, or computers embedded in their skin to become cyborgs. People with implanted chips in their hands have been seen to open workplace doors or even pay for a cup of coffee with a simple swipe of the wrist. What if you didn't have to carry your phone around with you all the time? Also on the horizon are gadgets that allow our brains to interface with computers directly. Google Glass-like, but without the spectacles. Dowd (2018) explored biohacking for women.

Women are typically regarded as the first biohackers (Figure 9.11). When it comes to our hormones, they have been in charge. Pregnancy has been naturally managed by women since thousands of years ago, employing methods that date back to (tracking cycles to the moon). Pills are used to control births in modern times. Women, on the whole, have a stronger sense of self-awareness. For Biohacking purposes, we're ideal candidates. When it comes to biohacking, men and women have quite different needs. Everything comes down to hormonal variations.

Mood swings are reduced when our hormones function in harmony with our moods rather than against us. We might expect to see improvements in our appetites, weariness, cramping, and overall energy levels. How can anyone say they don't desire greater control over their lives, greater control, and more time to devote to the activities they enjoy? To get started right away, here are five easy steps.

9.22 METHODS TO COMPREHEND BIOHACK YOUR HEALTH

9.22.1 HACK YOUR DIET HABITS

Our gut flora is directly influenced by the food we eat. This is the result of a colony of microbes residing in your digestive tract. Dysbiosis: an imbalance of beneficial and harmful microorganisms as a result of a diet based on Seasonal Affective Disorder. SAD is, to put it succinctly, an unhealthy way of eating. That means that SAD may be deficient in fiber, vitamins, and other essential elements. In this case, fiber is a crucial component. As a general rule, fiber feeds the beneficial microbes in your digestive system. Fresh, whole foods, such as vegetables, fruit, and pieces, are easy to consume when you eat them at every meal. Inflammatory foods like refined sugar and fast meals can't be given. When sugar enters the body, it alters gut flora, boosting the growth of dangerous bacteria and causing dysbiosis. Genetically Modified Organism gluten and dairy products, as well as sugary drinks and snacks, should be avoided. For a healthy body and reduced inflammation, you need high-quality nutrition.

FIGURE 9.11 Biohacking for better health.

Source: Reprinted from ParsleyHealth (2017). https://www.parsleyhealth.com/blog/biohacking-women-101/

A smoothie made with medical-grade protein powder and healthy fats is a great way to start the day. Green leafy vegetables like spinach, avocado, and kale fall under the category of "healthy foods." Lunch and dinner should include plenty of organic vegetables and a lean protein source. Organic starchy vegetables like sweet potatoes or gluten-free whole grains will be consumed as a source of carbohydrates. When it comes to dieting, it's not about keeping track of your calories. Nutrients and nutritional quality are figured out using this formula.

9.22.2 GET MORE (AND BETTER) SLEEP

Sleep apnea affects between 50 and 70 million people in the United States. The lack of sleep has been acknowledged as a significant public health concern. Lack of sleep can have a negative effect on the hormones involved in controlling blood sugar and hunger. We can gain more time by improving our sleep hygiene. The quantity of sleep you get will be enhanced by getting enough of it.

9.23 BIOHACKING FOR THE IMPROVEMENT OF BETTER FITNESS

Blue light from cell phones, computers, and LED lights should be kept to a minimum. The kind of light your brain associates with daylight is called "blue light." At night, it exposes your brain to dangerously high levels of blue light. Your body's natural melatonin production is affected by blue light. Try using a blue light filter like night shift duty on your smartphone or downloading flux to your PC to keep your electronics off for 2–3 hours before bedtime. All of this explains to your body what is going on. It's time to wind down and prepare for a brighter day the following morning.

9.23.1 LOWER ANXIETY LEVELS

It's not just guys who get stressed out; women are more likely to do so. It is natural for the body to release cortisol and adrenaline in response to a perceived threat. To prepare our bodies to deal with the stressor, these hormones activate the creation of glucose in the liver, which is crucial for the "fight or flight" response, as well. What can we do if stress is a given in our daily lives? Self-care should be treated as seriously as any other aspect of your health. Meditation in the morning, journaling before going to sleep, and a weekly massage all contribute to a healthy body and mind.

9.23.2 EXERCISE IN ACCORDANCE WITH THE MANUFACTURER'S INSTRUCTIONS

More than two-thirds of adults fail to fulfill physical activity guidelines for both aerobic and weight training. Consistent physical activity can help us lose weight and shrink our abdominal fat stores (fat). Inflammatory diseases and insulin resistance can be exacerbated by a large amount of fat in the body. Increased testosterone (which boosts sex desire and increases muscular mass) and decreased cortisol levels can be achieved by exercising. Consider low-intensity exercises like therapeutic yoga and walking to supplement your high-intensity workouts.

9.23.3 CONNECTION

Cortisol is a stress hormone that raises your level of anxiety. Oxytocin is a hormone that has been linked to stress in women. Those women who have a strong social network will produce more oxytocin than those who do not. Whenever life becomes tough, women turn to social support and contact for assistance. One of the most obvious hacks you may use is to combine the benefits of female friendship with some rate time!

9.24 FINAL THOUGHTS

Increasing numbers of women are turning to biohacking as a way to improve their health and well-being. People could start with simple tactics like eating a healthier diet and meditating to reduce stress. To help you get a better night's sleep, it reduces advertising to blue light. We must make modest adjustments to our routines in order to reduce the impact on our bodily and emotional well-being.

Hormones are the end outcome of this process. It's time to hack! Individuals' right to edit their DNA is now endorsed by high-profile biohackers, arguing that it will speed up the availability of life-saving gene therapies.

DIY (Do-it-yourself) gene therapy components, such as guide RNAs and CRISPR plasmids, are non-hazardous and can be obtained rather easily, according to biohackers.

9.25 NEWS ABOUT BIOHACKING

- **2010:** A 'Biosafety Level One' biotech laboratory is the first to be established outside of academia or business. In the lab, which is managed by Ellen Jorgensen, a former cancer researcher, workshops are held on a regular basis for locals of all ages.
- **2013:** In a crowdfunded effort to create *Arabidopsis thaliana* plants whose seeds glow in the dark, more than $5,00,000 has been raised for genetic engineering of the plants. It all comes to an end when the project fails because of the outcry over the unleashing of transgenic animals.
- **2015:** Liz Parrish, an American entrepreneur who flew to Colombia to dodge US laws, injects herself with an experimental anti-aging gene therapy. Although she is CEO of the biotechnology business developing the treatment, Parrish says that the unapproved and unmatched treatment is a success.
- **2016:** For $159.00, you can get a "do-it-yourself bacterial genetic engineering kit." Using this kit, anyone can produce *E. coli* that is able to resist growth on a variety of mediums.
- **2017:** Within a few months, two American biohackers experiment with gene therapy on themselves. The first is a breakthrough in HIV gene therapy, while the second uses CRISPR to target the myostatin gene, which has been linked to increased muscle mass in studies. Both men broadcast their surgeries live online.

9.26 BIOHACKS FOR HEALTH AND HAPPINESS

According to Dave Asprey, biohacking is "the art and science of altering your external and internal environments so that you have complete control over your own biology." Dave devised a solution after researching frequent fasting mistakes and everything wrong with the American diet. In Dave's opinion, the first biohack isn't about fasting. Forgiveness is the key to this situation. Grudges are just going to make you regret your actions in the long run (https://lewishowes.com/podcast/top-5-biohacks-for-happiness-health-with-dave-asprey/).

Forgiveness is a muscle that can be developed through practice, and if you can master it, you'll live longer and be freed from the burdens of resentment. That's exactly the point. Doing forgiveness and releasing resentment can be exhausting for me… "*Forgiveness is simpler when you raise your energy level*" – Dave Asprey. Dave argues that if you don't forgive, it will have a bad effect on your life. You will be wiped out by it. As a result, if you choose not to forgive yourself or others for their mistakes, you will feel resentful and depleted of energy. If you refuse to forgive, a gloomy cloud will always hang over your head, stealing your joy and health.

The second biohack is getting a good night's sleep. Good sleep can help you be more productive in your day-to-day activities. A good night's sleep can help you reach your peak performance. However, if you want to get a good night's sleep, you'll need to establish a quiet setting. Dimming the lights and blocking out the four different light frequencies that enter your eyes before you go to bed can make a significant difference. "This last night, I was able to get a good night's sleep with two hours in deep and change and two in Rapid Eye Movement sleep." Asprey, Dave. You'll get a better night's sleep if you wear special eyewear. To help you get a better night's sleep, Dave recommends wearing glasses that block out blue light. The final biohack is to get up and walk around a little bit. There are studies that indicate the importance of moving 20 minutes a day, which Dave mentions. High-intensity workouts such as squats, pushups, hard sprints, and so on should be used in this routine. High-intensity training pushes you to the limit.

In the fourth biohack, you'll learn how to maximize your body's ability to generate maximum power. Taking a cold shower may sound strange, but it's good for your health. When you take a cold shower, your sensors will be more attentive. When you're focusing on the head and chest, your body will respond more quickly. Neurofeedback, the fifth of Dave Asprey's biohacks, is akin to meditating. Learning relaxation techniques and cultivating a sense of well-being Neurofeedback has two main advantages.

"More than 1,000 entrepreneurs have gone through my neurofeedback program—I would say meditation. If you don't have access to neurofeedback, you can meditate," Asprey, Dave. This is a list of the biohacks that you can use. They can transform your life and help you avoid regret, sleeplessness, and sloth, all while enhancing your sense of awareness and bringing you peace of mind:

- Forgiveness is one of these biohacks;
- Sleep;
- Movement;
- Sensors on the body;
- Neurofeedback.

KEYWORDS

- **biohacking**
- **bioinspiration**
- **biomimicry**
- **health care**
- **medical devices**
- **parasitic wasp**
- **safety**

REFERENCES

Abaie, E., Xu, L., & Shen, Y. X., (2021). Bioinspired and biomimetic membranes for water purification and chemical separation: A review. *Frontiers of Environmental Science & Engineering*, *15*, Article number: 124.

Bhatia, R.K., Sakhuja, D., Mundhe, S., & Walia, A. (2022). Renewable energy products through bioremediation of wastewater. *Sustainability 12*(18), 7501. https://doi.org/10.3390/su12187501

Bhushan, B. (2018). *Biomimetics: Bioinspired Hierarchical-Structured Surfaces for Green Science and Technology* (3rd edn.). Springer International.

Bhushan, B., (2019). Bioinspired oil-water separation approaches for oil spill clean-up and water purification. *Phil. Trans. R. Soc. A, 377*, 20190120.

Bhushan, B., (2020). Bioinspired materials and surfaces for green science and technology (part 3). *Phil. Trans. Roy. Soc., 378*(2167). https://doi.org/10.1098/rsta.2019.0439.

Biomimicry 3.8. (2016). *Innovation Inspiration: What Bones Taught Air Bus About Optimizing Strength*. The site was designed by Six Pony Hitch.

Brown, P. S., & Bhushan, B., (2016). Bioinspired materials for water supply and management: Water collection, water purification, and separation of water from oil. *Phil. Trans. Royal Soc. A*. https://royalsocietypublishing.org/doi/full/10.1098/rsta.2016.0135 (accessed on 17 August 2023).

Celadyn, W., (2018). Inspirations and their modeling in architecture. *World Transactions on Engineering and Technology Education, 16*(3), 263–268.

Chen, H., Zhang, Y., Zhang, L., Ding, X., & Zhang, D. (2021). Applications of bioinspired approaches and challenges in medical devices. *Bio-Design and Manufacturing, 4*, 146–148.
Christopher, B., (2019). *What Do Lobster Shells Have to do with Body Armor?* COMSOL Blog. https://www.comsol.com/blogs/what-do-lobster-shells-have-to-do-with-body-armor/ (accessed on 17 August 2023).
Chuah, S. H. W., Rauschnabel, P. A., Krey, N., Nguyen, B., Ramayah, T., & Shwetak, L. S., (2016). Wearable technologies: The role of usefulness and visibility in smartwatch adoption. *Computers in Human Behavior, 65*, 276–284. https://doi.org/10.1016/j.chb.2016.07.047.
Chuan, Y. L., Kean, J. T. Y., Ben, L. Y., Domun, M. K. S., Naresh et al. (2017). Biomimicry-grasshoppers inspired engineering innovation. *Int Rob Auto J., 2*(2), 77–80. doi: 10.15406/iratj.2017.02.00020.
Coenen, C., (2017). Biohacking: New do-it-yourself practices as technoscientific work between freedom and necessity. *Proceedings, 1*, 256. doi: 10.3390/IS4SI-2017-04119.
De Melo, P. D., & Habibovic, P., (2018). Biomineralization-inspired material design for bone regeneration. *Adv. Healthc Mater., 7*(22), e1800700.
El-Rashidy, A. A., Roether, J. A., Harhaus, L., Kneser, U., & Boccaccini, A. R., (2017). Regenerating bone with bioactive glass scaffolds: A review of *in vivo* studies in bone defect models. *Acta Biomater., 62*, 1–28. doi: 10.1016/j.actbio.2017.08.030.
Filippi, M., Born, G., Chaaban, M., & Scherberich, A., (2020). Natural polymeric scaffolds in bone regeneration. *Front. Bioeng. Biotechnol.*, https://www.frontiersin.org/articles/10.3389/fbioe.2020.00474/full (accessed on 17 August 2023).
Gangadharabatla, H., (2020). Biohacking: An exploratory study to understand the factors influencing the adoption of embedded technologies within the human body. *Heliyon, 6*(5), e03931, https://doi.org/10.1016/j.heliyon.2020.e03931.
Gonzalez-Perez, A., & Persson, K. M., (2016). Bioinspired materials for water purification. *Materials, 9*(6), 447. https://doi.org/10.3390/ma9060447.
Helms, M., Vattam, S. S., & Goel, A. K., (2009). Biologically inspired design: Process and products. *Design Studies, 30*, 606–622. www.elsevier.com/locate/destud (accessed on 17 August 2023).
https://journals.co.za/doi/pdf/10.10520/AJA10195785_54 (accessed on 17 August 2023).
https://lewishowes.com/podcast/top-5-biohacks-for-happiness-health-with-dave-asprey/ (accessed on 17 August 2023).
https://medium.com/@amnaatharaliabbasi/benefits-of-ostrich-egg-de72115b92f8 (accessed on 17 August 2023).
https://www.newindianexpress.com/lifestyle/health/2021/sep/01/biohack-your-bod-2352506.html (accessed on 17 August 2023).
Knippers, J., & Speck, T., (2012). Design and construction principles in nature and architecture. *Bioinspiration & Biomimetics, 7*, 015002.
Koyama, T., & Tennyson, A., (2018). Ostrich is biologically highly interesting and useful for human life as a producer of antibiotics. *J. Vet. Med. Health, 2*, 107.
Lee, D., (2011). *Biomimicry Invention Inspired by Nature.* Kids Can Press limited, Canada. https://books.google.com/books/about/Biomimicry.html?id=Pnf2vUJjNCQC (accessed on 17 August 2023).
Norton, S., (2016). *The Beginners 'Guide to Biohacking*. iNews. https://inews.co.uk/essentials/lifestyle/wellbeing/beginners-guide-biohacking/ (accessed on 17 August 2023).

Rajshekhar, R., (2014). Biomimicry in architecture. *Intern. J. Adv. Res. Civil, Structural, Environmental and Infrastructure Engineering and Developing, 1*(3), 101–107 https://www.academia.edu/63197635/Biomimicry_in_Architecture (accessed on 17 August 2023).

Sheng, Z. Z., Liu, X., Min, L. L., Wang, H. L., Liu, W., Wang, M., Huang, L. Z., et al., (2017). Bioinspired approaches for medical devices. *Chinese Chemical Letters, 28*(6), 1131–1134.

Tang, Z., Kong, N., Zhang, X., Liu, Y., Hu, P., Mou, S., Liljeström, P., et al., (2020). A materials-science perspective on tackling COVID-19. *Nature Reviews Materials, 5*, 847–860.

Wei, S., Ma, J. X., & Xu, L., (2020). Biodegradable materials for bone defect repair. *Military Med. Res., 7*, 54. https://doi.org/10.1186/s40779-020-00280-6.

Whitesides, G. M., (2015). Bioinspiration: Something for everyone. *Interface Focus*. https://doi.org/10.1098/rsfs.2015.0031.

Wood, A. (2019). *Pufferfish-Inspired Pill Could Keep a Long-Term Eye on Your Gastric Health.* New Atlas https://newatlas.com/pufferfish-pill-stomach-gastric-health-mit/58267/ (accessed on 17 August 2023).

Wu, J., Qin, Z., Qu, L., Zhang, H., Deng, F., & Guo, M. (2019). Natural hydrogel in American lobster: A soft armor with high toughness and strength. *Acta Biomater., 88*, 102–110. doi: 10.1016/j.actbio.2019.01.067.

Young, M. (2018) How Women Are Bio-Hacking Their Health and Fertility, https://www.true-self.com/biohacks-for-womens-health-fertility-2569431099.html (accessed on 27 September 2023).

Yuk, H., Lin, S., Ma, C., Takaffoli, M., Fang, N. X., & Zhao, X. (2017). Hydraulic hydrogel actuators and robots are optically and sonically camouflaged in water. *Nat. Commun., 8*, 14230. https://doi.org/10.1038/ncomms14230.

CHAPTER 10

Nature-Inspired Business Lessons Create New Thought-Provoking Views

ABSTRACT

Leaders are accountable for producing results. Great leaders are educators. Great leaders help their teams grow. Great leaders create something that is significantly larger than themselves. All enterprises, organizations, and professional practices must take management and their desired outcomes into consideration. Both humans and animals exhibit leadership capabilities. Animals are excellent examples of leaders because they influence, set an example, guide, and communicate frequently with their followers. Biomimicry is an interesting new field of study. It investigates how nature functions and how we may apply lessons from nature to solve human issues. Innovation influenced by biology plays an important role in the management innovation process. The success of a business is heavily dependent on managerial abilities and talent. In this chapter, business skills applicable to the animal realm are outlined. Each animal's every action is dictated by the imperative to live at all costs. Each creature's ability to make the most of its resources and environment has been honed over the course of hundreds of thousands of years through trial and error as a result of evolution. The animal realm has much to teach us as humans. While animal experimentation may not be the most ethical way to do so, observing the basic management approaches each animal has experimented with over many years is highly informative. The following are some essential managerial lessons from the animal kingdom: Animals give a much-needed broad framework for the creation of a wide array of business techniques. Nature may teach us many things, including survival and discovery. However, few are aware of the ongoing marketing lessons nature has been teaching us. This chapter provides entrepreneurs,

Bio-Inspired Technologies for the Modern World: Emerging Trends and Applications.
R. Ramakrishna Reddy & T. Pullaiah (Authors)

executives, and investors with lessons from the animal kingdom. Packs of wolves and flocks of geese are both wonderful instances of teamwork in the animal kingdom. It is essential to their survival that they remain united. While teamwork may not be essential to our physical survival, it is essential to our professional success.

10.1 INTRODUCTION

Albert Einstein once remarked, "You will comprehend everything better if you contemplate nature in great detail." Nature is the ultimate classroom for all aspects of life, and the lessons are free to anybody who is observant and eager to learn. Numerous inventive businesses look to Mother Nature for solutions. Nature is also one of the greatest financial instructors. We have never considered a potential link between nature and your (future) business. Nonetheless, there is much we can learn from nature to improve our business practices. Nature is everywhere. Observation can provide an infinite source of inspiration. Therefore, one must study nature-inspired entrepreneurship, which involves drawing parallels between natural systems and human organizations (Benyus, 2002; Laroui, 2020; Friedman, 2015; Jennifer, 2013).

Biomimicry is a fascinating emerging field of study that investigates how nature functions and how we may learn from nature to solve human problems. Biomimicry is a relatively new term, although humans have been learning from other species for tens of thousands of years. The term was coined by Janine Benyus, the author of the book titled "Biomimicry." It derives from the Ancient Greek words *bios* (life) and *mimesis* (imitation). Janine is a biologist, author, and innovation consultant who describes herself as a "nature nerd." Her 1997 book, Biomimicry: Innovation Inspired by Nature, popularized the term biomimicry (Benyus, 2002). Her book is a collection of anecdotes describing the scientific process. Engineers, scientists, and innovators were transforming their fascinating findings of biological functional methods into novel technologies. The transformation toward sustainability necessitates concurrently addressing many challenges (Figure 10.1). Agriculture, architecture, manufacturing, transportation, healthcare, software, materials, and robots are just a few of the industries that have adopted biomimicry as a complete innovation strategy. Arizona and the Wyss Institute for Biologically Inspired Engineering at Harvard University were pioneers in putting biologists together (with experts in design, sustainability, Business, and engineering). Nature's 4-billion-year-old R&D laboratory provides an inexhaustible supply of energy-efficient, non-toxic, and time-tested innovations.

FIGURE 10.1 The Eastgate Centre has mimicked termite performance in passive cooling through the use of local material and porous building.

Source: Reprinted with permission from Jamei and Vrcelj, 2021.

Nature offers a small but largely undiscovered repository of technologies that process and manage data, materials, and energy. While the concept of going to nature for solutions has been increasingly accepted in innovation circles, the breadth and depth of the movement continue to surprise many. Despite the challenges of bringing innovation from inspiration to commercialization, innovative companies are increasingly abstracting ideas from nature to pave the way for technological breakthroughs and create profitable products that are often unattainable using conventional product design and development methods.

The current state of bioinspired innovation demonstrates that Biomimicry is influencing multiple industries, thereby expanding global markets. Biomimicry for Creative Innovation (BCI) embraces ecological thinking for radical transformation and is applied by collaborating professionals. This topic has created a set of Business Principles for the Future Firm. Innovation inspired by biology plays an essential part in the management field.

10.2 SUCCESS LESSONS FOR THE SMALL BUSINESS FROM THE ANIMAL KINGDOM

What lessons can your company learn from nature? The nature-learned insights that will make your organization better, stronger, and more efficient. Lessons in company management can be derived from the animal realm (Jennifer, 2013). In today's fast-paced business environment, launching a company is difficult because it demands a substantial amount of capital and expert managerial abilities. Management abilities will transform the business. Some management abilities from the animal realm can contribute to the success of this firm. We can acquire valuable lessons from the animal kingdom that are applicable to management. Animals have their own distinct advantages and disadvantages. The scientific community has pondered the features of the ideal leader. Those that create distinct regulations may confront a variety of challenging choices. The answer can be found by examining nature. The leadership and leadership lessons outlined by Warren Bennis (2009) are as follows:

- Leadership is the capacity to inspire people with a compelling vision and to pursue a common objective with fervor.
- Leadership is a function of understanding oneself, having a well-communicated vision, fostering trust among peers, and taking effective action to reach one's leadership potential.
- Leaders are developed, not born.
- Curiosity and willingness to take risks form a leader.
- A leader does what is right.

Bennis (2009) created innovative leadership theories that dismissed "command-and-control" as a killer of creativity and emphasized passion, honesty, curiosity, and seeing beyond quarterly figures as success drivers.

Biomimicry in the Workplace: The findings of this study reveal that while nature's principles may appear daunting to human businesses, they give a much-needed comprehensive foundation for developing a wide range of sustainable business strategies.

In order for businesses to "switch back on," fresh strategies are required. To satisfy this demand for innovative thought, one must comprehend an untapped and underutilized source of inspiration: nature. Compared to the hundreds of years of industrial innovation that we as a contemporary economy can claim, nature has conducted research and development for 3.8 billion years. As Janine Benyus notes, "failures are fossils." There is plenty to uncover if we search for it.

There is inspiration for businesses everywhere in nature (Farnsworth, 2020; Mashingaidze, 2014). There are numerous species in redwood forests, coral reefs, slow-growth forests, beehives, and plains. The goal for companies will be to draw inspiration from natural systems rather than single species. Working within the limits of a system is the new chapter that businesses must comprehend when developing growth plans for the new economy. The time has come for systems thinking.

Biomimicry is a potent driver of environmentally sustainable innovation. However, the reality is that many new ideas, whether biomimetic or not, fail to achieve market success. Some failures are noted because they will not be defined within the context of a clear, viable market need, while others are observed owing to a lack of resources required to reach the market.

10.2.1 BIOMIMICRY FOR CREATIVE INNOVATION (BCI)

Occasionally, innovators create their own ecosystems to promote their services. Biomimetic solutions are frequently disruptive innovations that can be seen as a threat to established competitors' interests. For these reasons, market entrants must build ways of interacting with industry participants and entry points into a market ecosystem that is mutually beneficial. Everything starts with a systems perspective.

10.3 BUSINESS LESSONS FROM ANIMAL KINGDOM

Survival is one of the most fundamental and effective management principles. If the very survival of an organization is in jeopardy, nothing else counts. Interestingly, management strategies are not the exclusive province of humans. In actuality, the animal kingdom is filled with instances of management styles and strategies that have enabled creatures to exist for thousands of years without human assistance (Srinivasan, 2015). Biomimicry for Creative Innovation (BCI), a consortium of experts applying ecological thinking to radical transformation, has formulated a set of Business Principles for The Future Corporation (https://thenatureofbusiness.org/2013/06/19/biomimicry-for-business/). "Biomimicry" or "biomimetics" is the concept of using nature to solve human issues (https://www.livescience.com/28873-cool-technologies-inspired-by-nature.html). All that is required is close observation of nature. Hence the adage, "turn to nature for solutions"—the concept of biomimicry. Biomimicry imitates nature in order to solve human problems and is applied to businesses ranging from energy to consumer goods. Imagine a day when bees supply recommendations for energy-efficient buildings and self-driving cars, sharks provide the promise of fewer hospital infections, and geckos provide suggestions for hanging large televisions.

According to Janine Benyus, co-founder of Biomimicry 3.8, "individuals invent by studying the natural environment" (https://sustainability-innovation.asu.edu/person/janine-benyus/). Janine and her colleagues merged the for-profit and non-profit branches of Biomimicry 3.8 into a single B-Corp in 2010. Nurturing and expanding this network is Biomimicry 3.8's true mission. The expanding topic of Biomimicry, in which scientists mimic nature to solve human issues, has attracted interest from a variety of businesses, including the energy and consumer products sectors. Geckos provide suggestions for mounting flat-screen televisions. Using the Biomimicry principle, Colgate-Palmolive, Levis, Nike, and Boeing reformulated their products (Low, 2010).

10.3.1 GECKOS

Scientists at Amherst College analyzed how geckos adhere to surfaces in order to design a fabric that can be reused for suspending large loads. Unique arrangement of tendons and skin is the gravity-defying secret. Geckos' throat pouches, also known as a dewlap, serve a variety of daily duties, but a swollen pouch may indicate a problem (https://animals.mom.com/why-do-geckos-have-a-neck-pouch-12305031.html). On a level surface, a piece the size of an index card could support anything from glass to a 42-inch TV screen or up to 700 pounds.

10.3.2 BEES

Bees are remarkable insects. They can teach us numerous leadership lessons. Bees have long been a simple source of inspiration. Thousands of honeybees function as a single organism (https://www.fastcompany.com/3031018/6-lessons-that-businesses-can-learn-from-bees). In their natural state, bees work concurrently. All of the bees' choices are governed by their natural understanding of design efficiency. They maintain focus on the objective. Each bee in a hive has a purpose and a job; they are aware of their function, carry it out, and communicate well. Equally, corporate leaders succeed when both communication and employee liberty are effective. Communication and sharing of information are fundamental to the team. The leader is, in fact, the servant of everyone. Future preparation is required. What occurs if the nectar source reduces or disappears? Never become complacent because of how rapidly things change. For the future success of the hive, we must be constantly adapting and innovating (company). Now, Regen Energy, located in Toronto, has employed "swarm logic" to produce software that enables heating and cooling units in big box stores to wirelessly collaborate to determine when and how often teams run to optimum energy efficiency. Regen Energy manages the energy at Target (TGT, Fortune 500) and Sears (SHLD, Fortune 500) stores.

10.3.3 LOTUS LEAF

The German Stucco and Painting Business (Sto AG) developed a paint inspired by the lotus leaf's ability to dry rapidly and remain clean during rainfall. Sto replicated the plant's microstructure, which resembles Lotusan paint, which is meant to remain cleaner and last twice as long as conventional paint.

10.3.4 MUSSELS

Researchers from Northwestern University examined the adhesion capacity of mussels and discovered that they had unique protein-based adhesives that are effective in moist settings. They then replicated this composition to develop a polymer surgical adhesive that, one day, may be used to mend embryonic membranes and so prevent premature labor.

10.3.5 TERMITES

African architect Michael Pearce was influenced by termite mounds. The architect aims to base the design of the Zimbabwe Eastgate shopping center on termite mounds. The Eastgate shopping area replicated the banks' natural thermal architecture by utilizing ducts and 48 enormous chimneys to exhaust the daytime heat. The building's unique design utilizes 60% less energy than conventional structures.

Nissan is investigating how schooling fish and bumblebees can cooperate together to reverse direction, going side by side at the same speed and intelligently navigating around obstacles. The objective is to imitate this mechanism in self-driving automobiles to prevent traffic congestion and collisions.

10.3.6 SHARKS

A scholar from the University of Florida revealed that the microscopic diamond-shaped patterns on a shark's skin protect the animal from marine microorganisms. Now, Sharklet Technologies in Aurora, Colorado, replicates these mathematical patterns to provide antibacterial surface textures for medical gadgets and hospital counters. The textures can reduce the size of microorganisms by between 90 and 99.9%.

10.3.7 FIREFLIES

Why do you use superior light bulbs in your home? Fireflies. Belgian, French, and Canadian scientists analyzed and replicated the firefly's jagged pattern on its reflector panels in order to develop a new coating for brighter LED lights. The outcome is a 55% increase in efficiency relative to conventional LED lighting. Researchers anticipate that the breakthrough will result in improved LEDs for use in cameras, televisions, automobiles, and household illumination.

From survival to self-discovery, nature has much to offer mankind. Nonetheless, few individuals are aware of the marketing teachings. Nature is continually instructive. In light of this, here are lessons from nature that every professional should know.

Evolve or lose to the competition.

- As Albert Einstein famously stated, "Just delve deeply into mother nature, and you will have a greater understanding of things." And he was correct. Nature may teach us a variety of lessons, including overcoming adversity and adjusting to multiple changes, among others. The human ego has a tendency to view natural phenomena as either beneficial or harmful in general. Observing nature in its whole teaches us that everything in the universe serves a purpose and that we should appreciate what each creature accomplishes, even if it does not contribute to our survival. Nature offers us many lessons, including survival, self-discovery, and generosity (https://www.feelgoodanyway.com/nature-inspired-marketing-lessons/, https://www.pickthebrain.com/4-business-lessons-from-mother-nature-to-accelerate-your-growth/). But have you ever considered the marketing lessons that nature has continuously offered us all?
- Consideration of nature-inspired marketing teachings for professionals.

- Adapt or fall behind the competitors.
- Capture the attention of the entire globe.
- Standalone.
- Never place all of your eggs in one basket.
- Adapt.
- Creativity cannot be compromised.
- Take your time.

Nature has always been a consistent educator who imparts new teachings to humans. The more you search for answers, the more likely you are to get them, and the responses provided can facilitate the design of marketing plans for your organization. Businesses may learn a great deal from the interdependence of various elements of the natural world.

10.4 IMPORTANT NATURE-INSPIRED MARKETING LESSONS

10.4.1 EVOLVE OR DIE

Let's be sincere, deadweight has no place in nature. If you cannot perform or evolve, you will inevitably fall. However, this "evolve or perish" dictum has driven all living things to become the best. Similarly, we must ensure that every component of our business is put to a rigorous test and is not allowed to become routine. One of the most important marketing lessons from nature is that dead weight has no place in nature.

A caged animal cannot survive in the wild for long. Instead of focusing on creating the ideal model, create a minimal viable model and test it in your market. You will encounter failures and unforeseen circumstances, but you must learn from them and adjust your business accordingly. Remember that you can only outperform your competitors if you adapt to market demands (https://www.pickthebrain.com/4-business-lessons-from-mother-nature-to-accelerate-your-growth/).

10.4.2 ONLY ROUTE IS PERFORMANCE

Fitness is the only criterion considered by mother nature. It is irrelevant if the animal is a lion or a deer. We must take a very important business lesson from the fact that there is no preferential treatment. At whatever level of your firm, you must favor and promote individuals on the basis of their performance, not their gender, caste, or relationship. Numerous exemplary businesses have suffered tremendously as a result of a family member taking on all the burden.

10.4.3 LOW WASTAGE

Humanity is the creator of waste. If you ever observe nature closely, you will see that everything has a specific purpose; for instance, birds and rats use fallen, dry tree leaves to create their nests. Likewise, you should work diligently to reduce waste in your workplace. Reducing waste can improve your company's standards and save you a lot of money. Utilizing the blank side of printed pages for rough drafts might conserve paper. In addition, you can count the number of superfluous workplace lights and turn them off. With these easy measures, you may reduce your waste without the need for a grand plan.

10.4.4 NOTHING IS MORE CRUCIAL THAN YOUR OWN SURVIVAL

Every animal views its own survival as the most important aspect of the world. Even if a predator is waiting, a deer will leave no stone unturned to protect its life. It imparts a valuable lesson. You must exert effort and make every effort to expand your firm. However, if someone attempts to suffocate or destroy your firm, you should do everything in your power to fight back and survive. Mother nature is the finest educator, and she has answers to every difficulty. You must have faith in her, observe her, and learn from her. John Muir was correct when he stated, "In every stroll through nature, one receives considerably more than he seeks."

We cannot throw tantrums every time our jobs or objectives shift from one hour to the next. We must learn to be adaptable and flexible in the face of ever-changing environments. You are one with the land, the water, and the air; nature does not belong to you; you are nature.

10.4.5 BIOMIMICRY: OBSERVABLE LEADERSHIP LESSONS IN NATURE

Nature is an extraordinary resource for exploring learning and leadership development. By examining nature in greater detail – how plants and animals adapt to their environments and the components of their bodies that enable them to flourish – we can be motivated to build and organize our surroundings to imitate what makes them successful.

Except for infrequent excursions to botanic gardens or summertime walks and camping, our opportunities to explore the natural world can be relatively limited. Yet with reruns of Earth Planet and even Netflix's Our Planet, people are able to discover and comprehend Nature's splendor for the first time, watching animals and plants in their original settings.

It is the subject of a field of study known as biomimicry. Biomimicry is the use of natural phenomena to human-centered goods, experiences, and design (https://www.leadershipdevelopmentlab.com/blog/2019/7/17/biomimicry-leadership-lessons-we-can-observe-in-nature). Rock climbing equipment has been influenced by the little hairs on gecko's feet that allow them to scale vertical surfaces. Consider Velcro; the tiny hooks found in bur fruits were the inspiration for this innovation. The study of birds was crucial in the development of the first airplanes. By examining nature in greater detail – how plants and animals, as well as the aspects of their bodies, enable them to survive – we might be inspired to build and organize our surroundings in a manner that mimics what makes them successful (http://smartblogs.com/leadership/2012/07/11/leadership-lessons-learned-nature/).

Wolves can serve as an outstanding example of leadership and teamwork. When considering a pack of wolves, we reverted to the concept of the wolfpack. It is the magic of many working together that enables achievement. Members of a wolf pack are often inherited. There are both young and old individuals in the same group. And, like many teams, wolves have a designated leader who encourages the others to continue and sets the pack's direction.

10.4.5.1 LESSONS IN LEADERSHIP OBTAINED FROM NATURE

Where, then is the pack's leader? It is logical to suppose that the leader of the wolfpack is in the front. Ultimately, they are setting the route and taking the initiative. Within teams, the leader takes command and communicates the vision from the front. What happens if a leader remains despite the fact that they no longer have many followers?

It may surprise you to hear that the wolfpack's oldest member often leads the pack. These people may have been leaders in the past, but they are now the elders. They are frequently slower than the rest and, as a result, set the pace. As quickly as they can move, the remainder of the group moves. Because of this, wolves travel in packs and move at the pace of the slowest individual so that they can stay together.

As you recall your team and leadership responsibilities, where do you stand? You either lead from the front or the back. How can we empower the newest and most experienced members of our team to set the pace while reserving our leadership roles for solely guiding, protecting, and advising? If you have the opportunity to explore nature this summer, take a closer look – what can you observe in our world that can inspire your leadership style? Hummingbirds construct their nests from lichen and spider webs, incubate their eggs, feed their extremely little young, and teach them to fly.

I've never seen something so small and yet so ferocious as that hummingbird mother. Despite her evident size disadvantage, if another bird entered her field of vision as she sat on the nest, she would immediately attack it. This bird can successfully ensure the safety of her family. Likewise, leaders must be relentless in their pursuit of the team's industries while safeguarding them from those who could disrupt their common objectives. Patient and determined: You're familiar with the phrase "pushing them out of the nest," Not applicable to the mother hummer. When it was time for the baby birds to fly away, they were encouraged patiently but repeatedly. She remained patiently by their side until they were ready to leap when it was awkward for them to climb out of the nest and onto the tree's limbs. As a leader, it is your responsibility to patiently encourage and coach the business team and to remain by their side when they are ready to fly solo.

10.5 BUSINESS CONCEPTS DERIVED FROM NATURE: THE RELATIONSHIP BETWEEN BIOMIMICRY AND MANAGEMENT THEORIES

As a novel method of creating technology, biomimicry offers new insights and breakthroughs for business strategy. Since the 1990s, biomimicry has been utilized in various sectors, including architecture, product design, engineering, and manufacturing processes. Despite its extensive history, biomimicry and management concepts have only been examined in a handful of research. Celep et al. (2017) noted that biomimicry and managerial principles work hand in hand and give biomimicry a managerial perspective. Regarding innovation, biomimicry is an effective catalyst for achieving and sustaining innovative processes. Biomimicry, on the other hand, is a relevant notion as an open system of enterprises and a crucial strategy formulation tool. In terms of being organic, lean, and self-replicating, organization structure is correlated with biomimicry.

The Future of Business—adheres to the effective principles, tactics, and practices discovered in Nature. Nature has had over 3 billion years to figure out what works and what does not; evolution eliminates ineffective processes. In the past, enterprises were operated as machines. Our developed components are designed to support the manufacturing process and the production component. We are currently changing from growth based on consumption to growth based on knowledge and from value pulling to value sharing (https://thenatureofbusiness.org/tag/business-inspired-by-nature/, https://thenatureofbusiness.org/).

Nevertheless, the rate of change is fairly rapid. For success, we must be nimble, inventive, vigilant, spontaneous, and receptive, acting in wholly novel ways. We want our organizations to be living learning systems, not regimented and segregated entities that inhibit learning. We expect our organizations to be decentralized, bottom-up, and interdependent. It should also consist of multifunctional, self-organizing components as opposed to centralized, hierarchical, top-down monoliths.

Therefore, today's quickly changing business climate necessitates businesses that survive in such situations. A firm inspired by nature is more comparable to dynamic, adaptable living entities and should grow within ever-changing business and socioeconomic systems. It is resilient and able to withstand unforeseen interruptions. A business inspired by nature functions similarly to nature, where waste is food, form follows function, consumption increases value and productivity, and self-interest improves the system.

It endeavors to establish symbiotic ties with the biosphere. Historically, organizations have viewed themselves as distinct from nature. Nature was a location to obtain and guard against resources, a place to observe from a distance, and a place to dispose of garbage. This transition is accompanied by a shift in consciousness from taking from nature to learning from nature to being a part of nature.

10.5.1 WHAT LESSONS CAN BE LEARNED FROM NATURE?

Nature will teach us that fearlessness is a trait of all successful leaders. They are accountable for confronting difficulties head-on and accomplishing results. Great leaders instruct (Hyacinth, 2015). Great leaders aid in the growth and development of their teams. Nature can offer us unique opportunities to acquire leadership lessons. It can contribute to an organization's health. Nature teaches us to give without discrimination or expectation of recompense. When someone we have helped forgets our kindness, we feel remorse and regret for our assistance. It causes us pain and sorrow. We learn about the Circle of Life from nature. Nature teaches us about origins and beginnings, development and growth, death and endings, and rebirth and rejuvenation. Being in nature, or even just witnessing natural scenes, reduces anger, fear, and stress while increasing good emotions (https://answeregy.com/what/what-lessons-can-you-learn-from-nature.php). Exposure to nature improves not only your mood but also your physical health by reducing blood pressure, heart rate, muscular tension, and the production of stress-related hormones.

10.5.2 LESSONS FROM THE ANIMAL KINGDOM ENTREPRENEURS, EXECUTIVES, AND INVESTORS

There are many beautiful voices, but not all of them are human. Even if animals are incapable of speech, they can nonetheless teach us vital life lessons (https://www.goodnet.org/articles/5-life-lessons-we-learn-from-animals). These programs are designed for business owners, executives, professionals, and students. Feed the chicken to the eagles. This is how the world operates. Those who perform in sales, marketing, or startups (and are also great team players) are rewarded in the career success race, while others are removed. In addition to capability, other aspects, such as diplomacy, interpersonal skills, and even a degree of luck, exist (Dumbre, 2021).

10.5.2.1 BE LIKE ANTS AT WORK

Our social structure has evolved into a more collaborative superorganism, where none of us can honestly thrive on our own, as ants can. Examining ants closely reveals that each performs a unique role and is dependent on the others for survival. Human communities possess similar undertones. One did not create all the clothes he wore, all the food he ate, or his phone. People are intrinsically interconnected and dependent on one another for survival. This is a significant departure from our nearest relatives.

People appreciate the following Jim Ron quotation: I believe that everyone should study ants. They possess a four-part concept that is superior. Never give up, look forward, keep an optimistic attitude, and do everything you can to have insight, perseverance, and tenacity. Nature constantly favors perceptive, persistent, and industrious individuals.

10.5.2.2 BE CAUTIOUS OF LONE WOLVES

Occasionally, or more frequently, you will encounter someone who desires growth at the expense of others. Respect them for their determination, attention, and shrewdness, but be sure they are not harming us.

A blind squirrel may still locate a nut. It is comparable to saying that a broken clock is accurate twice a day. Everyone can occasionally experience good fortune even more so if one had a strategy and plan.

It is OK to ride a cow until a horse can be found. While everyone must strive for the greatest in life, achieving the best requires time, patience, planning, intelligence, difficult effort, some luck, and much more. In the interim, it is acceptable to pursue alternatives that may not be ideal. Depending on the availability of reputable venture capitalists as investors, startup founders also collaborate with angel investors. Before you may aspire to become CEO, you must hold a variety of other positions.

A person can be a bull or a bear, but you do not want to be a baker (scapegoat). Mani, one of India's top intellectual property specialists, used to joke, Mahesh, that one need not be a bull or a bear (as in the stock market), but one must avoid becoming a "bakra" (a poor goat that is sacrificed). The globe is a highly competitive place. Given the restricted growth chances for a large talent pool, individuals will occasionally resort to humorous strategies to advance. It is solely your responsibility to be aware of tricks and traps established by others and to avoid falling victim. Sharks are present in the water. Animals are unable to communicate verbally, yet they may teach humans invaluable lessons in other ways.

Dumbre (2021) provides the following animal kingdom lessons for entrepreneurs, executives, and investors (https://starthub.london.edu/10-lessons-from-the-animal-kingdom-for-entrepreneurs-executives-and-investors/).

A lion is never disturbed by the opinions of sheep. There is no need to be concerned about people who are small-minded and insignificant. There are no experts on butterflies among caterpillars. If you are attempting to achieve something novel or unique, it is possible that others will not understand or appreciate it. Entrepreneurs: The sweet, obedient, and adorable dogs have an ulterior motive. They are fed and cared for by humans, who are unaffected by any other species. So, what then? They attack criminals and wag their tails to ensure that we continue to love them.

10.5.2.3 BIG LESSON LEARNED

A prominent businessperson may offer assistance out of friendship or as a favor. Learn how to obtain the greatest assistance here while maintaining your credibility and meeting expectations. Additionally, he is never dependent on them. Nobody can expect something to always be free (Small-scale industries and e-commerce merchants).

The cuckoo lays its eggs in a crow's nest, as Cuckoo Bird teaches. As long as the corvid is oblivious, it will be all right. There is no way to prevent it if, one day, all the crows decide to roll the cuckoo's eggs down the tree. There is no other method for the cuckoo to continue its species!

Important Lesson Learned: Such a parasite start-up may be effective and fraught with danger. Avoid it unless there is no other option (Entrepreneurs that create things from scraps fall into this group! A hazardous start-up!).

Work hard, like ants. For success in whatever sphere of life, one must work hard. There are many animals linked with hard work, but the ant stands out. There are 12,000 known species of ants. These diverse species utilize essentially the same castle system. This structure is comprised of queens, soldier ants, worker ants, and fertile males. Each class has a designated function inside the colony. The queen lays eggs, and soldier and worker ants execute various activities, including caring for the larvae, cleaning the eggs, and seeking food.

Moreover, according to a study conducted by the University of Regensburg, we share a striking similarity with ants in that we both appreciate rewards acquired through effort. The research was straightforward. Ants were positioned in front of two routes in order to obtain a particular reward. There was one method that required only horizontal travel and another that was much more difficult. Regardless of the incentive at the end of the path, the results demonstrated that ants always took the difficult option. This displays a willingness to work hard, as well as the fact that hard-earned benefits feel and, in this case, taste better.

Lead, like lions. Since childhood, we have been informed and instructed that the lion is the ruler of the forest. Consider well-known films such as The Lion King, children's literature, and National Geographic documentaries. We can learn a great deal about leadership from these creatures by observing their behavior individually or collectively (DeNio, 2020). For instance, the characteristics that are commonly associated with lions, such as courage, strength, and decisiveness, are also the characteristics that will set you apart and make you a great leader. Moreover, lions are gregarious creatures who live in groups of 10 to 15 individuals. These groupings are comprised of lionesses, whose function is to hunt, cubs, and lions whose purpose is to defend the pride.

Occasionally, silence speaks better than words. According to Kashyap (2017), Nature is silent. To prove his claim, one must hear a story.

Through the forest traveled a bunch of frogs. Two of them plunged into an inexhaustible pit. When the other frogs saw how deep the hole was, they informed the two that they were effectively doomed. The two frogs disregarded the comments and attempted to climb out of the hole with all their strength. The other frogs repeatedly warned them to stop, stating that they were already dead. Eventually, one of the frogs paid attention to what the other frogs were saying and gave up. The frog plunged to its death. The second frog continued to leap as vigorously as possible. Once more, the crowd of frogs cried at him to cease his suffering. He jumped even harder and eventually escaped. When he emerged, his fellow frogs asked, "Did you not hear us?" The amphibian explained that he was deaf. He believed they were always encouraging him (https://parenting.flinto.in/child-development/moral-stories-for-children). Reason that frog was deaf and never heard a word; all he could perceive was others urging him to work harder. Keep up the good work. As a great leader, you may gain more from your team if you are optimistic rather than pessimistic.

Instead of telling your team what to do, set an example for them to follow. Speak in silence. The mouth holds the power of life and death. A word of encouragement to someone who is feeling depressed can help them get through the day. Therefore, watch what you say. Communicate the significance of life to those who cross their path. Occasionally, the power of words is difficult to comprehend. It can be a word of encouragement that goes so far.

10.5.3 EAGLES WILL TEACH POSITIVE LEADERSHIP QUALITIES

Eagles, seemingly larger-than-life birds, have intrigued and inspired us for millennia. We learn numerous things, including their exceptional leadership qualities. When we think of eagles, we often envision a massive hunter soaring above vast open regions with enormous wings. Indeed, eagles are among the world's known raptor species. They are revered as living emblems of strength, liberty, and greatness. In certain religions, these beings are believed to communicate directly with God. According to legend, the Aztecs valued birds so much that they constructed their city, Tenochtitlan, on a cactus where an eagle nested. Eagles have been a symbol of beauty, bravery, courage, honor, pride, resolve, and grace for millennia. Eagles are significant to humanity because of the features of this bird. Here are five key characteristics of the eagle (Janus, 2020).

10.5.3.1 EVERY INTELLIGENT LEADER CULTIVATES EAGLE-LIKE QUALITIES

Eagles have the sharpest vision of all birds, and their eyes are designed for long-distance clarity and focus (http://easeandglam.blogspot.com/2011/10/monday-motivation-fly-like-eagle.html). These Birds can detect another eagle flying at a distance of 50 miles. Abraham Lincoln, the 16th president of the United States, led his nation through the most catastrophic event in its history, the American Civil War. He wishes to save the union and be liberated from slavery. Numerous historians regarded him as a distinguished American president. You must have a perspective that guides and directs your team toward organizational or social objectives. The image must be large and sharp. A large, thoughtful notion will generate exhaustive results.

Throughout history, people have revered and studied eagles for their leadership qualities and abilities. Some countries have sports teams with eagles as their names or mascots, and they consider eagles to be a symbol of their country. This longing is engendered by the eagles' distinct personality traits. Eagles are an excellent example of leadership and success because of their vision, boldness, and love for their eaglets (Figure 10.2). Have a clear picture of where you want to go – the image of the eagle's leadership lesson is stunning. Because of its exceptional clarity, this camera can capture objects up to five kilometers away in the air. It is only when an eagle spots prey that it narrows its focus and continues searching until the game is over. Have a goal in mind and stick to it no matter what comes your way. A company's top executive should Identifying and disseminate a clear vision is one of the primary responsibilities of a business manager.

It's the ability to turn a dream into a reality that defines leadership. As cited by Warren Bennis: the Eagles have a fearless nature. No matter how large or strong its prey may be, an eagle will always put up a battle to keep its territory. Try to see the Golden Eagle's extraordinary hunting tactic, in which it throws goats considerably larger in size than itself from the cliff face in order to prey on them. Fearlessness and a crystal-clear vision are hallmarks of great leaders. When it comes to developing his team, he needs to be bold and unafraid to take some chances. Without regard for your own safety, you would shoot that individual regardless of their size or the type of weapon they were wielding. You wouldn't even consider being terrified since your nature is to safeguard everything you value.

It is estimated that the eagle's eye is four to eight times more powerful than the ordinary human. Even from 3.2 kilometers away, an eagle's keen eyes can pick out a rabbit. The eyes of an eagle are about the same size as those of a human being. The muscles in an eagle's eyes constantly change the curvature of its eyeballs to maintain keen focus as it descends to attack its prey. By weight, their eyes are far larger than their brains. Color vision, great resolution, and transparency are some of the most notable characteristics of eagles' eyes.

FIGURE 10.2 The Bald **Eagle**.

Image courtesy: Carole Robertson. https://creativecommons.org/licenses/by-sa/3.0

The eagle has much to teach us about life (https://www.expords.com/amazing-life-lessons-from-the-eagle/).

- **Lesson 1:** Eagles are rarely seen in groups of smaller birds because they prefer to fly alone at high altitudes.

  Get comfortable being by yourself so you can surround yourself with like-minded individuals if you want to achieve greatness.

- **Lesson 2:** The eagle's vision is so sharp that it can see through any obstruction to its prey until the sun goes down. Keep your attention and avoid becoming sidetracked from your goals if you want to achieve them.
- **Lesson 3:** Unlike vultures, eagles may feed on both live and dead prey, as demonstrated in Lesson 4.
- **Lesson 4:** Get excited about the storms in your life. Challenges bring opportunities. We could see the opportunities, wrapped in our challenges if we weren't so busy, complaining and trying to avoid them. Embrace our challenges and learn from them. That's how we learn new skills, grow and get stronger. They teach us the importance of individuality and need to the distance of ourselves from negative influences. There's always an opportunity lurking behind every challenge; just be bold enough to go through it. Embrace our challenges and learn from them. That's how we learn new skills, grow and get stronger.
- **Lesson 5:** Before mating, the female eagle conducts a thorough examination of the male. Before entrusting someone with your time and resources, verify their sincerity and devotion.
- **Lesson 6:** Young eagles are taught to fly by being lowered from mountain peaks. Don't give up hope if you fall from a tremendous height. Until you regain your ability to fly, practice makes perfect.
- **Lesson 7:** Old and tired eagles retire in solitude at faraway places to rejuvenate and robot. When the stress of life drags you down, spend alone time and give your mind and body rest and space to rejuvenate.

There are many lessons to be learned from leadership (https://neilpatel.com/blog/10-lessons.html). Developing one's leadership skills is a lifetime endeavor. There is no end to it. There is always room for improvement in the leadership of a great person. In order to be a leader, you must have both strategy and character. Integrity is the defining characteristic of a great leader. Entrepreneurs may find it particularly difficult to hold positions of leadership (i.e., products, investors, clients, and so on.).

1. **Work as a Team:** The most effective leaders, it seemed to me, never use the word "I." Moreover, this is not due to the fact that they have learned to refrain from using the word "I." They're not thinking 'I.' They think 'we' and 'team' are synonymous. They understand that their role is to help the team work together. 'We' get the credit, but they should take responsibility and not avoid it. In order to succeed, you need to be able to rely on each other.

2. **Help Others Grow:** Good bosses like watching their employees flourish. It's the day you fail as a leader when you're worried that someone else will be better than you.
3. **Instill a Sense of Self-Worth in Employees:** They go above and beyond to help others feel good about themselves. When people believe in themselves, they may accomplish remarkable things.
4. **"Really with Technology":** "It's about the people, acquiring the finest people, maintaining them, cultivating a creative environment, and helping to find a way to develop," he says.
5. **Coordination is Essential in Order to Achieve a Goal:** Everyone on the team must work toward a common goal, according to an expert in this industry.
6. **Allow Yourself the Freedom to be Misunderstood:** You must be willing to put yourself out there for long periods of time if you want to invent and pioneer.
7. **Achieve a Large Following:** Leadership is earned through "followership," a term coined by Anne Mulcahy.

   "In my opinion, we tend to forget that we're not actually in charge. It's up to us to get it. It's imperative that we have people who can rely on and follow us. That, in my opinion, is what separates great leadership from merely adequate leadership."

   The ability to inspire others to work for you even when they are not required is a hallmark of leadership.
8. **Recognize the Truth and Offer Encouragement:** The leader's responsibility is to define reality and provide hope…

As indicated by the expert, any statements made in this lesson are mine alone and not those of the original author(s). The thought of becoming a pariah like Napoleon disgusts him. For someone in a position of authority, it might be challenging to be real and to express that authenticity. If you're dealing with business difficulties or other people with differing viewpoints, you have a lot of opposing forces working against you.

If you want to be a good leader, you must be a good listener.

Listening is critical to becoming a good leader. In the most improbable of settings, you should keep your ears open for some smart advice. It's up to the leaders to make sure the goals are met. Great leaders pass on their knowledge to their followers. Great leaders help their team progress and expand. We may learn about leadership in a new way by paying attention to nature.

It has the potential to improve the health of a company. True bosses don't exist in the actual world. They change with the people they serve.

People in the company or society grow to their fullest potential because of their perseverance. It's as if they're imitating the mother eagle in every way. They're never done presenting obstacles, but they're also never done motivating and directing. When a team member is ready to fly on their own, a good leader is there to guide and encourage them. To help people evolve into full-fledged leaders, they also create an environment that empowers and enables them to take on more responsibility. If you can stand the pressure and wait, you'll shine. One must be reminded of the priceless diamond whenever the word "pressure" is used. Is it possible that something so gleaming, translucent, and delectable could be made from nothing but trash? For several years, the leaves and other forest waste have been subjected to extremely high pressures.

Waste can be repurposed and turned into priceless treasures. The same is true for us. Be brave enough to tackle anything life throws at you. Sooner rather than later, you, too, will be as brilliant as a diamond!

Be on time and do your job to the letter. Imagine a day without the sun; it's impossible. Will you no longer refer to it as a 'day'? Is the sun too lazy to rise early in the morning and rest for a while? We will not be able to survive if the sun doesn't rise at the correct time. Without sunshine, photosynthesis will fail, and the plants will perish. Flowers won't grow, animals will be harmed, water will not evaporate, and there won't be rain if there is no sun. The Earth will go into a state of slumber. No more life on this planet for us to enjoy. We've learned the most important lesson from nature as a result of this. You must be on time and execute your job. Do your part to the best of your ability if you expect others to be perfect.

Many great leaders must have imparted their wisdom to you. It's time to find out how some of the world's greatest leaders found their inspiration. Nature has so much to teach us. To do this, you must learn to decipher the language of nature. Nature's actions have a deeper meaning to us than we realize. Each and every one of history's greatest leaders is a lifelong student. Exceptional leaders are adept at picking up new concepts by studying examples. Seeing the moon prompted a guy to invent the wheel. It was a huge leap from nil to one. Nature, on the other hand, has never been more important than it is today as a source of inspiration for modern man.

10.5.3.2 NATURE HAS A FEW WORDS TO SAY ABOUT LEADERSHIP

Each and every one of history's greatest leaders is a lifelong student. A short lesson in leadership from nature (https://scottamabry.medium.com/

e79544264f01) by Scot Mabry (2016) demonstrates that great leaders are adept at picking up new skills by observing the examples set before them.

- The season shifts to winter as the weather gets colder.
- The trees aren't bothered by the fact that their leaves have fallen off.
- The winter season marks the beginning of a new chapter.
- The birds seem to like the journey.

Give a song as a gift.

"*In nature, we may learn a lot about how to be a leader.*
When things are going well, it's only a matter of time before things get tough.
Relationships flourish and fail.
Nature has the ability to change and adapt.
Practical adaptability is at the heart of effective leadership.
Awareness, reactivity, readiness, and self-assurance are all necessary for successful adaptation.
In order to be aware, you must pay attention to what you are doing."

10.6 BUSINESS FUNDAMENTALS LEARNED FROM TREES

Without trees, there would be no life on Earth. As one of the most intellectual species on Earth, we owe our existence in large part to them. They are not only essential to our survival, but they also teach us valuable lessons like how to deal with difficulties while remaining calm, patient, and flexible; how to face challenges with courage; how to steal happiness in every season; how to endure pain and loneliness; and how to live life happily no matter what comes your way.

Acceptance and non-resistance are taught as the seasons change. When fall arrives, a green leaf can't help but become red. When winter comes, trees can't resist losing their leaves. Their vulnerability is exposed, and their branches extend wide, submitting to the Universe (https://workingabetterlife.com/lessons-from-trees-about-the-business-principles-that-every-entrepreneur-needs/), and they stand firmly in the earth.

Intense roots: The tree's roots were well-developed and extensive. The tree was healthy as long as the seeds were healthy. If an organization wants to remain stable, its leadership must be based on a strong set of principles and ethics. In order to keep things simple, keep it to the point: A tree can grow and live as long as it has access to the resources it requires (such as water, sunlight, food, and space). They don't appear to necessitate a lot of time and effort in the planning stage. It's easy to get things done if you offer them what they need.

10.6.1 HAVE A SOLID FOUNDATION

Branches and leaves are supported by the tree's trunk, which serves as a conduit for nutrients to flow from the crown down into its stems and leaves. It is imperative that an organization's top executives serve as a conduit for life. If you want your team to succeed, you need to make it easy for them to share knowledge. Leaders and followers must work together to ensure the long-term success of the company.

Trees have both strength and flexibility in their living structure, which is why it's important to be flexible as well. Rather than resisting the wind, trees are able to withstand storms without snapping. There is no reason to be apprehensive about the turbulences of conflict or change that leaders may face. Both the leader and the organization benefit from having a better energy position.

People, bugs, and other creatures all coexist with them as they try to adapt and survive the harsh conditions of their environment. The best companies do this as well (well, maybe not bugs and animals, but you get the point). Everyone has a role to play in the success of a good leader. As a leader, you must possess vision, courage, honesty, and empathy.

When a tree works as a unit, it does its task admirably. The health and growth of a tree are dependent on the efforts of all its parts. All kinds of good things happen when everyone in an organization is working on what they are good at and executing their job to the best of their abilities. Groups of individuals working together to achieve a common goal are what produce successful businesses. Being a lone leader and trying to "shine" can be perilous, and it typically ends in failure. Involving others in your ideas is essential; leaving them out is not. In numbers, there is power.

Encourage others to rest in your shade by being welcoming and non-judgmental. In contrast to humans, trees do not weep or fret. Similarly, a good leader doesn't dwell on the shortcomings of his or her employees. The company is open to suggestions and new ideas.

Trees provide good health by absorbing CO_2 from the air and exchanging it for oxygen. A company's top executives are supposed to help, not to take and hoard resources. Keep your feet firmly on the ground, but keep your head held high to see the bigger picture. It's up to the leaders to make sure the goals are met. Great leaders pass on their knowledge to their followers.

Exceptional leaders aid in the growth and development of their staff. We may learn about leadership in a new way by paying attention to nature.

Please pay attention: Anyone who has learnt to monitor and listen to animals for what they need, such as food, and a potty break/playtime, is an excellent social worker.

A chance to develop empathy and the ability to pay attention to other people can be provided by having a pet. Children's empathy and social abilities improve when they have a close relationship with a pet, according to research.

Negative attitudes toward animals, such as brutality and low levels of empathy among youngsters who own pets, are all linked to lower levels of social competence. The hearing talents of animals are significantly superior to those of humans. We tend to neglect the importance of our hearing even though we've evolved to be less reliant on it for day-to-day survival. Animals first observe and then respond, which is a skill that we might all learn from.

Be like elephants: Today's recruiters are looking for applicants who have the correct attitude and personality, as well as the necessary soft skills. They make it a point to blend in with the crowd and do their best job as part of a group. There are few animals more empathetic, calm, united, and determined than the elephant. Since elephants have been around for a long time, we know that they can be trained to act at public exhibitions or carry out specific everyday chores. An additional example of elephant intelligence is their ability to remember and use tools in order to accomplish their goals.

10.7 THE HUMMINGBIRD'S WISDOM

Every day is like the last for a hummingbird because they only have enough energy to last one day. Because of this, they are fantastic professional eaters and also very fast (http://mba-blog.audencia.com/2017/05/02/lessons-from-a-hummingbird/), thus, they constantly eat buds. The message is the same whether you use the carpe diem or yolo idiom: don't be afraid to take chances. You can't achieve greatness by slouching through your job. Missy, it's time to put yourself out there.

Every time you spread happiness to others, the effect is multiplied. A crow is the best example of this. When negative things are shared, their impact is reduced by half.

It doesn't matter how little food they find—they share it with their companions first, and then they consume it. They even use their beaks to carry food for their offspring. They're never greedy or envious, and that's a good thing. They consume and supply even more cheerfully while sharing with all of their other friends' beings. If your company receives assistance, you should split the cost equally.

10.8 THE GIRAFFE IS A MANAGEMENT EXPERT

The birth of a giraffe cub is a major event. A giraffe calf emerges from its mother's womb eight feet in the air (Gazzara, 2022). It collapses to the ground, unable to move. She bends her neck to kiss the young giraffe with tenderness. Unbelievable things happen, and she kicks the young giraffe with her lengthy leg. It sends it hurtling through the air before crashing to the ground. The mother kicks the infant many times as it lies curled up. Tense, the baby giraffe pushes its legs in an effort to keep moving. For the first time, she is able to walk unassisted. The mother giraffe is delighted to see her youngster standing on its own two feet, so she rushes over and gives it one more stomp. Giraffes are known for their quickness and agility. Mama Giraffe is ecstatic about the new addition. He or she realizes that her infant has learned a vital lesson (https://wildtrails.in/mother-giraffe-kicks-their-newborns.html).

- **Lesson:** Don't worry about how far you've fallen. Always remember to get back up and keep moving forward.

In the end, the giraffe prevailed.

Figure 10.3 shows that the best time to grow is when you're in decline, as a giraffe informs us today. When you're about to fall, that's the perfect time to get your feet under you. A clear correlation between human and animal behavior has long been shown. It's the straightforwardness of these lessons that makes them so effective. These creatures serve as a reminder that we can take on even the most challenging situations.

A hungry giraffe is always on the lookout for something new to eat. It's time for giraffes to face up to the fact that they're running out of food. For them, it is only natural and advantageous to relocate to locations with more resources. "The only thing that is constant is change itself," stated Heraclitus. Having just one item to keep us going is not enough; we need a diversity of things. Patience is a strategy that can easily be outperformed by one that is more proactive.

Remember to get back up and keep pushing forward no matter how far you fall. This behavior by the giraffe's mom is baffling. Giraffe meat is a favorite of lions and leopards. Consequently, the baby giraffe has no chance of surviving unless it quickly learns to stand and gallop with the rest of the herd. Giraffe calves, on the other hand, have it far better than the rest of us. We're never taught how to get back up after falling down by anyone. We tend to give up on ourselves after a setback or difficulty.

We forget how to live and achieve because we aren't forced to leave our comfort zones by anyone else. Getting back on our feet is something we need to work on. Successful people have a reoccurring pattern in their lives. Were they always able to achieve their goals? No. How quickly did they see success? Their ability to get back up after falling down is the same thread that runs through all of their lives. The giraffe cub wields incredible influence. Accomplishing Tasks Quickly and Easily: Agility.

FIGURE 10.3 South African Giraffes:

Photo Credit: D. Gordon E. Robertson. https://creativecommons.org/licenses/by-sa/3.0.

In the animal realm, business agility may be learned. An organization's agility can be measured by how quickly it responds to changing circumstances.

You must:

- Be nimble in responding to market shifts–both internally and internationally;
- Be responsive to client needs with speed and flexibility without sacrificing quality;

- Be constantly ahead of the competition.

The leopard is one of the most agile animals on the planet. It has the ability to alter its behavior according to the situation. Stealth, speed, and agility all play a role in the pursuits of this vehicle. Prey can be tracked on the ground and climbed into trees by this predator. ' Silently moving through the woodland, it may also unleash a burst of speed when necessary.

Being nimble and flexible in today's fast-paced business world is essential for success. People and companies that are unable to adapt to an ever-changing business environment are doomed. According to Jack Welch, "The rate of change on the outside exceeds then the rate of change on the inside reaches the end."

Remember that the change inside a firm must be greater than the change outside. One who is even more wise is at risk of failing. TCS is one of the most agile companies in the IT industry when it comes to agility. It has undergone radical transformations in terms of its internal structure and procedures. It employs tactics that have helped it become one of the most valuable corporations in the world's history. IT service providers around the world are experiencing the same problem.

Leopards thrive in delivering unique performance on demand because of their rapidity and skillfulness, making them exceptional at fulfilling their needs. It hunts in perfect stealth, using its speed and agility to its advantage. Leopards are used to evading their prey by climbing trees or scurrying over the ground. Today's fast-paced business environment necessitates proactive customer service and a firm grasp of the company's long-term objectives in order to succeed. Customers are always looking for one-of-a-kind performances that are also adroit. The transformation on the inside of commerce must be greater than the external change. If you don't, you run the risk of failing.

10.9 LIFE LESSONS THAT YOU CAN APPLY TO YOUR OWN LIFE CREATURES HAVE A LOT TO TEACH US

Everything from a gnat on the sidewalk to a lion prowling the African savannah can teach us something important about life. We can learn a lot from other creatures if we take the time to pay attention and learn from their experiences. We can learn a lot about ourselves and the world around us just by casually engaging with other animals. Each and every day, you may learn something new from your own pet. Because we live in such an intriguing cosmos, there are always new things to discover (The power of positive,

2nd of August 2015, https://www.powerofpositivity.com/10-things-animals-teach-us-matters/).

Social rules, codes of conduct, and modes of communication are all present in animals. Certain behavioral tendencies can be recognized in animals that differ from others. Animals can teach us a number of great lessons that we may use to improve our own lives and those around us. Make sure to follow your gut: In other words, follow your instincts, as the saying goes. That small voice in your head may be referred to as it by certain people. As soon as we see it, we know something is wrong. Every circumstance may be dealt with by following this advice. A more pertinent issue is: How often do we pay attention to significant signals about our surroundings—events, circumstances, and people? It is instinctual for animals to trust their senses and respond to the world around them.

Bring out your inner beast: Take a mental vacation! Listen to your inner ideas instead of dissecting a situation from 20 different aspects, and use the innate survival skills that all people possess.

10.9.1 RESPECT YOUR ELDERS

Older people's wisdom and knowledge can be beneficial to the younger generation. Experience and new ideas have accrued over time for those who have been around longer. Listening to and learning from elders is a worthwhile investment of time. Let the animal kingdom teach you some life lessons that you may use in your own life. Animals can teach us a lot about life, and here are five of the most important ones.

Elephants are well aware of this. Until they die, the matriarchs of African elephant tribes continue to be the leaders. A wealth of wisdom and experience can be expected from the elderly: Visiting an older relative or neighbor can be a rewarding experience. If you listen carefully, you may be able to take something valuable away from the exchange.

10.9.2 ACHIEVE YOUR OBJECTIVES AND KEEP AT IT

Even if you're just starting out in a new career or preparing for a major exam, setting objectives can be intimidating. Keep your chin up! In order to return to the spot where they were born, salmon must journey thousands of kilometers upstream against all odds in order to reproduce properly.

Friends are there to help and support you: You are the source of all that is good in the world. With a little help and encouragement from your pals, you can put your intentions into action. Check out this proactive list of five suggestions for achieving your goals if you're looking for inspiration.

10.9.3 MAINTAIN A LONG-TERM PERSPECTIVE ON LIFE

It is up to us to ensure that we are aware of our carbon footprint – keeping track of how much we buy and discard on a daily basis.

Reduce and Recycle: You may recycle and decrease trash in small ways, but if you want to go the extra mile, follow these seven simple suggestions for a waste-free existence.

10.9.4 DO WHAT IS RIGHT

Developing the ability to play fair as a child entails learning to share and wait one's turn. This set of abilities serves as the foundation for healthy human connection and serves as a reminder to keep others in mind at all times, even when having a good time. Canids, including wolves, jackals, foxes, and domestic dogs, have a unique code of play that emphasizes playing by the rules, communicating clearly, and apologizing when required.

Give and share: Be aware of how you connect with people and whether or not you are willing to share and wait your turn in line. Take a hint from these Spanish schoolchildren who took part in a social experiment to see how giving and sharing can be practiced at any age.

- Prepare your cash flows and put away some funds in case of a shortage.
- To ensure that they have enough food for the winter, squirrels gather and store nuts in October. That frugality helps the squirrel, just as careful cost control helps your company.
- Never let a chance to close a deal pass you by without being prepared.
- A spider patiently spins a web to catch prey. When an insect becomes entangled in the web for any reason. In response to the spider's sense of motion, it rushes to locate the source of the vibration. Spiders can cover a larger area with their senses thanks to the web.
- Motivate your employees to work together toward a common objective.
- There is a large herd of wild horses guarding the alpha mare, and they are well-versed in the area and its resources. When the horses go between different locations, they plan the route ahead of time.

- A decentralized response and rapid distribution of food is used by ant colonies to successfully and safely distribute food in the aftermath of famine.
- A society is made up of people. In a well-ordered community, everyone has a job to do. There are two ways in which a society might be organized. One person or a small group of people serves as the "control center" of a centralized society. The power to plan and control society is in the hands of this command-and-control facility. They are the ones who take the lead in making decisions.
- Finally, this information is shared around the community. In this approach, there is ample opportunity for collective decision-making and task assignment (Figure 10.4).

On the other hand, there is no command-and-control structure in a decentralized society. Instead, individuals or small groups make their own decisions. Individuals can respond more quickly because they don't have to wait for information from a centralized source.

FIGURE 10.4 Black ants.

Source: Rakesh Kumar Dogra, https://creativecommons.org/licenses/by-sa/3.0.

10.9.4.1 THE PURPOSE

Decentralized societies, such as ant colonies, can be seen in action. An ant's decisions can be influenced by its interactions with other ants or the environment in which it lives. Ants, for example, may quickly direct other ants to a source of food. First, they tell us to make sure we're in sync with the 'control center.' Pheromones are chemical signals that ants generate from their bodies to let other ants know where the food is. This signal will be followed by other ants, guiding them to the source of food. Ants are able to respond swiftly and with resiliency because of their decentralized response. In the face of adversity, it persists and recovers. When faced with food shortages or famine, an ant shows that he or she can adapt. An ant's behavior can be influenced by its relationships with other ants in the area or by cues from the natural environment. Stormy weather can induce food shortages in ants for a number of reasons. If ants find food, they will communicate with each other. Within the confines of the nest, the ants communicate and immediately distribute the information to other ants. Famine-fighting ants can wander about the nest freely to help disperse food. Those ants who are gathering food dig a little deeper into their perch. Then, the Ants rush to feed everyone. Taking care of the offspring at the bottom of the nest is a sign of good quality in ants (Chen, 2020).

During a famine, ants must make do with lower-quality food in order to find and spread it rapidly. Contain substances that can be dangerous on occasion. It's unlikely that ants would ever eat these things naturally, but providing them with a diverse diet can decrease their exposure to potentially dangerous foods. During a famine, many foragers would leave the nest in search of food in various locations. Diverse dietary sources help to reduce the impact of contaminants from a single source. Using living "silos" where an ant stores and checks food for dangerous compounds is another option. Even if the ant dies, the colony will be spared from a potentially dangerous food source. Individual ants are rapid and well-coordinated in their responses in all of these instances. These actions work together to help a colony withstand starvation.

10.9.5 NATURE HAS A LOT TO TEACH US ABOUT MANAGING CHANGE

Learning from nature: Six lessons in change management (Forbes India): information technology (IT) company Infosys CEO Narayanan Sampath, whose thoughts and opinions are shared here:

1. **Prepare for Predictable Change:** Nature's greatest asset is the degree to which it can be predicted. Predictable occurrences, such as days and nights Seasons, come and go according to a predetermined schedule. A hurricane, earthquake, or tsunami are all examples of unpredictability, but they're exceptions, not the rule. Adapting a company's operations to new circumstances requires foresight and planning.
2. **Preparing People for Change:** Birds begin returning to their nests as dusk falls, lights are turned on in residences, headlamps are turned on in automobiles, and the ground is progressively illuminated as a day-night cricket match continues. Understanding the nature of change and its repercussions is well-known. People are aware of the necessary alterations. Everyone in the workplace has to know what to expect from change in order to better prepare themselves. With careful planning and communication, changes can be welcomed by those who have been affected by them.
3. **Create a Sense of Excitement for the Future by Building Anticipation:** As the days grow shorter and the nights longer, many people look forward to the colder weather, time to unwind, and the opportunity to spend time with friends and family. Business changes that are presented in an attractive way, with clear benefits that are anticipated and widely known, will lead to more success in adapting and succeeding.
4. **Choose a Speed That's Right for You:** Everyone is at ease with the transition from day to night, which occurs in an effortless and unhurried manner. Introducing too much change too quickly can produce a sense of uncertainty and worry, but slowing down the pace of change can have the opposite effect.
5. **Change in a Subtle and Unobtrusive Manner:** Evening falls silently and unobtrusively. Successful change projects, on the other hand, tend to be more cautious. It is not necessary to accompany a change management exercise with a high-volume performance. Despite the need for communication, change leaders must maintain a cool demeanor. They'd feel more at ease and less stressed out as a result of the shift.
6. **Take into Account the Impact on All Stakeholders of the Change:** The natural world ensures that the change will be accepted by all those who will be impacted by it. Customers, partners, employees, and shareholders are just a few of the groups who are impacted as

things change in the business world. Others are more or less affected. In cross-cultural and global contexts, it is particularly prevalent. Change managers must be aware of the impact on all stakeholders and find ways to make the change most acceptable to everyone.

7. **Intuition:** Nature, like life, is all about syncing things up. Without the moonlight, there is no sunlight. Without winter, you can't see the rainbow, and rain is frequently required to see a rainbow. Without the other, you can't have one. Nature teaches us to appreciate the good times as well as the bad and to learn to cherish both. People can look forward to the brighter days to come.
8. **The Ability to Accept Responsibility for One's Actions:** We've become accustomed to taking control of most or all aspects of our life. We must, however, come to terms with the fact that we have no control over them. For example, death and life are inextricably intertwined and beyond our control. Nature also teaches us to love and accept one another. Nature has such a devoted following. It's a safe haven where you may take a deep breath and forget about the world around you. It would be a game-changer if we could do the same for others.
9. **Self-Worth:** Everything in the natural world has intrinsic value, from animals and plants to worms and rocks. In ecology, it serves a purpose. Changing even the tiniest detail by force could have disastrous consequences. The same thing occurs to us. Each of us brings something unique to the table in our homes, workplaces, and communities. Everyone is unique and valuable in their own right.
10. **Gratitude is the Key:** How do you enjoy spending time outdoors? When the sun rises, I'm particularly interested in the little beams of light that peek through the horizon. Gratitude is the key. How do you like to spend your free time outside? During sunrise, the author of the lesson said, I'm particularly fascinated by the stray rays of light that poke their way through the haze. It could be that a cool breeze or a breath of fresh air can help. Is that the sound of birds singing or the breeze through the trees? Everything comes down to the tiniest details. For those of us who don't spend a lot of time outside, these moments mean a lot to us. The exercise would be equally valid if performed in reverse. For example, you'd discover just how much you missed the little things like being able to sleep in your own bed and having a home-cooked meal.

11. **Happiness is a Positive Emotion:** It's impossible not to feel happy when you're out in nature. We can experience contentment based on how much we do in our daily lives. It's interesting how much happier and more at ease we feel when we're not surrounded by the things we're accustomed to in the comfort of our own homes. It's possible that happiness isn't tied to material possessions but rather to savoring the little pleasures in life. For the most part, these are things we take for granted. Take a break from your work and go for a walk. Going on an adventure can broaden your outlook while also providing you with priceless experiences.

    When you take the time to go outside and pay attention to nature, it may be a tremendous teacher. Nature is trying to teach you something (https://nature-mentor.com/21-things-nature-teaches-you/).
12. **The Ability to Observe:** This is a skill that hardly everyone possesses. Spending time in nature is a great way to improve one's capacity for observation.

    Learn to identify all the trees, plants, and birds in your area's bioregion. You'll be able to navigate the area like a pro.

    Unexpected discoveries can be made by venturing out into nature and pursuing your curiosity to new and exciting locations. As a traveler, it's exciting to see the world from a new perspective.

    Use your senses to their fullest extent; they have a lot of potential. Spend some quiet time in nature with your eyes closed, and hear what you can hear. The amount more we observe when we slow down is astounding.
13. **Intelligence:** Our thoughts contain vast amounts of previously untapped intelligence. With an attitude of curiosity, we're able to discover remarkable things and learn practical problem-solving skills when we get out into nature.

In terms of comfort, the outdoors can be hit or miss. It's possible to feel at ease even in the harshest, most lonely wilderness if you can discover a deeper sense of security within yourself.

Nature has always served as a source of inspiration for artists, musicians, thinkers, and writers. Nature has inspired some of the most innovative architectural concepts.

Listen to the dawn chorus of a bird singing in the midst of a storm to learn the true art of happiness. Emotional resilience can be learned from animals.

Richard Louv invented the phrase "nature deficit disorder" to describe what happens when people don't get enough outdoors in their lives, and the outcomes aren't always pleasant. You will learn how to be healthy, happy, and connected by spending time in nature (How to cure nature deficit disorder).

The past can be accessed through nature, which acts as a window into the present. It's inscribed in the sands of time and space. Try to envision what it would have been like to live in your ancestors' time and see how it compares to your own.

When you're out in nature, it's easy to lose your bearings. Nature will teach you how to keep track of where you're going by allowing you to visualize what you're doing (Natural navigation techniques).

Every living individual already possesses peace within themselves. This freedom and joy can only be found when we are able to relax and let go of our thoughts while enjoying the natural world around us.

Learning to be grateful for all the wonderful things we have in our lives comes from experiencing the gift of life through our senses in nature. Take a moment to savor what you've got.

Our decisions in life have natural repercussions, as nature repeatedly demonstrates. We learn to keep an eye on the weather and take steps to avert unpleasant outcomes.

It's important to be quick-witted so that you can take advantage of the opportunities that nature presents you. If you don't have a raincoat, you'll be drenched!

The survival of our world is a constant reminder of the resilience of nature. We can overcome any obstacle if we commit to growth and progress (How to survive in the wild with nothing).

It takes more than a little effort to climb the mountain. As you work to fulfill your want for a closer relationship with nature, you'll develop the ability to persevere.

While storms and trees may fall, there will always be a clearing and a return of sunlight after they have died down. Everything in life is the same.

Naturalists are always on the hunt for new things to learn about the world around them. What will tickle your curiosity, and where will it lead you next?

There are many more people on this planet than just you and me. A worldwide community of plants, animals, people, ideas, and dreams includes all of us in a very small way.

To be humble is to recognize that not all questions have a corresponding solution. Every turn brings with it a fresh set of problems. We can only let go and keep going.

KEYWORDS

- **animal kingdom**
- **biomimicry**
- **Eagles' positive leadership qualities**
- **leadership lessons**
- **life lessons**
- **management theories**
- **nature-inspired business**

REFERENCES

Bennis, W., (2009). *On Becoming a Leader*. Perseus Books Group.

Benyus, J. M., (1997). *Biomimicry.* HarperCollins, New York.

Benyus, J. M., (2002). *Biomimicry: Innovations Inspired by Nature.* New York, Perennial., Business lessons from nature #2. https://www.academia.edu/38300413/Janine_M_Benyus_Biomimicry_Innovation_Inspired_by_Nature_2002_Harper_Perennial_1_ (accessed on 17 August 2023).

Celep, S., Tunç, A. Ö., & Düren, A. Z., (2017). Can biomimicry and managerial concepts come together? *Global Business and Management Research, 9*(3), 31–45.

Chen, N. Ask Nature collections. *Ant Colonies Respond, Quickly to Distribute Food, Rock Ant.* https://asknature.org/strategy/ant-colonies-respond-quickly-to-distribute-food/ (accessed on 17 August 2023).

DeNio, M., (2020). *10 Lessons Every Entrepreneur Can Learn from The Lion King*. https://www.linkedin.com/pulse/10-lessons-every-entrepreneur-can-learn-from-lion-king-michelle-denio (accessed on 17 August 2023).

Dumbre, M., (2021). *10 Lessons from the Animal Kingdom for Entrepreneurs, Executives, and Investors.* https://starthub.london.edu/10-lessons-from-the-animal-kingdom-for-entrepreneurs-executives-and-investors (accessed on 17 August 2023).

Farnsworth, M. (2020). *Biomimicry and Business, How Companies Are Using Nature's Strategies to Succeed.* Routledge.

Friedman, S., (2015). *Life Lessons, We Can Learn from Animals*. Good net. https://www.goodnet.org/articles/5-life-lessons-we-learn-from-animals-list (accessed on 17 August 2023).

Gazzara, K., (2022). *Leadership Lessons We Can All Learn from Giraffes*. https://www.lifehack.org/articles/work/the-10-leadership-lessons-can-all-learn-from-giraffes.html (accessed on 17 August 2023).

https://animals.mom.com/why-do-geckos-have-a-neck-pouch-12305031.html (accessed on 17 August 2023).

https://asknature.org/strategy/ant-colonies-respond-quickly-to-distribute-food/ (accessed on 17 August 2023).

https://beleaderly.com/11-leadership-lessons-learned-2/ (accessed on 17 August 2023).

https://businessconnectindia.in/what-ants-can-teach-us-about-business/ (accessed on 17 August 2023).

https://eightymphmom.com/6-important-life-lessons-we-can-learn-from-nature/ (accessed on 28 September 2023).

https://innovatureinc.com/what-is-agile-business-process/ (accessed on 27 September 2023).

https://neilpatel.com/blog/10-lessons-on-leadership/?lang_geo=in (accessed on 17 August 2023).

https://parenting.flinto.in/child-development/moral-stories-kids (accessed on 17 August 2023).

https://scottamabry.medium.com/a-short-leadership-lesson-from-nature-e79544264f01 (accessed on 17 August 2023).

https://starthub.london.edu/10- -lessons-from-the-animal-kingdom-for-entrepreneurs-executives--and-investors/ (accessed on 17 August 2023).

https://starthub.london.edu/10-lessons-from-the-animal-kingdom-for-entrepreneurs-executives-and-investors/ (accessed on 17 August 2023).

https://thenatureofbusiness.org/2013/06/19/biomimicry-for-business/,https://www.livescience.com/28873-cool-technologies-inspired-by-nature.html (accessed on 17 August 2023).

https://wildtrails.in/category/business-lessons-from-our-wildlife/ (accessed on 17 August 2023).

https://worldineyes.com/5-leadership-traits-we-can-learn-from-an-eagle/ (accessed on 17 August 2023).

https://www.2ascribe.com/articles/business-2/leadership-lessons-learned-from-nature (accessed on 17 August 2023).

https://www.eleanorgoold.com/7-life-lessons-you-can-learn-from-trees/ (accessed on 27 September 2023).

https://www.exoplatform.com/blog/6-leadership-and-career-lessons-we-can-learn-from-animals-exo-platform/ (accessed on 17 August 2023).

https://www.expords.com/amazing-life-lessons-from-the-eagle/ (accessed on 17 August 2023).

https://www.fastcompany.com/3031018/6-lessons-that-businesses-can-learn-from-bees (accessed on 17 August 2023).

https://www.feelgoodanyway.com/nature-inspired-marketing-lessons/ (accessed on 17 August 2023).

https://www.forbesindia.com/blog/business-strategy/learning-from-nature-six-lessons-in-change-management/ (accessed on 27 September 2023).

https://www.leadershipdevelopmentlab.com/blog/2019/7/17/biomimicry-leadership-lessons-we-can-observe-in-nature (accessed on 17 August 2023).

https://www.linkedin.com/pulse/10-lessons-every-entrepreneur-can-learn-from-lion-king-michelle-denio (accessed on 17 August 2023).

https://www.linkedin.com/pulse/5-powerful-management-lessons-from-animal-kingdom-srinivasan-r (accessed on 27 September 2023).

https://www.mindfueldaily.com/livewell/5-life-lessons-from-a-hummingbird/ (accessed on 28 September 2023).

https://www.successfolks.com/motivation/inspiration-must-read-get-back-up-and-be-prepared/ (accessed on 17 August 2023).

Hyacinth, B., (2015). *10 Leadership Lessons from Nature*. Linked In. https://www.linkedin.com/pulse/10-leadership-lessons-from-nature-brigette-hyacinth (accessed on 17 August 2023).

Jamei, E., & Vrcelj, Z. (2021) Biomimicry and the built environment, learning from nature's solutions. *Appl. Sci. 11*(16), 7514; https://doi.org/10.3390/app11167514

Janus, A., (2020). *5 Leadership Traits We Can Learn from an Eagle*. https://worldineyes.com/5-leadership-traits-we-can-learn-from-an-eagle/ (accessed on 17 August 2023).

Jennifer, A., (2013). *Business Lessons from the Birds and the Bees*. Science, Fortune. https://fortune.com/2013/03/28/8-business-lessons-from-the-birds-and-the-bees/ (accessed on 17 August 2023).

Kashyap, V. (2017). *Leadership Lessons From Nature*. https://medium.com/thrive-global/leadership-lessons-from-nature-hint-its-not-survival-of-the-fittest-23070b9be9b6 (accessed on 17 August 2023).

Laroui, F., (2020). *Leadership and Career Lessons We Can Learn from Animals 2020*–Tips & Tricks.

Life Lessons We Can Learn from Animals, (2015). *Let the Animal Kingdom Teach You Some Valuable Life Lessons.* Goodnet. https://www.goodnet.org/articles/5-life-lessons-we-learn-from-animals-list (accessed on 17 August 2023).

Low, P. K. C., (2010). *Leadership Lessons from the Animal Kingdom*. E-Leader Singapore, http://www.g-casa.com/conferences/singapore/papers_in_pdf/tue/Low-second.pdf (accessed on 17 August 2023).

Mabry, S., (2016). *A Short Leadership Lesson from Nature*. https://scottamabry.medium.com/a-short-leadership-lesson-from-nature-e79544264f01 (accessed on 17 August 2023).

Mashingaidze, S., (2014). Biomimicry: Descriptive analysis of biodiversity strategy adoption for business sustainable performance. *J. Governance and Regulation, 3*(2), 18–28. https://doi.org/10.22495/jgr_v3_i2_c1_p2.

Power of Positivity, (2015). https://www.powerofpositivity.com/10-things-animals-teach-us-matters/ (accessed on 17 August 2023).

Srinivasan, R., (2015). *Powerful Management Lessons from the Animal Kingdom*. https://www.linkedin.com/pulse/5-powerful-management-lessons-from-animal-kingdom-srinivasan-r (accessed on 17 August 2023).

CHAPTER 11

Biofouling and Bio-Purification of Water

ABSTRACT

Microorganisms, plants, algae, and small animals can build up on moist surfaces that have a mechanical function, producing structural or functional problems. Biofouling is the buildup of these organisms. Microfouling and macrofouling, or soft and hard fouling, are two forms of these organisms. The buildup of biofouling on ship hulls is a major issue. This chapter discusses biofouling issues, membrane fouling, and biofouling mitigation in membrane water treatment systems. Biomimetic nanotechnology and ultrasonic waves inspired by a sea creature were used in a variety of ways by scientists. Existing methods of water filtration may be supplanted by new water purification technology. Living water treatment and purification procedures can be mitigated by the generation sector. Water is just as essential to survival as food and shelter. Toxic compounds and disease-causing bacteria can be found in water. In impoverished countries, waterborne infections claim many lives each year. The development of water treatment technology that can reduce dangerous contaminants to drinkable standards is essential. There are a variety of water treatment systems that use low-cost, simple technologies. Disinfection via boiling, chemicals, sun radiation and filtration, coagulation, flocculation, and desalination are just a few of the ways available. Biomimetic membranes, water filtration inspired by non-flowering trees, and gymnosperm sapwood engineering and characterization are all covered in this chapter. Advanced chemical or physical treatment technologies are more expensive and energy-intensive, but they are more robust over operating conditions and water quality. Water filters that are simple, unique, and low-cost are being developed by researchers.

Bio-Inspired Technologies for the Modern World: Emerging Trends and Applications.
R. Ramakrishna Reddy & T. Pullaiah (Authors)

11.1 BIO-FOULING PROBLEMS

Biofouling is the term used to describe the attachment and development of organisms on artificial surfaces. Ships and offshore platforms are protected by antifouling coatings containing organometallic or inorganic hazardous species, and cooling water systems are protected by biocides injected into the water column. When there was no strong, practical knowledge available in the past, certain recognized practices grew and proliferated, resulting in even more confusion regarding which approaches to use. Due to biofouling and its control, reverse osmosis (RO) plants have long been plagued by operational difficulties (Bixler & Bhushan, 2012).

Filtration operations using membranes are plagued by the problem of membrane fouling. Two approaches to preventing and controlling membrane biofouling were proposed in the late 1990s. The physical removal of germs from the feedwater by means of micro- or ultrafiltration (UF) is addressed by biocide and UV radiation. Some of the bacteria's long-term adaptive skills, such as rapid genetic adaptability to harsh environments, were basically neglected by the extraction versus killing strategy.

11.2 FOULING TYPES OF MEMBRANE

Fouling is the accumulation of foreign materials on the surface of a device (Nguyen et al., 2012). In addition to increasing hydraulic resistance, increasing energy consumption, and potentially causing system components to malfunction, it can also cause significant damage. In a feed stream, fouling can occur in the form of particulate/colloid, biological/microbial, scaling/precipitation, and organic fouling, as mentioned below. Either a simple cleaning can restore the membrane's function, or fouling becomes permanent and cannot be undone. The latter scenario occurs when foulants chemically connect with a membrane and permanently impair its performance.

Membrane-based solutions for water treatment offer high energy efficiency, compactness, and low space needs, as well as operational simplicity and automation. Membrane desalination techniques like respiratory osmosis, membrane distillation, microfiltration, and nanofiltration are among the most well-established (Al-Sawaftah, 2021).

11.2.1 FOULING BY PARTICLES AND COLLOIDS

Membrane pore clogging or membrane surface adhesion of paused solids and colloidal debris is known as particulate fouling. Water cannot pass through the membrane pores because of the buildup of particles on the membrane, which results in symptoms such as an increase in pressure differential measurements and more energy being used up by the membrane. A stream's feed water is likely to contain non-biological and inorganic particles (such as silt or clay) when it originates from a body of surface water. Particle fouling can be prevented with proper coagulation and filtration upstream. In order to estimate the probability of particulate/colloidal fouling in a feed stream, water treatment specialists usually use the Silt Density Index (SDI). RO systems have the smallest membrane filtration holes, necessitating SDI measurements.

11.2.2 BIOFOULING AND MICROORGANISMS

It is a process in which microbes, plants, algae, and other biological pollutants develop and enter membranes and pores. Warm, low-flow conditions are ideal for biological and microbiological foulants. As they grow, they release an extracellular polymetric substance known as EPS, which acts as a kind of scavenger (EPS). A biofilm is a thick covering of bacteria and EPS. Standard cleaning methods like backwashing or using biocides like chlorine can't remove biofilm because of its chemical qualities. Even though it isn't considered a foulant, free chlorine assault causes membrane oxidation. It's irreversible and can't be undone. Membranes that have become infected with microorganisms pose maintenance issues and, in some situations, may need to be replaced. A biofouling membrane will impede water flow over time. Pressure demand has increased, and the cost of energy has gone up, as well as a drop in membrane flow.

11.2.3 SCALING OR PRECIPITATION FOULING

Scaling is a type of inorganic fouling caused by salts, oxides, and hydroxides crystallizing in the feed solution. Scaling in the membrane happens when dissolved components precipitate or aggregate in its pores. Crystallization of ionic components occurs when precipitation fouling leads a solution to become saturated. When the feed stream contains significant quantities of calcium or magnesium, scaling is a serious challenge for RO/NF systems with high conversion rates. An acid injection, softening, or the use of chemical scale inhibitors can be utilized to avoid the formation of inorganic fouling in water. Even if chemical treatments are effective, they should not be employed because the membrane material is incompatible with them.

11.2.4 FOULING MADE UP OF ORGANIC MATERIALS

When a carbon-based material infiltrates a filter membrane, an organic fouling results, in soil, groundwater, and surface water, decomposition products of natural organic matter can be discovered. Depending on the membrane material, organic materials can offer a significant foulant danger. Organic fouling concerns can be reduced in facilities with natural water treatment.

11.3 BIOFOULING IN THE OCEANS: WHY GROWTH IS THWARTED BY ANTIFOULING SYSTEMS?

Fouling is the deposit of organic debris in submerged things such as ship hulls and dock pylons (Bixler & Bhushan, 2012). Sessile types of macroalgae and microbial slimes are among the roughly 4,000 marine species recognized as biofouling organisms. In the US Navy alone, the problem costs an estimated $1 billion per year. Roughness in a ship's hull indicates increased hydrodynamic drag, which is caused by biofouling. The expenses of biofouling include increased fuel consumption, hull cleaning, paint removal, repainting, and environmental compliance requirements. All kinds of environments are at risk for biofouling. There are two types of fouling: biofouling and inorganic fouling, both caused by nonliving particles. Fouling can be controlled in a variety of ways thanks to the diverse solutions provided by nature. Some examples are sloughing, wettability (the ability to repel and attract water), microtexture, low drag, and low stickiness.

The flora and fauna of the natural world are contemplating the development of new antifouling technologies that mimic those used by engineers. Coatings and cleaning methods are just a few of the antifouling measures now in use. Marine invasive species introduced by ships into new habitats pose a serious danger to ocean biodiversity and the health of the world's oceans. It's possible that marine animals transported by ships in their ballast water or on their hulls could establish a population in the host environment, outcompete local species, and grow into pest proportions (Schultz, 2007). Over time, marine life accumulates in the ship's hull due to biofouling. Because of their invasive nature, they can have an adverse effect on ship operations and the maritime environment. As an alternative, antifouling is an option. Antifouling is used on the ship's hull to keep these organisms from growing.

The marine economy and human health are jeopardized when various kinds of marine organisms attach themselves to the surface of marine facilities of all sizes (Abioye et al., 2019). To make matters worse, marine pollution (Tian et al., 2019) prevents humanity from taking full advantage of the ocean's resources. Bacteria, diatoms, and Ulva spores were only some of the thousands of microorganisms and large fouling critters that were detected. A vast range of fouling organisms find reproductive grounds in ships, oil platforms, and aquatic breeding facilities. Barnacles and other invertebrates attach themselves to the ship's hull via a biological glue. Anti-corrosion layer on the hull surface will be eroded by the sharp edges of the barnacle as it grows, causing hull degradation (Figure 11.1).

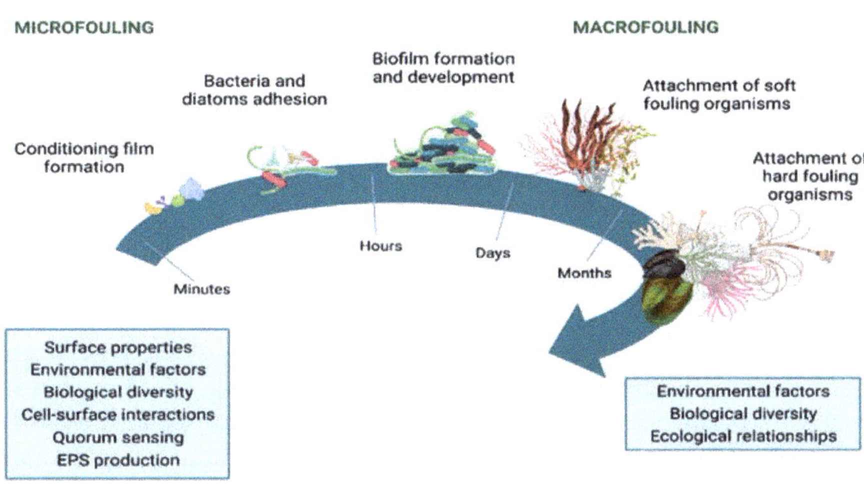

FIGURE 11.1 Representation of the marine biofouling (mainly bacteria, cyanobacteria, and diatoms) process and the main parameters/factors that affect microfouling and macrofouling events.

Source: Reprinted with permission from Romeu and Mergulhão, 2023. Copyright © 2023 by the authors. Licensee MDPI, Basel, Switzerland. https://creativecommons.org/licenses/by/4.0/.

11.3.1 SCENES INSIDE TUBES

The coolant lines in a ship's engine become clogged with biofouling. Barnacles can impede the flow of coolant via your intake pipes. A container shipping company's chief engineer tells the New York Times that "unless you clean it out, everything will be halted." The primary approach for removing clogged engine tubes is overengineering. If the cooling system is overengineered or constructed with more pipes than the engine needs to cool at peak speeds, it can afford to lose efficiency due to barnacles choking off pipes. As an example, Raines points to the ship's cooling system. Cooling is one of those things that only matter when you're running out of it. Copper plates, which release ions and sodium hypochlorite (often referred to as bleach) formed from oxygen and sodium chloride seawater, can be used to combat fouling organisms in coolant pipes.

11.4 UNDERSTANDING MATINE FOULING

Plankton, the primary source of nutrients for marine fouling, grows in estuaries where sewage disposal and fertilizer runoff are prevalent. Animal or plant waste can accumulate on the submerged portions of the hull and along the waterline. Polyps that grow in dark locations under the hull are mistaken for grasses, but they are actually tiny tentacled creatures that capture animals. However, for the boat owner, a foul hull means a slower boat and less efficient use of fuel. Years of combat have been waged by its battle boat crews (Schultz, 2007).

11.5 TYPES OF ANTIFOULING

In the ocean, biofouling is a major concern for all ships of any type. It's not just the exterior of the ship that is affected by biofouling, but also the water intake. Antifouling is a term used to describe a technique used to keep underwater structures (such as ship hulls) clean. Typical antifouling paints contain water-soluble biocides that are gradually released to lower the growth rate of marine life, which is why they are commonly used. Because they include biocides, antifouling paints containing biocides are considered pesticides and subject to the same stringent regulations as pesticides. Biofouling prevention has been achieved using nanocoatings (Kumar et al., 2021). Nature serves as an inspiration for the antifouling properties of nano- and microscale surface patterns (Richards et al., 2020). It is important to consider the impacts on non-target organisms, buildup in the food chain, and application safety when evaluating antifoulings' safety. Antifoul purchased in one jurisdiction may not be lawfully applied or used in another, especially in situations that are considered sensitive, in some nations, even inside the EU. As a result, you must make certain that the product you intend to use is legal in both your home country and any other country to which you intend to go. Copper oxide (CuO_2) is the most important biocide in modern antifoulings. For example, it is more easily dissolved in water than plain copper. It is possible to achieve brighter colors with copper thiocyanate (CuSCN), but it is less efficient against animal fouling when used alone. As soon as slime begins to form, it becomes a breeding ground for other species that can attack the ship's shell. Additionally, organic biocides are sometimes used in products to avoid slime buildup, which copper oxide isn't very effective at. Optimizing organic biocides can make a significant impact on performance, and manufacturers often put a lot of time and attention into this aspect of their products (https://www.marineinsight.com/tech/4-types-of-anti-fouling-systems-used-on-board-ships-to-prevent-marine-growth/).

11.5.1 APPLICATION OF ANTIFOULING

On a range of factors, antifouling needs to be done again and again. Included in this list are the type of vessel, intended use, environmental circumstances (such as pollution), and the expected time spent at sea. In order to maintain a clean bottom throughout the year, many race boats are dry sailed, and for good cause. Day sailers and powerboats up to 9 meters in length can also benefit from stack facilities. For some boaters, this may not be an ideal answer. The best antifouling is used by the majority of competitive racing fleets that stay afloat throughout the season, which includes a twice-weekly scrubbing regimen. Many boat owners may not need to be as meticulous in their approach, but the extent of the problem they are dealing with can be measured by this. In the spring, cruising boats have traditionally had two coats of paint done before they are launched. A few weeks afloat is usually plenty for most, but those in high-fouling locations may begin to notice slime within weeks. Alternatively, a single coat may suffice for a three- or four-month-long commissioning period in areas with minimal fouling.

An interesting natural phenomenon known as Biofouling can be seen in the form of micro- and macrofouling on surfaces. As the name suggests, microfouling is caused by tiny creatures, whereas macrofouling is caused by larger, more persistent species (e.g., invertebrates). Microorganisms can colonize both natural and man-made surfaces, according to this study. Microbial biofilms are formed by the accumulation of enormous amounts of microbial metabolites. Biofouling protection measures were explored by Abioye et al. (2019). Anti-biofouling compounds are electrolytically deposited in paints. Biofouling can be combated in the maritime industry with nanomaterials and composites containing anti-biofouling natural ingredients. Microbial biofilms and high concentrations of metabolites on seashore rocks contribute to their slippery properties.

Environmental hazards are discovered when antifouling substances are applied in the traditional method, which is a regular practice. Antifouling that is long-lasting, broad-spectrum, and ecologically friendly is the attention of engineers and researchers. Antifouling technologies based on low toxicity and non-toxicity antifouling chemicals and materials have recently been developed, and they have a wide range of application possibilities. The latest eco-friendly antifouling technologies have been reviewed by Lian et al. (2018).

11.6 BIOFOULING REMOVAL FROM SHIPS USING IN-WATER CLEANING AND COLLECTION: A PRELIMINARY ASSESSMENT OF EFFICACY AND SAFETY FOR THE ENVIRONMENT

Ship biofouling is increasingly being used to spread non-indigenous marine species around the world. Historically, biofouling on ships has been a problem since it can reduce ship performance, increase fuel consumption, and lead to increased pollution from the ship's engines. The removal and collection of macrofouling organisms and associated antifouling coating chemicals by ship in-water cleaning and capture (IWCC) systems is becoming a feasible solution to both concerns. As a condition for IWCC licenses and permits, as well as to assist design of future biosecurity standards, independent and thorough evaluations of their effectiveness and environmental safety are necessary; Tamburri et al. (2020) developed an approved protocol for testing an IWCC system in order to measure biofouling removal and capture efficiency on two ships. Through ship biofouling, non-indigenous marine species are spreading over the planet. Since biofouling can increase drag, fuel consumption, and exhaust emissions, it's been a problem for ships for a long time. Using ship IWCC systems to remove macrofouling organisms and their accompanying antifouling coating chemicals is becoming a viable solution to both challenges. For IWCC licenses and permits, as well as to help develop future biosecurity standards, independent and comprehensive assessments of their effectiveness and environmental safety are necessary. An IWCC system evaluation technique was developed by Tamburri et al. (2020) to test biofouling removal and capture efficiency on two ships.

11.7 MEMBRANE BIOFOULING MITIGATION FOR WATER

External deposition of contaminants and microbes is common with membranes, the fundamental instrument in membrane separation procedures. Membrane water and wastewater treatment is plagued by the problem of biofouling, which seriously reduces the effectiveness of the system (Aslam et al., 2018). Current research efforts are yielding new insights on how to combat the biofouling problem in membranes (Pichardo-Romero et al., 2020). Fouling management approaches for MBR membrane biofouling include uncoupling, and quorum quenching, and others have their own particular advantages, according to Cui et al. (2021). In the previous few decades, engineering science has made remarkable progress. Membrane technology includes UF. For membrane-based technology, separation strategies have been anticipated. A membrane is vital in the process of separating different types of molecules. Membranes are classified according to the materials from which they are constructed. Their polymer-based construction makes them both organic and inorganic (e.g., glass, ceramic, silica, graphene, and metal oxide). Membrane-based technologies are organized by the driving force for pressure-driven membrane processes.

Drinking water is produced using UF membrane technology. Suspended particles, colloids, and bacteria can all be intercepted by UF membranes because the pore diameter is smaller than the typical pollutants. Membrane bioreactor fouling is a big problem. Membrane fouling is the subject of extensive study. Various types and configurations of membranes have become an important part of the separation toolkit. The presence of macrofouling indicates the presence of larger, more conspicuous organisms. Deposits on the membrane's exterior and internal surfaces are what lead to fouling. The higher hydraulic resistance and the fouling-enhanced polarization caused by fouling lead to an irreversible decrease in performance. Because the deposit (biofilm) does not travel with the meal, befouling differs from other forms of fouling in that it develops on-site, utilizing nutrients from the feed. Fouling and the resulting effects must be given special attention in reducing the problem. In wastewater treatment, the MBR method (Cui et al., 2021; Anjum et al., 2021) has gotten a lot of attention (Figure 11.2) due to its excellent effluent quality and low sludge production, both of which are advantages. In spite of this, membrane fouling is an inevitable obstacle to the widespread implementation of MBR technology.

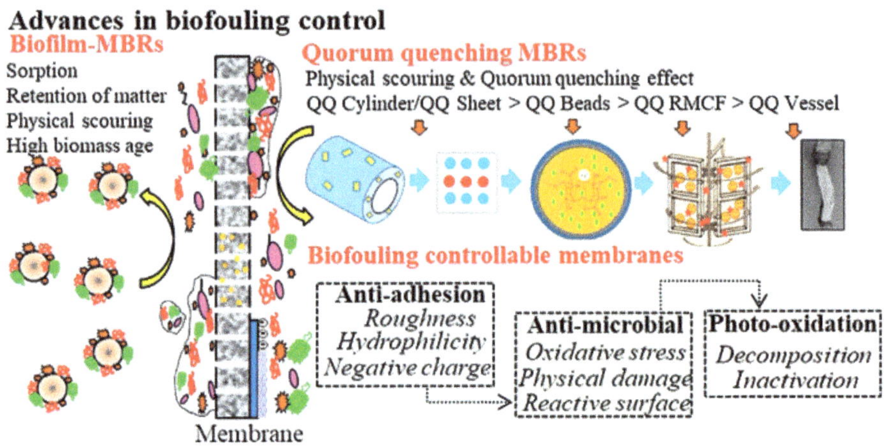

FIGURE 11.2 In particle-sparsed MBRs, biofouling is controlled by functionalized medium.
Source: Reprinted with permission from Aslam et al. (2018). Copyright © 2018 Elsevier B.V.

Compared to other fouling, membrane biofouling has made little progress. Many new methods for controlling membrane biofouling have been created in the last few years. There has been a rise in the use of both physical and biological methods in the fight against biofouling (Cui et al., 2021; Anjum et al., 2021).

11.7.1 THE USE OF ULTRASONIC WAVES TO PURIFY WATER: USE AND OBSTACLES TO OVERCOMING

When we talk about water quality, we're talking about whether or not a body of water has the chemical, physical, and biological attributes necessary for its intended use. Water is essential for a wide range of human activities, including recreation, drinking, fishing, farming, and industry. A sufficient supply of high-quality water is required to support all human activities. Water pollution affects drinking water, rivers, lakes, and oceans around the world, as well as the natural environment. When we talk about water quality, we're referring to its chemical, physical, and biological qualities, which are all related to whether or not it's suitable for the intended use, whether it's for drinking, fishing, farming, or industrial. In order to maintain all human activity, water must be available in the right quantity and in the right quality. A vast range of water bodies around the world are affected by water pollution. In some cases, water treatment methods are used to improve public health and water quality. There are many ways to get rid of bacteria, however, chlorine is the most commonly used. Public health advancements such as water filtration and the addition of chlorine are possibly the most important advances in human history.

Patients with impaired immune systems, such as those undergoing chemotherapy, may be more vulnerable to contaminants in drinking water. Many diseases can be transmitted by drinking contaminated water, such as diarrhea, cholera, dysentery, typhoid, and polio. It doesn't matter if you're using it for drinking, cooking, or farming; everyone needs access to clean, safe water. There is a direct link between water supply and sanitation and economic growth and poverty reduction. Ultrasonic waves are utilized to improve the water treatment process. The filtration of water has never been easier thanks to this approach. Ion exchange, carbon adsorption, membrane processes, and disinfection are some of the most common methods for removing contaminants from water. Fetyan and Attia discussed water purification methods (2020). Researchers have developed a method for efficiently removing pollutants from water in the wake of Actinia. A wide range of scientists and researchers is experimenting with new ways to ensure that drinking water is safe to ingest. A considerable role is played by ultrasound waves.

A sea creature came up with the idea for a water purification breakthrough. Menachem Elimelech of Yale University and Huazhang Zhao of Peking University used nanocoagulants in their research. The discovery has the potential to revolutionize the centuries-old use of coagulants by eliminating a wide variety of pollutants in a single operation (Science News, 2018; Liu et al., 2019). Using common coagulants like aluminum sulfate and other metallic salts, larger particles in water can be removed from the supply and flushed down the drain. It can be time-consuming, costly, and resource-intensive to implement a wide range of water treatment systems. A gadget that removes all contaminants from water must be developed to solve global water shortages.

For the first time, researchers have developed an ultra-stable nanocoagulant that is structurally and functionally distinct from ordinary coagulants. This nanocoagulant eliminates tiny, dissolved pollutants as well as suspended particles. Actinia is a sea anemone with a spherical body with retractable and extendable tentacles. The coagulant was created using this marine predator as a model. The nanocoagulant, like *Actinia*, has a core-shell structure that flips over when submerged in liquid. Larger suspended particles are destabilized and ensnared by the shell, while smaller, dissolved ones are captured by the exposed core. Trace micropollutants and bigger particles are all effectively removed. When it was discovered that typical water-coagulants had no power to remove nitrate, it was quite a surprise. Existing water treatment systems might be replaced, and water treatment costs could be drastically reduced if this effort is successful. Additionally, it paves the way for the development of 'smart' materials that are able to change their shape and function in reaction to their surroundings. In addition, nitrate poisoning is linked to the 'blue baby' syndrome, which affects young children in several regions of the world, and water treatment is crucial to preventing this.

11.7.2 BIOMIMETIC NANOTECHNOLOGY ENHANCING WATER PURIFICATION

Wastewater treatment, pharmaceuticals, and the food sector are just a few of the industries that benefit from water purification. As a result of the industrial revolution, pollution, and the changing climate, freshwater sources are no longer fit for drinking. Various water treatment methods have been developed in an effort to help with this issue. However, the majority of them are located in difficult-to-reach, expensive, and time-consuming areas.

11.7.3 NANOTECHNOLOGY IN WATER PURIFICATION

In addition to providing remedies for water purification problems, Richard Feynman's scientifically prophetic phrase, "There's plenty of room at the bottom," helped usher in the nanotechnology revolution. Water purification systems have been improved because of the use of nanotechnology, which was inspired by biological organisms and their great productivity. The removal of molecular contaminants using a membrane-based system saves both energy and waste. 'High water throughput, high filter efficiency, and high molecular loading capacity are required while designing water membrane purification.' Polymer membranes can be chemically altered, but the use of multi-layer architecture, which mimics nature, has been around for a million years. All contaminants are removed from the soil by rain, and we have pure groundwater in wells and tanks for our consumption. Much like the organelles seen in living systems, organisms and cellular organelles are constantly collecting, transporting, and recycling water. Similarly, many organisms and cellular organelles collect, transport, and recycle water while removing pollutants and salts along the way.

11.7.4 USING NANOTECHNOLOGY TO PRODUCE WATER THAT IS SAFE AND LONG-TERM

Nanotechnology-enabled water treatment (NWT) is beginning to show promise as a solution that can cut costs while simultaneously enhancing performance in the water treatment industry. Advances in fundamental nanoscience have made it possible to utilize catalysis and energy from all throughout the electromagnetic spectrum in innovative ways to purify drinking water, handle industrial waste, and access new supplies of water. Before the industry can broadly employ NWT devices, it is imperative that they be safe and long-lasting in order to gain the approval of consumers, governments, and non-governmental organizations (Nagar & Pradeep, 2020; Qu et al., 2013). The identification of new pollutants necessitates the development of cost- and energy-efficient, as well as water treatment and remediation-friendly, technology (Falinski et al., 2020). In order to achieve these goals, sophisticated nanomaterials are being used. The safe and sustainable use of nanoparticles (NPs) is a major concern in every therapy application (https://www.mdpi.com/2079-4991/12/2/263).

When it comes to nanotechnology, altering and regulating matter at the atomic level is the goal. When it comes to water purification, nanomembranes are employed to remove biological, chemical, and physical particles and molecules while softening the water. Nanotechnology can be used to remove pollutants from water using magnetically active NPs, a new method of water filtration currently being researched. Magnetic NPs can be used to remove contaminants from drinking water, which are attracted by the approach. Furthermore, this method could have a good impact on both human health and the environment. The breakdown of microplastics using NPs is another option in nanotech. As a result of nanotechnology's modular, scalable, and high-efficiency capabilities, it is possible to upgrade outmoded infrastructure and develop high-performance products.

11.7.5 INCORPORATED BIOMIMETIC MEMBRANES INTO AQUAPORIN

Biomimetic block copolymer membranes mimic biological membranes, according to Neal et al. (2016). Chemically synthesized membranes can't compete with Aquaporins, which are naturally occurring water channels. Because of their high-water flux, aquaporins containing carbon nanotubes are useful in water filtration, but their cost is a deterrent. Aquaporin-embedded biomimetic membranes, according to Neal et al. (2016), have strong architectures. These types of studies serve as a catalyst for additional work aimed at advancing technology.

A water filter based on nanochemistry for pesticide removal was developed by Neal et al. (2016). Low-cost point-of-use water filtration systems can now utilize chitosan derived from shrimp and other crustaceans' chitin shells by means of a novel family of nanocrystalline metal oxyhydroxide–chitosan granular composite materials. One study estimates that more than 7.5 million people have been helped by the clean water solutions developed with the help of nanotechnology. According to Dwivedi (2019), the biomimetic approach has been improved. The multilayer nanoporous membrane structure she created filters out our molecular contaminants. Biomineralization and protein self-assembly techniques are employed to make nanoporous materials. Aquaporin, a transmembrane protein, has received a lot of recent interest. For the synthetic biomimetic membrane to be developed and overcome the challenges that stood in its way, a considerable deal of work was required. Efficiencies and costs in desalination are expected to be improved.

11.8 WATER PURIFICATION–INNOVATIVE ATTEMPTS

Over the course of history, water has had a significant impact on the lives of people and the development of civilization. Because water is so crucial to our bodily functions, we are completely reliant on it as living organisms. Water pollution has increased in tandem with human activity. In order to eliminate the contamination, water filters have arisen (http://sciencehowstuffworks.com/10-innovations-water-purification.htm), which have been shown to be a success.

To meet people's high expectations for water quality, low-quality technologies were developed. There are a number of factors that contribute to water scarcity:

- Population increase has led to a decrease in groundwater supplies.

- Agriculture irrigation, energy production, and other water-intensive businesses are all sources of waste.
- Poor sanitation and tainted water sources (affecting almost 1 billion people). More than half a million people die each year from diarrhea alone, according to the World Health Organization (WHO).

By removing impurities from the drinking water supply, water purification technologies work (https://www.herox.com/blog/633-ten-awe-inspiring-water-purification-inventions).

Many new technologies have emerged in recent years to supply water to those who need it the most (https://www.theguardian.com/sustainable-business/new-water-technologies-save-planet).

- Nanotechnology in filtration;
- Membrane chemistry;
- Seawater desalination;
- Smart monitoring.

India's Institute of Science won first place at Google Pitch Fest in Switzerland recently with a revolutionary wastewater purification technique. People in disaster-affected areas that have access to water can utilize these devices, but the water they produce is unfit for human consumption. No membranes or chemicals are used, and no water is wasted during the procedure.

Future water treatment technology must meet the following criteria: The technology must have broad applicability, be cost-competitive at large scale, meet regulatory standards, and have a high degree of dependability in comparison to existing technologies.

11.9 NON-FLOWERING TREES INSPIRED LOW-COST WATER FILTRATION

Human health and survival depend on the availability of clean drinking water. Drinking contaminated water can cause diarrhea and the spread of disease (Ask Nature, 2021). Wastewater treatment systems are prohibitively expensive for many countries to install and require specialized parts that must be imported from all over the world.

To cleanse drinking water, MIT engineers used tree branch filters. As part of their research, the MIT team is looking into methods that individuals may be encouraged to regularly practice water purification by giving replacement filters in economical, pay-as-you-go packages that can be produced at bigger sizes utilizing locally accessible resources. New xylem filters have been developed to eliminate pathogens like *E. coli* and rotavirus from contaminated springs, taps, and groundwater in laboratory studies. Non-flowering trees like pine and ginkgo are used to make the water filtering system. The xylem, a thin tissue that serves as a sieve, is found in the sapwood of these branches. Pathogens and bacteria are eliminated from the water as it passes through the xylem. One liter of clean water can be produced every hour using a natural filter formed from sapwood branch cross-sections peeled apart. *E. coli* and rotavirus, two of the most prevalent causes of diarrhea, can also be eliminated by this method.

11.9.1 GYMNOSPERM SAPWOOD ENGINEERING AND CHARACTERIZATION FOR WATER FILTRATION DEVICE DESIGN

Natural membranes found in gymnosperm sapwood xylem tissue can be exploited to produce filters that provide safe drinking water to those living in resource-constrained places. There is, however, little information on the filtering properties of the substance, and issues such as a limited shelf life are not addressed. It is discovered to self-block in this investigation of xylem filters due to a high non-linear relationship between flow resistance and thickness after drying. Gravity-operated xylem filters with shelf lives of more than two years were devised and built by scientists. More than three logs of *E. coli*, MS-2 phage, and rotavirus were removed from synthetic test waters, as were coliform bacteria from contaminated spring, tap, and groundwater, thanks to the filters' ability to remove them from the synthetic fluids. Water filters with daily to weekly-replaceable xylem filters were built using a user-centric approach to discover the social acceptance of xylem as a natural water filter. With this innovative low-cost, biodegradable xylem filtering technology, the world may soon benefit.

11.10 HYDRATION MAY BE TRANSFORMED BY EDIBLE WATER BLOBS IN THE FUTURE

A biodegradable, edible water bottle (water ball) and a ball name "Ooho" were invented by Skipping Rocks Lab. Packaging constructed of seaweed, the water ball is completely natural and biodegradable. You can eat the edible container by itself. To maintain the peripheral clean for consumption, a layer of edible containers can be unwrapped from the main body. There will be no need to waste away millions of single-use plastic bottles when the water ball is introduced in the near future (Schneider, 2017).

First exposed to the public in 2013, the edible "water blob" (http://www.mentalfloss.com/article/68911/edible-water-blob-could-change-hydration-forever) generated news. Three members of London's Skipping Rocks Lab designed the orb: Rodrigo Garca González, Guillaume Couche, and Pierre Paslier (which Fast Company once described as looking like a silicone implant). Keeping a bottle or blob of water with you at all times may not be possible for everyone, but a new device is making it easier than ever to do so.

It's manufactured by layering a seaweed extract membrane on top of a frozen water ball (Oohowater Admin, 2019; Patel, 2019). Because the gelatinous covering is edible, this method is a spin on the culinary technique known as spherification. The water blobs produced by Ooho! since obtaining an EU sustainability prize of $22,500 in 2015 appear to be ready for sale. Ooho! has a lot of possibilities when it comes to conservation: Spherical Ooho! Packaging could one day replace petroleum-based plastic in the United States, where 50 billion plastic bottles are thrown away every year. As long as it isn't a problem for you to bite into something that isn't food, don't worry. You have the option of not eating the water if you don't want to.

One portion of the water is contained in a clear chloride and seaweed sachet. It's biodegradable and even edible, so it's safe to throw it away. The fact that it can be consumed, on the other hand, proves just how natural it is. To be honest, the tap would be more convenient in such circumstances because it's so easy to make at home. People are overjoyed at the prospect of developing a packaging material that is safe enough to eat (Figure 11.3).

An additional patent application (EP 2759212 A1) was submitted by Le Labogroup SAS, which deals with the conveyance of food and drinkables in biodegradable water bottles. When generating a water ball, a membrane forms around a frozen liquid when it is dipped into various calcium chloride and brown algal solutions. For the Soho! bubble, Skipping Rocks developed an edible seaweed extract film that encases drinking water. You can dispose of the flexible bubble-like packaging in 4–6 weeks, just like you would a piece of fruit. As a result of people using single-use bottles, more non-renewable resources are used. Skipping Rocks Lab was founded by master's degree holders in innovation design engineering from the Royal College of Art and Imperial College London. A new production technique has been developed and refined by the design team over the last few years. Plastic bottles are more expensive to create than bubbles. A crowdfunding campaign for an "edible water bottle" that aims to replace the millions of plastic bottles thrown away each year has garnered more than £5,00,000. A natural and biodegradable membrane called Ooho can be consumed and digested in the same manner that regular water can be. Using seaweed extract, which has no flavor on its own but can be flavored, the product is created.

FIGURE 11.3 Edible water bottle.

Source: Reprinted with permission from National Biotechnology. *Copyright © 2017, Springer Nature America, Inc.*

The use of sodium alginate gel creates a water container that resembles an edible water bottle. Using brown algae and calcium chloride, the container's creators created a barrier that traps some of the water. Single-serving plastic bottles can be replaced by biodegradable blobs produced by Skipping Rocks Labs. In exchange for allowing anyone to use the recipe, a Creative Commons license is granted, which safeguards the process of manufacturing.

11.10.1 A BENGALURU-BASED START-UP IS CREATING EDIBLE SEAWEED WATER PODS!

Each water pod costs roughly Rs 2 for 100 ml, which is both cost-effective and environmentally friendly. Workbench Projects is already in touch with a lot of hotels and cafes, so it's even better (https://www.thebetterindia.com/180797/). Bengaluru's efforts to decrease plastic waste may be found here (Patel, Bengaluru, Environment, 2019). Workbench Projects members had a fascinating debate about the Skipping Rocks Lab water pod when one of them inquired about the following. These water pods, which were recently utilized at the London Marathon, are edible packets that contain water or a sports drink in order to limit the usage of disposable bottles. At the time, Richard Gomes, a former Manipal Institute of Technology student, was engaged in a vermicomposting project. Apparently, he was looking at a variety of ways to make a water capsule that could be eaten (Mistry & Kharate, 2020).

"To think that no one in India has looked into how simple the process was is astonishing. It was a no-brainer for me to create water pods as a biohacker since I believe that nature can be harnessed to create valuable tools for the human race." Richard tells us about a book called The Better India.

When it comes to making edible pouches, it is how they are made (Mistry & Kharate, 2020, https://www.neliti.com/media/publications/336039, a study of edible water balls in India.

- An Alginate polymer and water mixture is frozen in an ice tray at–10°C.
- The ice is then dipped into a biochemical combination once it has frozen. This mixture is safe for human ingestion because it contains only a little amount of each of these substances, says Richard.
- As a result of the biochemical mixture, the alginate hydrogel can form a pouch around the ice. The liquid pouch is ready once the ice has melted.
  We heat it till the alginate forms a membrane around the cubes, air bubbles appear, and they round out in shape, Richard told the Bangalore Mirror.
- Each pod is then submerged in a clean glass of water to remove any remaining gel.

Currently, the team is looking into fruit juice, medicine, and sports drinks as suitable water substitutes. Because they are so delicate, the pods are difficult to use. In the end, plastic bottles have mostly taken the place of glass or clay containers. This does not, however, mean that the water pods are limited in variety. It is possible that India will follow London's lead and add them to their races. Richard mentioned that it might also be used to analyze liquid samples. Using water pods during social or informal gatherings can also significantly reduce the quantity of plastic used.

11.11 POTABLE WATER CAN BE PRODUCED USING MICROBIAL BIOTECHNOLOGY

Water, which makes up around 70% of all land and water, is unquestionably the most valuable natural resource on the planet. It contains both hydrogen and oxygen. Water is a complicated system of chemical species when it is found in nature. Water is an excellent solvent because of its high heat absorption capability compared to other liquids. Because of water's high surface tension, it is an excellent solvent for dissolving a wide range of substances, including minerals. A solvent's solid state must be lighter than its liquid state in order to support aquatic life. The typical person requires 7.5 liters of water every day to maintain their health (WHO recommends). It is advised that each person drink a minimum of 20 liters of water each day to maintain appropriate sanitary conditions. That's enough freshwater for the whole population of 7.5 billion people per day. In wealthy countries, people tend to overconsume safe water, despite the fact that there are only limited supplies in other parts of the world. Global water needs can be met by freshwater, but infrastructure in some locations is lacking to ensure safe water production and distribution. UN SDG 6 aspires to make the world a global supply of safe drinking water for everyone. As a result, this goal is strongly linked to the goals of SDG 13 (climate change) and SDG 12 (sustainably consuming water and reducing pollution).

Sustainable Development Goal 1 aims to ensure that everyone has access to safe drinking water. The grade of safe drinking water is such that it poses no major harm to health when consumed over the course of a lifetime. It is possible that the pathogenic burden during water treatment will be reduced by using physical and chemical disinfection procedures in addition to these. Pathogens are less likely to infiltrate a supply chain when a product has been made physiologically stable through the use of biofiltration methods like those described by Fowler & Smets (2017). Biological water treatment technologies can eliminate a wide variety of chemical pollutants. Compared to modern chemical or physical treatment procedures, they are less expensive and less energy-intensive and can be used in a wide range of circumstances and water quality. Fowler & Smets (2017) argued that increasing biological process exploitation is essential to this goal because of their cheap economic and energetic costs.

11.12 MEMBRANE WATER FILTRATION WITH SOLAR POWER

Water, which covers 75% of the Earth's surface, is the most plentiful natural resource on the planet. The amount of free-flowing fresh water in the world's ecosystems remains relatively constant. The amount of water that humans are generating greatly exceeds the amount of water that is actually available. Since the industrial revolution, population growth, rising living standards, and global warming, the freshwater market has grown at an alarming rate. Freshwater supply is a growing concern in many places these days. By eliminating them from the water supply, water purification can eliminate undesired chemicals, biological poisons, suspended sediments, and gases. Water shortages sparked a public outcry. Lack of access to safe drinking water is one of the most important worldwide issues today. Water treatment improvements and rising energy costs could restrict the future availability of safe drinking water. Treatment systems for drinking water are standard in more developed nations like the United States. Because of the abundance of water supplies, thermal and membrane-based processes are dependent.

It's impossible for many countries and economies, notably those in Asia and Africa, to be sustainable or thrive without appropriate water resources. Membrane-based technologies can be used alone or in conjunction with other membrane-based approaches to purify water and generate sustainable energy. There are a variety of ways that renewable energy sources like hydropower, geothermal heat pumps, and wind turbines can be used to generate electricity. Industrial settings can benefit from this procedure because it is less expensive and easier to regulate (Zewdie et al., 2021). When considering water purification methods, RO and electrodialysis should be taken into account.

Due to their utilization of low-cost heat sources, membrane-based water purification systems are critical for water purification. Because their services were both cost-effective and technically practicable, small, medium, and large-scale applications could all benefit from them. Water purification systems may be powered by solar electricity in the future. Solar energy can provide the potable water needed in the future. A reliable supply of safe drinking water is a major issue for many rural populations in semi-arid and arid parts of Asia and Africa. For the purpose of producing safe drinking water, it is imperative that we learn more about how solar energy can be used in conjunction with water treatment equipment. In semi-arid and dry environments, solar-aided membrane-based water filtration systems may be feasible options. A solar water treatment system can be powered by solar thermal collectors or photovoltaic panels. In this investigation, solar-powered membrane-based water filtration technologies and precise energy consumption are employed. Membrane distillation and hydrolysis, RO and electrodialysis, and hybrid systems were all used in this investigation (Zewdie et al., 2021). This work will also improve solar-aided membrane-based water filtration technology, giving researchers a fuller understanding of other researchers' efforts. Solar-powered membrane-based water filtration systems have better energy efficiency. RO, a solar-powered membrane-based water purification system, is the most extensively investigated water purification approach for small, medium, and large-scale water treatment systems. Recent years have seen a resurgence in the efficiency of solar electricity generation when combined with the most prominent new separation technology, such as the MD. Membrane-based water purification systems are less expensive than solar-powered non-membrane approaches.

According to Zewdie et al. (2021), we need more theoretical and experimental research in this area. Membrane-based separation technology advancements will undoubtedly improve solar-powered water treatment and desalination. In the future, membrane-based water filtration systems will be proved to be economically feasible and have the potential to become increasingly energy efficient, cost effective, and environmentally benign.

Fir trees could play an important role in solving the clean water crisis. This technology could prove a lifesaver for many people around the world who lack access to clean water. In the absence of clean water production from trees, it is possible that they can provide a simple and inexpensive filtration method for the water. Researchers have used tree branches to filter water in the lab and out in the wild. An inexpensive, biodegradable, and low-tech alternative to more expensive methods of supplying impoverished regions with safe drinking water may be found in this device. In a recent study, researchers found that the natural structure of wood from non-flowering plants like pine and cedar can eliminate *E. coli,* rotavirus, and other bacteria from water. This method was successfully tested in a lab in 2014 and was shown to be effective. Researchers have discovered that the tree-based purifying system works just as well in the field as it does in India, where the villagers are willing to use it.

Thin straw-like tubes called xylem are used to build branches in the sapwood of gymnosperms or non-flowering plants. According to Ramchander et al. (2021), a graduate student at MIT and the study's primary author, the specific shape of the xylem tissue in branches allows it to function as a water filter since these channels connect through porous membranes (Chu, 2014). Water filter membranes consisting of synthetic materials, such as those in hiking or camping water filtration systems, can "entrap" pollutants. Particulates, bacteria, and protozoa can't get past the membrane pores, which are only a few nanometers to a few 100 nanometers wide. For example, Ramchander noted that a 4-cm x 1-cm wood disc can be built to work like a standard gravity-based water filter. In a vessel, water is poured into a tube that hangs below. Water drips into a container at the bottom of the setup at a rate of one liter per hour after passing through the wood xylem filter at the end of the tube (which is still quite a bit slower than other options on the market). Many other designs are feasible, and replacing the filter disc is a simple and inexpensive process, as demonstrated by Ramchander.

Using a hot water and ethanol mixture as a protective coating, the researchers were able to keep dry wood filters fresh for up to two years. A two-star "complete protection" rating from the WHO classifies the wood xylem filters as the second-highest level of domestic water treatment in the WHO's classification system. Around 88% of the deaths are due to hazardous drinking water and sanitation difficulties, which is why the researchers decided to focus their efforts on India, Ramchander explains. Over a million Indians lack access to clean water, making it the most dangerous country in the world in terms of waterborne sickness mortality, according to the WHO. As it turned out, the natural chir pinewood of India was the perfect material for the job of securely filtering water in the actual world. In real-world assessments, the effectiveness of the technology is heavily influenced by factors such as usability and accessibility.

To make this potentially life-saving technology accessible to those who need it, the filter's construction has been made open source. You can find detailed instructions on how to build your own wood water filter on their website. According to Rohit Karnik, a mechanical engineering professor at MIT and research author, "We have had an enormous level of interest from students globally and non-profits and entrepreneurs." People from underdeveloped countries want to use readily available resources to make water purification more accessible, says the company's CEO. "We believe that our work encourages such persons to further develop and market xylem water purifiers customized to local demands for the benefit of international communities." It may take some time before xylem filters are widely used, but the issues of logistics, quality control standards, and production are being tackled, adds Karnik. An NGO is helping us to build up a pilot production factory, develop the business model, and conduct pilot studies of filter use.

11.13 MIT AND IIM-A RESEARCHERS ARE WORKING ON NEW, LOW-COST WATER FILTER TECHNOLOGIES

Innovation in water filtration is needed for low-income consumers. Comprehensive Initiative on Technology Evaluation (CITE), an Ahmedabad-based organization that tests the efficacy of water filters in homes, worked with the Indian Institute of Management Ahmedabad (IIM-A) and the Massachusetts

Institute of Technology (MIT) on a study to discover new, low-cost water filter options. Nine main brands in Ahmedabad, India, had their water filters tested by CITE experts, who looked at more than 100 different models. They gathered information from nearly 400 people, including shoppers, merchants, distributors, and producers. Cloth/jali mesh filters, gravity non-electric filters, and RO filters were the three main divisions of these devices.

RO is widely regarded as one of the most effective water filtration processes. RO systems can be used in a variety of settings, including whole-house filtration, faucet filtration, aquarium filtration, and restaurant water filtration. An RO system is likely to be able to handle any type of water you start with. A system that uses RO "For every 26 liters of water they filter, they use 74 liters of water. Those in water-strapped areas like Ahmedabad and India as a whole are unable to sustain them. Additionally, environmentally friendly RO filters for individual homes must be developed." The findings of the CITE research for low-income consumers there is room for innovation in the market for residential water filters.

KEYWORDS

- **biofouling**
- **bio-purification of water**
- **edible water blobs**
- **membranes**
- **sustainable water**
- **water purification**

REFERENCES

Abioye, O. P., Loto, C. A., & Fayomi, O. S. I., (2019). Evaluation of anti-biofouling progresses in marine application. *J. Bio Tribo Corros*, *5*, 22. https://doi.org/10.1007/s40735-018-0213-5.

AlSawaftah, N., Abuwatfa, W., Darwish, N., & Husseini, G. (2021). A Comprehensive review on membrane fouling: Mathematical modeling, prediction, diagnosis, and mitigation. *Water, 13*, 1327. https://doi.org/10.3390/w13091327.

Anjum, F., M. Khan, I. M., Kim, J., Aslam, M., Blandin, G., Heran, M., & Lesage, G., (2021). Trends and progress in AnMBR for domestic wastewater treatment and their impacts on process efficiency and membrane fouling. *Environmental Technology & Innovation, 21*, 101204. https://doi.org/10.1016/j.eti.2020.101204.

Ask Nature, (2021). *Simple Filtration System from MIT Contains an Abundantly Available Resource, Xylem Tissue from Native Trees, That Naturally Filters Pathogens from Water.* MIT, https://asknature.org/innovation/low-cost-water-filtration-inspired-by-non-flowering-trees/ (accessed on 17 August 2023).

Aslam, M., Ahmad, R., & Kim, J., (2018). Recent developments in biofouling control in membrane bioreactors for domestic wastewater treatment. *Separation and Purification Technology, 206*, 297–315. https://doi.org/10.1016/j.seppur.2018.06.004.

Barath, H., (2015). *Innovative Water Purification Technology Wins Google Award.* https://indiabioscience.org/news/2015/innovative-water-purification-technology-wins-google-award (accessed on 17 August 2023).

Bixler, G. D., & Bhushan, B., (2012). Biofouling: Lessons from nature. *Philosophical Transactions of The Royal Society a Mathematical Physical and Engineering Sciences, 370*, 2381–2417.

Chu, J., (2014). MIT news office. *MIT Engineers Make Filters from Tree Branches to Purify Drinking Water.* https://news.mit.edu/2021/filters-sapwood-purify-water-0325 (accessed on 17 August 2023).

Cui, Y., Gao, H., Yu, R., Gao, L., & Zhan, M. (2021). Biological-based control strategies for MBR membrane biofouling: A review. *Water Sci Technol., 83*(11), 2597–2614. doi: 10.2166/wst.2021.168.

Dwivedi, R., (2019). *Biomimetic Nanotechnology Enhancing Water Purification.* Azo Nano. https://www.azonano.com/article.aspx?ArticleID=5213 (accessed on 17 August 2023).

Eat a Water Bottle or Two. *Nat Biotechnol* **35,** *495* (2017). https://doi.org/10.1038/nbt0617-495a

Falinski, M. M., Turley, R. S., Kidd, J., Lounsbury, A. W., Lanzarini-Lopes, M., Backhaus, A., Rudel, H. E., Lane, M. K. M., Fausey, C. L., Barrios, A. C., et al., (2020). Doing nano-enabled water treatment right: Sustainability considerations from design and research through development and implementation. *Environ. Sci.: Nano, 7,* 3255–3278.

Fetyan, N. A. H., & Attia, T. M. S., (2020). Water purification using ultrasound waves: Application and challenges. *Arab J. Basic Appl. Sci., 27*(1), 194–207. doi: 10.1080/25765299.2020.1762294.

Fowler, S. J., & Smets, B. F., (2017). Microbial biotechnologies for potable water production. *Microbial Biotechnology, 10*(5), 1094–1097. https://doi.org/10.1111/1751-7915.12837.

Gu, J. D., (2005). Chapter 9–Biofouling and prevention: Corrosion, biodeterioration, and biodegradation of materials. In: Myer, K., (ed.), *Handbook of Environmental Degradation of Materials* (pp. 179–206). William Andrew Publishing. https://doi.org/10.1016/B978-081551500-5.50011-2.

Harkness, B. *Ten Awesome Water Purification Inventions.* https://www.herox.com/blog/633-ten-awesome-water-purification-invention (accessed on 17 August 2023).

https://media.neliti.com/media/publications/336039-a-study-of-edible-water-balls-in-indian-a347d0fa.pdf (accessed on 17 August 2023).

https://www.marineinsight.com/tech/4-types-of-anti-fouling-systems-used-on-board-ships-to-prevent-marine-growth/ (accessed on 17 August 2023).

https://www.mentalfloss.com/article/68911/edible-water-blob-could-change-hydration-forever (accessed on 17 August 2023).

https://www.surfertoday.com/environment/what-are-edible-water-bottles (accessed on 17 August 2023).

Kiger, P. J. *10 Innovations in Water Purification.* How stuffworks, https://science.howstuffworks.com/environmental/green-tech/sustainable/10-innovations-water-purification.htm (accessed on 17 August 2023).

Kumar, S., Ye, F., Dobretsov, S., & Dutta, J., (2021). Nanocoating is a new way for biofouling prevention. *Front. Nanotechnol.* https://doi.org/10.3389/fnano.2021.771098.

Lian, X., Shi, D., Ma, J., Cai, X., & Zhongwei Gu, Z., (2018). Peptide dendrimer-crosslinked inorganic-organic hybrid supramolecular hydrogel for efficient anti-biofouling. *Chinese Chemical Letters, 29*(3), 501–504. https://doi.org/10.1016/j.cclet.2017.08.014.

Ling, S., Qin, Z., Huang, W., Cao, S., Kaplan, D. L., & Buehler, M. J., (2017). Design and function of biomimetic multilayer water purification membranes. *Sci. Adv., 3*(4), e1601939. doi: 10.1126/sciadv.1601939.

Liu, J., Cheng, S., Cao, N., Geng, C., He, C., Shi, Q., Xu, C., et al., (2019). *Actinia*-like multifunctional nanocoagulant for single-step removal of water contaminants. *Nat. Nanotechnol., 14*(2), 64–71. doi: 10.1038/s41565-019-0378-1.

Mistry, U., & Kharate, S. P., (2021). A study of edible water balls in the Indian market context. *Journal NX – A Multidisciplinary Peer Reviewed Journal*, 51–55.

Nagar, A., & Pradeep, T. (2020). Clean water through nanotechnology: Needs, gaps, and fulfillment. *ACS Nano, 14*(6), 6420–6435. doi: 10.1021/acsnano.9b01730.

Neal, C. T. S., Wang, H., Sun, G., Duong, H. H. P., & Zhong, P. S., (2016). Aquaporin-incorporated biomimetic membranes. In: Drioli, E., & Giorno, L., (eds.), *Encyclopedia of Membranes.* Springer, Berlin, Heidelberg. https://doi.org/10.1007/978-3-662-44324-8_812.

Nguyen, T., Roddick, F. A., & Fan, L., (2012). Biofouling of water treatment membranes: A review of the underlying causes, monitoring techniques, and control measures. *Membranes (Basel), 2*(4), 804–840. doi: 10.3390/membranes2040804.

Oohowater Admin, (2019). Ooho Water, The edible bottle. OohoWater.com.

Patel, T., (2019). *Exclusive: Meet the Bengaluru Startup Making Edible Water Pods from Seaweed!* https://www.thebetterindia.com/180797/bengaluru-innovation-edible-water-pod-seaweed-plastic-free-india/ (accessed on 17 August 2023).

Pichardo-Romero, D., Garcia-Arce, Z. P., Zavala-Ramírez, A., & Castro-Muñoz, R., (2020). Current advances in biofouling mitigation in membranes for water treatment: An overview. *Processes, 8*(2), 182. https://doi.org/10.3390/pr8020182.

Qu, X., Brame, J., Qilin, L. Q., & Alvarez, P. J. J., (2013). Nanotechnology for a safe and sustainable water supply: Enabling integrated water treatment and reuse. *Accounts of Chemical Res., 46*(3), 834–843. doi: 10.1021/ar300029v.

Ramchander, K., Hegde, M., Antony, A. P., Wang, L., Leith, K., Smith, A., & Karnik, R., (2021). *Engineering* and *characterization* of *gymnosperm sapwood toward enabling* the *design of water filtration devices. Nature Communications, 12*, Article No. 1871. https://www.nature.com/articles/s41467-021-22055.

Richards, C., Slaimi, A., O'Connor, N. E., Barrett, A., Kwiatkowska, S., & Regan, F. (2020). Bio-inspired surface texture modification as a viable feature of future aquatic antifouling Strategies: A review. *Intern. J. Mol. Sci., 21*(14), 5063. https://doi.org/10.3390/ijms21145063.

Romeu, M.J., & Mergulhão, F. (2023). Development of antifouling strategies for marine applications. *Microorganisms 11*(6), 1568; https://doi.org/10.3390/microorganisms11061568

Schneider, C., (2017). *This Edible Water Blob Could Change Hydration Forever*. https://www.mentalfloss.com/article/68911/edible-water-blob-could-change-hydration-forever (accessed on 17 August 2023).

Schultz, M. P., (2007). Effects of coating roughness and biofouling on ship resistance and powering. *Biofouling, 23*(5), 331-341. doi: 10.1080/08927010701461974.

Science News, (2018). *A Water Treatment Breakthrough, Inspired by a Sea Creature*. Yale University. https://www.sciencedaily.com/releases/2018/11/181126142856.htm (accessed on 17 August 2023).

Tamburri, M. N., Davidson, I. C., First, M. R., Scianni, C., Newcomer, K., Inglis, G. J., Georgiades, E. T., et al., (2020). In-water cleaning and capture to remove ship biofouling: An initial evaluation of efficacy and environmental safety. *Frontiers in Marine Science*, 437. https://doi.org/10.3389/fmars.2020.00437.

Tian, L., Yin, Y., Bing, W., & Jin, E., (2021). Antifouling technology trends in marine environmental protection. *J. Bionic Engineering, 18*, 239–263. https://doi.org/10.1007/s42235-021-0017-z.

Zewdie, T. M., Habtu, N. G., Dutta, A., & Van, D. B. B., (2021). Solar-assisted membrane technology for water purification: A review. *J Water Reuse and Desalination, 11*(1), 1–32.

CHAPTER 12

Nature Inspires Healthcare and Modern Medical Technologies

ABSTRACT

Henry David Thoreau mentioned that Nature is doing her best in each moment to make us well by her. Nature has another name called health. Contact with Nature is tied to health in an abundance of studies. It's safe to say that healthcare is the most crucial area to profit from the use of new technology. Nature-inspired medical gadgets have been around for a while, but their most important attribute is well-known. Nature-inspired medical gadgets have been identified by scientists as being of great importance. Bio-inspired breakthroughs influencing the development of medical technologies are described in this chapter. Biosensors and needles are inspired by parasitic wasps, peacocks, and electric eels. Inspire by parasitic worms as a power source for artificial organ construction. Following the laws of nature produces Theragrippers. In the next few decades, biomimetics will be transformed by AI and machine learning approaches. It's been shown that biomimicry and artificial intelligence (AI) have the ability to significantly improve human health. With the help of recent technology advancements, complex biomedical systems have been developed in the fields of biomedical informatics and molecular medicine. Computational intelligence is incorporated into innovative clinical and computer-based decision support systems, knowledge acquisition and management, medical imaging, and healthcare organizations. Doctors and patients can benefit greatly from AI in terms of productivity, efficiency, workflow, accuracy, and speed. A wide range of advantages and benefits can be gained from various types of healthcare. However, the methodologies employed to support the majority of AI and healthcare technologies differ substantially. The use of various IT tools and

Bio-Inspired Technologies for the Modern World: Emerging Trends and Applications.
R. Ramakrishna Reddy & T. Pullaiah (Authors)

software by hospitals has grown in importance in recent years. Telemedicine makes it easier to connect with medical experts. Virtually eliminating the need for patients and healthcare providers to travel to the doctor's office is the major goal of telemedicine. Supplementing with vitamin D3 has been shown to have numerous advantages for the elderly. There has been a slew of medicinal uses proposed for nanotechnology in recent years. IoMT aims to connect medical devices and applications to healthcare IT systems via internet computer networks (IoMT). The Internet of Things can be used to monitor and diagnose the health of the target structure of interest. It's easy to see how AI is assisting in the eradication of long-standing healthcare issues.

12.1 NATURE INSPIRES HEALTHCARE

Cutting-edge technology may consider sea worms, jellyfish, geckos, and spiders' odd inspirations. Medical advancements, such as novel adhesives, diagnostics, and needles, have been spurred in part by these critters and are now beginning to make their way into clinical settings. Nature has figured out through millions of years how to create a wide variety of materials and methods for carrying out its tasks. There have been several attempts, dating back to ancient civilizations, to heal and treat the human body by drawing inspiration from nature. A variety of natural resources have been used to offer raw materials for the fabrication of prosthetic devices that can be used *in vivo*. Nature is a source of technical inspiration for architects, engineers, and researchers in the medical field. There is a belief that human ingenuity is the source of future innovation and that nature inspires healthcare (https://www.ultracleansystems.com/how-nature/). Identifying research and addressing scientific issues are becoming increasingly difficult, despite our technological prowess. Research and experimenting necessitates careful consideration of time, resources, prices, and personnel. Nature has a hand in it. It's called biomimicry (*bio*= life + *mimicry* =to resemble), and it's happening all around us. Native Americans have traditionally relied on the medicinal powers of numerous flowers, tree bark, and plants. For example, the poppy flower can be used to alleviate both anxiety and pain. Every day, scientists uncover new species that could hold the key to medical advancements that are desperately required.

Humans have studied nature for thousands of years, collecting everything from animal teeth and spears to fish fins and paddles out of pure curiosity and the desire to learn from nature in order to create new things. Observing the morphology of creatures in nature is becoming increasingly outmoded as new technologies allow us to better understand their intricate compositions and how they carry out their various roles. As a result of this, people are now using natural inspiration to create products like shark skin swimwear, buzzard flaps for aircraft, and self-cleaning apparel inspired by lotus leaf wettability. Examples like these are endless. Modern technological design and engineering challenges can be met with the help of bio-inspired strategies that draw inspiration from biological processes and natural techniques. Through enhancing the body's immune system, nature has a good impact on health. Spending time in nature has been shown to have a positive impact on one's well-being. Nature has a wide range of health consequences, including depression and anxiety problems, diabetes mellitus, and attention deficit/hyperactivity disorder (ADHD). Patients can now benefit from more precise diagnostics and treatment options due to the growing usage of medical technology.

Biosensors, which aim to identify and prevent disease before it has a chance to take hold of the body, are the most interesting area for new health-care solutions. The advantages of reducing hospitalization include that fewer medications are required. Products that originate from nature's biosystems, such as sensors, are the most common. Electrical signals are converted and interpreted by a biosensor, which is a type of sensor. Using a computer or a brain, they can decipher this information. Millions of biosensors are found in all living things, including humans.

Less-invasive medical techniques are becoming more popular. We're seeing a new generation of medical technologies that are inspired by natural phenomena. Japanese micro-engineers have created a painless injection process after analyzing mosquito bites. Blood sampling and drug administration are about to become much easier for doctors, thanks to the development of a needle that looks like a mosquito's proboscis. A mosquito's beak vibrates while it injects, allowing its maxillae (serrated parts) to slide smoothly into the skin. Medical issues can be addressed through the use of animals and plants.

Traditional remedies, such as heart disease, dementia, and malaria, have been utilized to treat a wide range of global health challenges using pharmaceuticals developed from or inspired by plants and fungi. Biodiversity has been lost as a result of the global desire for medicinal plants and mushrooms. We pause to consider the evolving role of plants and fungi in global healthcare as new risks to human health and biodiversity emerge. Nature-based therapies' prospects are brightening as new information about human health and biodiversity emerges. As a result of scientific advancements, many plants and fungi from around the world may now be studied for their therapeutic capabilities, which have shown the value in protecting biodiversity. This has helped to illustrate the value of natural capital as a motivation for biodiversity conservation. As fresh data on human health and biodiversity emerges, the outlook for nature-based medicines is looking up. There are currently many plants and fungi from throughout the world that can be examined for their therapeutic properties because of scientific developments. This has demonstrated the need to protect biodiversity for the sake of our natural capital. Emerging technologies, on the other hand, give hope for the future preservation of life-saving medications. Interactions between humans, plants, and fungi reveal the importance of natural resources to our health and well-being.

It's important to note that the natural and healthcare industries are interdependent. The advancement of cutting-edge medical technologies and treatments in the 21st century has made it possible for people to live longer, healthier, and happier lives because of the better tissue healing, faster disease detection, and improved rehabilitation tools made possible by these advancements. Medical and healthcare sectors can benefit greatly from biomimicry's proven economic success areas, such as wound healing and cosmetics. To develop effective treatment and prevention techniques in a variety of health categories, healthcare researchers are actively drawing inspiration from the natural defense mechanisms of an organism. These solutions can be natural, bioderived, or synthetic. What can the Internet, social networks, artificial intelligence (AI), and even biomimicry contribute to support long-term economic, social, and environmental progress?

12.2 ADVANCES IN HEALTHCARE SCIENCE AND TECHNOLOGY

Various IT technologies and software are used by the healthcare industry to increase hospital and organization productivity, generate new ideas for drugs and treatments, and enhance the standard of care. The "Insurtech" industry provides more people with access to quality healthcare. Hospitals are properly utilized, and patients don't have to wait much longer.

Technology use in healthcare has improved patient diagnoses and treatment over time (Jenkins, 2022). As a result, it helped people live longer and healthier lives while also saving countless lives. Information technology (IT) can be used to keep track of and access medical records for patients, according to the ied.eu blog. It also makes it more convenient for patients to access information in a clear and understandable way. As a result, it is less likely that medication errors may occur. The specter of COVID-19 can be felt around the world today. The global epidemic has had the greatest impact on the healthcare business. The healthcare business has developed in new ways to maintain the same high level of service quality in light of substantial technological breakthroughs and the growing digitalization of protected health information (Figure 12.1).

FIGURE 12.1 Medical banner with doctor working on laptop:

Source: Reprinted from Freepik. https://www.freepik.com/free-photo/medical-banner-with-doctor-working-laptop_30555907.htm#query=healthcare%20system&position=0&from_view=keyword&track=ais&uuid=5693bed8-9a5f-4c87-bfdb-6c6bf5826ce7".

One of our society's most talked-about and critical topics is how to better provide healthcare and keep people healthy. New and improved ways to deal with today's most important health challenges can be found through the use of cutting-edge technology in the healthcare industry. A world without technology is unfathomable to most of us. Regardless of our physical appearance, we all rely on technology in our daily lives, from cell phones to laptops. It happened the same way in medicine. As technology continues to progress, it is only a matter of time before it has a profound impact on the medical industry as a whole. Currently, the field of healthcare is one of the fastest growing. In scientific study, health science has reached the point where it can evaluate the health safety of any emerging or developed country. As a result, it helped people live longer and healthier lives while also saving countless lives.

Continuous monitoring of one's health status is made possible through the use of personalized healthcare (Xiao et al., 2021). Wearable Bioelectronics is a rapidly expanding interdisciplinary study topic that involves the integration of biomaterials with electronics in a variety of biodevices, including biofuel cells, biosensors, ingestible, and implantable. Because of its portability, low cost, and simplicity of use, flexible bioelectronics have the potential to revolutionize personalized healthcare in the near future. Deep learning based on neural networks is becoming increasingly crucial in healthcare as AI advances, raising the bar for medical devices' performance needs. As it happens, a wide range of biological species are able to teach us important lessons about how to develop wearable bioelectronics capable of completing previously inconceivable tasks. The human epidermis served as an inspiration for academics working on overlapping receptive fields. It is clear that bio-inspired approaches can be used to create artificial peripheral nerve systems with extraordinarily high sensitivity. Biomimicry is a useful method for reimagining technology in the field of biomedical engineering (https://www.asme.org/topics-resources/content/nine-bioinspired-medical-technologies). Natural events are inspiring a new generation of medicinal technology. Innovators in the medical field are being influenced by bioinspired discoveries (Sakes et al., 2020).

12.2.1 NEEDLE INSPIRED BY PARASITIC WASPS

Laying eggs inside a caterpillar is no problem for the parasitic wasp's female ovipositor, which is shaped like a needle. The larvae thrive in the caterpillar's body because of the ideal environment it provides. When the larvae mature, they break through the caterpillar's body, resulting in its death. To create their ovipositor-like needle, scientists Paul Breedveld and Johan van Leeuwen of Delft University of Technology and Wageningen University researched this terrible process. Seven different rods are used in the needle, one of which is permanently attached to the needle head (see also: https://www.advanced-medical-technology.org/content/9). One rod at a time can be moved to bend the needle into a S shape with very little external force. It is perfect for delivering drugs and performing delicate treatments in difficult and delicate areas of the body.

12.2.2 BIOSENSORS INSPIRED BY THE PEACOCK'S TAIL

A variety of elements influence the coloration of a butterfly, beetle, or even a peacock's wings, which can range from blue to green to purple.

Scientists have been inspired by the biomimicry of butterfly and peacock feathers. Sensors based on a novel opal-like substance inspired by peacock and butterfly feathers could be on the way (see: https://tribune.in.com). Photonic crystals that change color in response to temperature, chemical changes, strain, and other stimuli, including light, have been developed by engineers at the Universities of Surrey and Sussex. Medical applications for the crystals include biosensors for the detection of viruses in the respiratory system, biomonitoring of physical performance, and even healthcare safety to determine the cleanliness of the surface.

12.2.3 CATHETER INSPIRED BY ELECTRIC FISH

Inspired by the electrolocalization behavior of weakly electric fish (Figure 12.2), a novel catheter guidance system for interventional vascular procedures was introduced in health fields.

FIGURE 12.2 An electric blue ram cichlid.

Photo credit: BlueRam92. https://creativecommons.org/licenses/by-sa/4.0/

The Johns Hopkins University catheter eliminates the need for fluorescent dyes and radiation. Bioinspired medical solutions are inspired by the ability of a fish to generate an electrical field that can travel beyond space. In addition, it uses electroreceptors to detect prey's electric fields. Similar to the Johns Hopkins catheter, the electrodes on the tip of the catheter transmit information about its location in a blood vessel. Engineers can utilize this information to direct the catheter to the precise spot where it is needed. With this new technique, patients who can't have fluoroscopy operations will have more options.

12.2.4 ARTIFICIAL ORGANS TO FEATURE ELECTRIC EEL-INSPIRED POWER SOURCE

The electric eel, or knife fish, likewise uses electricity to shock and stun its prey by generating fast pulses. Electrolytes, the charged cells that make up electric organs, were explored by engineers at the University of Michigan in Ann Arbor. An eel's shock can be as high as 600 volts thanks to the energy that can be gathered from thousands of these focused cells. Engineers created a synthetic organ that can generate up to 100 volts using a combination of hydrogels and salt. It's just enough juice to keep pacemakers and other minor medical devices powered up without the need for additional power sources. It could lead to a device powered by the body's own natural activities, such as respiration.

12.2.5 "THERAGRIPPERS" INSPIRED BY THE PARASITIC WORM

Inspired by parasites, an inventor came up with an idea for an implantable microdevice. When a hookworm gets inside an animal's digestive system, it latches onto the gut and begins feeding on its blood. For the purpose of developing an intestinal-embedding microdevice, Johns Hopkins researchers David Gracias and Florin Selaru researched this parasitic worm. Smaller than a dust particle, the theragripper is a star-shaped contraption constructed of a thin metal sheet covered in paraffin wax. After entering the digestive system, wax melts, and the device is able to remain in the digestive tract for a long amount of time, allowing the drug to be delivered gradually. Current drugs don't have time to take action before being flushed from the body, but theragrippers may provide a remedy for many GI disorders that are difficult to treat.

12.2.6 LONG-TAIL GLASSWING BUTTERFLY

To distinguish between male and female butterflies, the bright colors of a butterfly are used as a form of camouflage to evade predators. Known as "little mirrors," long tail glasswing butterflies can be found in Central and South America. As a result of their transparent wings, they are virtually invisible to predators. Would you be interested in learning that their wings have inspired researchers to develop new treatments for glaucoma? These transparent wing portions sparked the interest of scientists at the California Institute of Technology (Caltech), who discovered something unusual. A human hair is around 50–100 times smaller than the distance between these tiny pillars, which are each about 100 nanometers in diameter. Ocular pressure should be checked more regularly in patients with glaucoma in order to prevent further damage. It is possible to monitor glaucoma patients' intra-eye pressure using an implanted sensor that is put into the eye. It is possible for patients to view data from their implants on their cellphones, giving clinicians the ability to respond to problems as they develop.

12.2.7 HORSESHOE CRAB

Around 450 million years have passed since the first horseshoe crabs appeared on Earth. The blue blood is what really draws people to these creatures, not their 10 eyes or seven pairs of legs. The red hue of our blood is caused by the iron in our blood, not the copper it utilizes to bind oxygen. There is so much money to be made from their blood that a single quart of it might fetch up to $15,000. Their blood is extremely pricey because of this. The blood of horseshoe crabs is extremely sensitive to external substances. When *E. coli,* fungi, or bacteria come into contact with horseshoe crab blood, a substance known as Limulus amebocyte lysate (LAL) forms clots quickly. Within three days of being caught, about 30% of the fish's blood is drained and returned to the ocean, where it can be replenished, however, some fish do not survive the bleeding. The FDA mandates that all new medicine and surgical implant tests be free of bacterial contamination before they can be released into the market. Horseshoe crabs can be spared by utilizing a synthetic counterpart that has the same impact as the real thing (https://www.asme.org/topics-resources/content/9-bioinspired-medical-technologies).

12.2.8 SANDCASTLE WORM

The three-inch-long sandcastle worm lives off the coast of California. They secrete an adhesive and stick fragments of sand and broken seashells together, creating a tube-like shelter to protect them from severe underwater conditions. It was revealed by bioengineers that sandcastle worm glue is denser than water, does not dissolve in water, and is as strong as superglue. An insoluble synthetic water-based glue has been discovered by researchers that can adhere to wet materials even in moist environments. Small bone pieces can be permanently glued into place, removing the requirement for external fixation devices such as plates and screws.

12.2.9 MEDICAL TECHNOLOGY INSPIRED BY NATURE

New medical and high-tech advancements are being sparked by animal inspiration, and these concepts include painless injections, feigning foes in military conflict, and improved body armor, among other things. Squirrels, for example, inspired researchers at the Georgia Institute of Technology to build robots that can mislead each other. While other animals are distracted, squirrels visit bare burrows in an attempt to grab food from them. Predator robots are lured to the wrong areas by these robots, delaying their discovery of protected resources.

It is impossible to get spider webs wet. To speed up the healing process of wounds, researchers have developed a glue that replicates the natural stickiness found in barnacles. Spider webs were often employed by ancient Greek and Roman medics to produce bandages for their patients (https://www.dailymail.co.uk/health/article-9941011/The-medical-breakthroughs-inspired-animal-magic.html).

12.2.10 BIO-ADHESIVES IN HEALTHCARE

Humanity has mastered the analysis and exploitation of a wide variety of natural resources. Adhesives for medical use are strong and simple to use. Structure, non-structure, and pressure sensitivity are the three types of adhesives. Adhesives must be able to adhere to the substrates and the specific needs of the application. Adhesives, backing substrates, and coating techniques should all be carefully considered. The final product is almost always made with synthetic or biological formulations. Examples of common medical adhesive applications include surgical instruments, biosensors, electrode labeling, catheters, implanted devices, and wearable devices. The use of medical adhesives to encapsulate wet tissues during surgery is simple. Mechanical fixations like sutures and staples, which are intrusive, may be replaced with these devices. Bioadhesives inspired by mussels are becoming increasingly popular in the healthcare industry. In wound healing and hemostasis, bio-adhesives are commonly employed. Due to their biocompatibility, bio-adhesives are of economic importance. Wound closure with biomedical adhesives is expected to be less unpleasant. Topical bioadhesives are in high demand due to their wide range of uses across a wide range of sectors. Bio-adhesives are committed to redefining the surgical procedure itself. From the perspective of engineering, Pandey et al. (2020) examined bioadhesion and adhesives.

Nanoparticles (NPs) have lately been explored as prospective possibilities for wound treatment, and researchers have developed astonishing, enhanced nanotechnology-based sticky hydrogels. The creation of a bio-based hydrogel adhesive for wound closure must strike a balance between excellent mechanical characteristics and muscular adhesion strength. Injured tissues or organs can be reconnected using tissue adhesives. The employment of these devices promises a reduction in the complexity of the standard wound closure method. Tissue adhesives are preferred because they are simpler, faster, and do not require the use of sutures. A catechol-based biomaterial has also been suggested as a viable biomaterial for clinical use.

12.2.11 BIOMEDICAL ADHESIVES AND COATINGS BASED ON BIOLOGICAL POLYPHENOLS

Plant and animal tissues contain polyphenols, which perform a wide range of functions, including adhesion, structural support, and coloring. When it comes to animal examples of adhesion proteins that include high quantities of amino acids thought to be vital for adherence, mussels' adhesive proteins stand out. They contain 3,4-dihydroxy-L-alanine (DOPA). Benzenediol (catechol) and benzenetriol (gallon) functional groups are found in phenolic and polyphenolic chemicals found in plants (Messersmith, 2016).

12.2.12 SHARKSKIN – THE LATEST CRAZE IN CATHETERS

As a result of their oily skin and antibodies used to treat cancer, sharks make an excellent role model. Nature is made use of by industrial firms for their own convenience. For its antifouling and drag-reducing characteristics, shark skin serves as a model. An animal's immune system is currently being studied by professor of biotechnology Johannes Buchner at Germany's Technische Universität München. In order to diagnose and cure cancer, medical antibodies are used. However, the organism quickly breaks down these antibodies. In order to solve this dilemma, Mr. Buchner went to nature. He required a creature that was far enough removed from humans to have an immune system that was distinct from ours. A shark's immune system is remarkably resilient despite the fact that we've been separated from it by 500 million years. The German research team tweaked human antibodies to more closely match shark antibodies after discovering the power of shark antibodies.

Fouling (the collection of undesired material on solid surfaces, in this case, algae and barnacles) is a problem for ships and submarines in the US Navy. Dr. Anthony Brennan noticed that sharks are the only marine species that do not foul when he hired him to research antifouling technologies (meaning they stay clean). A diamond-shaped repeating pattern of denticles on shark skin prevented germs from developing. They discovered that they could duplicate shark skin and build coatings that prevent bacteria not just on ships but also on medical devices (https://www.worldanimalprotection-us/blogs/five-animal-inspired technologies). The Sharklet pattern is now used by Sharklet Technologies to create medical equipment and consumer products.

A hospital's primary concern is the spread of infectious diseases. Hospital-acquired infections are the leading cause of death for around 1,00,000 Americans each year. It's been more than 100 million years since sharks have been completely squeaky-clean. And thanks to them, infections may soon be extinct. A shark's body is clean and free of algae and slime, like those found on other large marine species. Tony Brennan, an engineer working on a better coating for Navy ships, was fascinated by the occurrence when he learnt about it in 2003. He discovered that a shark's entire body is covered by small, rough scales, like a carpet of tiny teeth, by additional investigation of the skin. When it comes to the likes of *E. coli* and *Staphylococcus aureus,* they can't get a foothold. After learning about Brennan's work, a startup called Sharklet set out to figure out if the idea of using sharkskin to create a germ-repelling coating might actually work. When it comes to protecting hospital surfaces, the company is now testing a sharkskin-inspired plastic cover (light switches, monitors, handles). Further examination of the skin revealed that a shark's entire body is coated in minute, rough scales, like a carpet of tiny teeth. They have little chance against bacteria like *E. coli* and *Staphylococcus aureus.*

12.2.13 PARTICLE FILTERS DESIGNED INSPIRED BY ANIMAL NOSES

Airborne disease transmission and particulate pollution can be reduced using passive filtering, which can be applied to any scale. The development of high-performance filters with low flow resistance but high virus or particle blocking efficiency is not a simple challenge, as is evident with industrial filtering systems and face masks (Yuk et al., 2022). Next-generation industrial filters and masks must be able to hold small droplets and aerosols while minimizing resistance. The complicated nasal architecture of animals served as inspiration for a new 3D-printable particle filter developed by Yuk et al. (2022). This new design utilizes airflow routes that are both uniformly spaced and tortuous, as opposed to the usual random-media-based filters. Due to their sinuous routes and local flow instability, both of these techniques lower resistance and increase the possibility of particle trapping. Pressure drop and particle filtration effectiveness are tested by the structures at various airflow rates. In addition, numerical simulations have been used to cross-validate the efficiency data. Particles larger than 10 micrometers were found to be more easily captured by the proposed filters, which had a lower pressure drop than commercial masks and filters. Animal noses could be used to suggest a new and scalable filter design. These scientists developed new masks and air filters based on the nasal architecture of nine different kinds of animals. Mice were found to be 50 times more sensitive to smell than humans, dogs were found to be 200 times more sensitive, and pigs were found to be 700 times more sensitive.

12.2.14 WOODPECKERS

By any account, woodpeckers are remarkable birds. The pileated and ivory-billed woodpeckers of North America are as large as crows, whereas the piculets of South America and Asia are much smaller. They have zygodactyl feet, with toes pointing both forward and backward to flit up and down vertical tree trunks effortlessly. They have shockingly long barbed tongues to extract insects tunneling in decaying wood. Even their raucous calls are strange and wonderful. But perhaps their most exciting adaptation is their namesake behavior: hammering away with their beaks into hardwood to search for their insect prey and to excavate nesting cavities. Woodpeckers can chisel away using specialized muscles and shock-absorbing skulls and necks (https://www.huffpost.com/entry/nature-technology_b_841987) using specialized muscles at a rapid-fire 20 pecks per second without suffering any impact-related injuries!

Shock-absorbent materials and helmets have already been developed based on the woodpecker's skull skeleton structure. Given that woodpeckers' anatomy reduces the amount of shock absorption, it's not a good idea to wear a safety helmet around them. Electric endodontic motor (ECOM) Woodpecker is a Cordless Handpiece. E-COM plus cordless endomotor is mainly used in Endodontic treatment during root canal preparation procedure, and it is used to mold and clean the root canal. E-COM plus can be connected to the matched Apex locator for the measurement of accurate working length. By connecting the E-Combine to the matched Apex locator, the position of the file tip inside the canal can be monitored during the procedure, and many automatic functions, such as Auto Apical Reverse, can be activated. Woodpeckers are able to sustain a ferocious pecking attack without suffering any damage to their brains. The jaw apparatus of woodpeckers plays an important part in the impact-resistant process while being actively involved in the pecking process. For example, the quadrate bone and joints of the jaw worked as a cushion during pecking, which lowered impact strain on the brain by changing the type of load (Xu et al., 2021).

12.2.15 NATURE-INSPIRED WEARABLE TECHNOLOGY

In order to protect athletes from damage, an Imperial alumnus has developed wearable technology based on the protective structures seen in animals' scales (https://www.imperial.ac.uk/news/202873/next-generation-cast-inspired-by-animal-scales-could/). Flexible, nature-inspired scales could be used to prevent and treat sports-related injuries as well as increase athletic performance. There is a lot of evidence that joint injuries are recurrent and require a long and expensive rehabilitation process to recover motion, strength, and function, which can lead to long-term immobility. As a result of current protective and supporting garments (such as a cast or brace), the affected area suffers from weakened muscular strength and stiffness.

12.2.16 SKIN GRAFT ADHESIVE

Skin grafting is used to treat big wounds, replace skin that has been lost, and improve the appearance of scarred skin. The vast majority of skin graft operations are successful (Figure 12.3).

Do you believe that all worms are parasites? Also, they might be used as a source of inspiration in the sciences. *Pomphorhynchus laevis*, a parasitic spiny-headed worm, provides the basis for a new surgical approach for skin grafts—transplants to cure wounds or burns. As soon as it has penetrated the host's intestinal wall with its razor-like spine, this parasite puffer-like worm expands inside the tissue to stay connected (https://didyouknowscience.com/amazing-inventions-inspired-by-animals-and-plants). These cone-shaped needles, which inflate when they come into contact with water, have been used to make patches that maintain grafts in place. A minimum amount of force and pressure would be applied to the tissues as the needles were inserted. In comparison to skin staples and sutures frequently employed by surgeons, a skin transplant would cause less stress to tissue and nerves after the adhesive has been removed. It would also pose a much lower risk of infection than current methods.

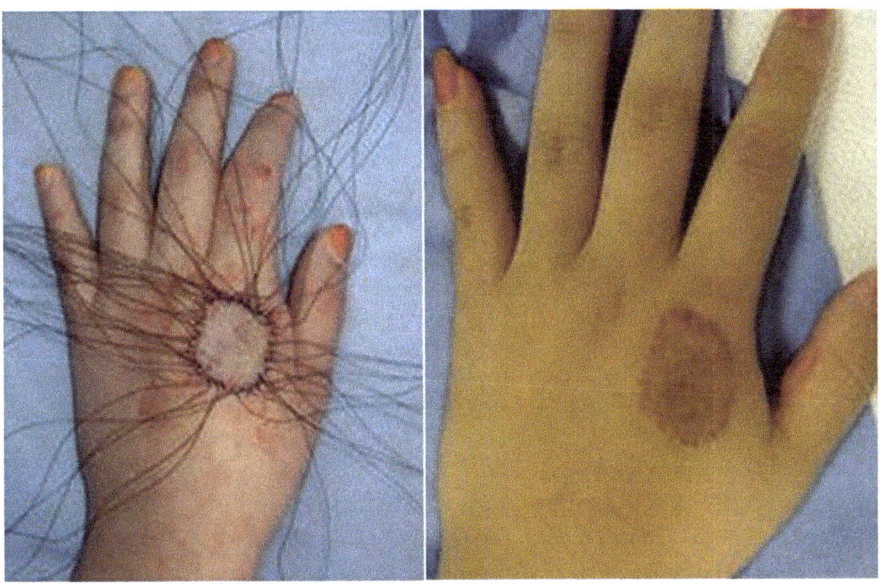

FIGURE 12.3 Split-thickness skin grafting was performed using sutures, and tie-over dressing was applied (right).

Source: Han et al., (2016). http://creativecommons.org/licenses/by/4.0/

12.2.17 BASED ON TRIBOELECTRIFICATION, A BIONIC MEMBRANE SENSOR

To monitor self-powered physiological and behavioral data, such as local internal body pressures, Yang et al. (2015) developed the first bionic membrane sensor to use triboelectrification reporting. With the sensor, it's feasible to do multimodal biometric authentication and high-accuracy self-powered anti-interference speech recording. Self-powered medical and health monitoring gadgets, interactive input/control devices, and less intrusive biometric authentication systems could all benefit from this technology.

12.2.18 DETECTION OF CORONAVIRUS ILLNESS USING A NATURAL APPROACH

Coughing, sneezing, and exhaling are the primary means of transmitting Coronavirus, according to Qureshi et al. (2021). Global solutions are needed to address the many facets of this pandemic's impact on health and well-being. COVID-19 can be controlled and treated with a variety of medical and technological approaches. A mix of machine learning and nature-inspired algorithms, such as Machine Learning and Nature Inspired Model for Coronavirus, was used to process brain MRI scans of COVID-19 patients (MLNI-COVID-19). The suggested model improves MRI image categorization as well as optimization. The model's overall performance will be improved. The area of images of the brain has been left out of this study due to a dearth of relevant datasets. The patient's brain is severely damaged by COVID-19.

12.2.19 BIOMIMICRY AND ITS HEALTH APPLICATIONS

Biomimicry has played a significant role in healthcare innovation. Studies of natural processes have produced everything from antibiotics to injections that don't cause pain. It is essential to understand the path of new diseases like the novel coronavirus (https://biomimicry.org/nature-reveals-solutions-for-supporting-public-health/). Because of the broad application of biomimicry in engineering, everything from Velcro to airplane design has been impacted by the forms and functions of living things. Biomimicry in healthcare refers to the imitation of characteristics found in plants and animals in order to give protection and treatment against infections and bacteria as well as disease management. Even on a small basis, scientists are continuously refining their ability to mimic nature. For example, nanorobotics can be used to fend off harmful infections in the body or restore specific functions by mimicking the functions of cells, tissues, proteins, and more. Even the treatment of disorders like Alzheimer's is included in this broad definition.

A biomimetic nanocarrier receives the proper medications into the brain's receptors, preventing the signaling. One of the many successes in the fight against Alzheimer's disease shows how biomimicry may make a genuine difference in healthcare. Technology, tools, and even nanorobots can be used to mimic nature's behavior in order to treat and save lives. Our institutions should use nature's power in innovative educational programs, especially in the healthcare industry, where advances like this hold such enormous promise.

12.2.20 BIOPHILIA AND ENVIRONMENT

The affiliation of humans to nature and the natural environment is innate and genetic; hence it has a solid correlation to and quality of life—this inherent affiliation toward nature was coined as Biophilia (Figure 12.4).

FIGURE 12.4 The "Regeneration Pods" provide a sense of shelter while workers access the views outside. Photo by Tom Donald for Aldworth James & Bond. Courtesy Dae Wha Kang Design.

Source: Reprinted with permission from Andreucci et al, 2021. Copyright: © 2021 by the authors. Licensee MDPI, Basel, Switzerland. (http://creativecommons.org/licenses /by/4.0/).

Human well-being can be improved by interacting with nature, natural elements, and the surrounding natural environment, all of which go under the umbrella term "biophilia." Additionally, the Biophilic design can help realize sustainable constructed form and environment through its design considerations. This can be done through the use of stabilized moss as a creative ecological background in the form of a wall. Architects and designers have just realized the benefits of Biophilia as a design co-conspirator, but greater attention is needed. In the future, nature-inspired design must be included in manmade surroundings. Patients will benefit from improved outcomes and speedier recovery if healthcare facilities incorporate cutting-edge technology into their design, according to Abbas & Jawaid (2017).

12.2.21 BASED ON NATURAL PRINCIPLES, ARTIFICIAL INTELLIGENCE (AI)

Scientists have drawn inspiration from natural systems, particularly social colonies, for the development of novel algorithms useful in the study of AI. Machines known as intelligent agents or living systems, which are capable of participating in behaviors that are perceived by humans as intelligent by AI, are the subject of this discipline of computer science. Studying how biological systems like our brains plan, learn, and make decisions can be done with AI. Algorithms are based on social-colony behavior. AI research draws on a wide range of fields and disciplines, including computer science, psychology, philosophy, neuroscience, ontology, operations research, economics, and probability and optimization theory.

For decades, the term "artificial intelligence" (Rothman, 2020) has been bandied around. There were times in the past when computers were nothing more than sophisticated devices capable of sifting through vast amounts of data. Only instructions that are given explicitly could be followed by a computer. To move from computing to thinking, today's engineers and computer scientists are making great strides. Some progress has been made. The BlueGene L supercomputer was used in 2008 to simulate a mouse's brain. Even a simulated mouse's brain, despite its seeming simplicity, is highly intricate. There were just too many variables to keep track of for even the most powerful computer to handle. The motions of a pendulum can now be studied using a computer program developed by Cornell University researchers in 2009. A genetic algorithm was used to deduce the fundamental physics rules from a set of data. One day, machines may be able to solve even the most difficult engineering challenges. Even more powerful machines may be designed by computers in the future.

For example, AI can automate fraud detection and maintenance schedules for aircraft, automobiles, and other physical assets. From capital project management to customer retention and new product launch strategies, it improves human decision-making. Everywhere around the globe, scientists and corporations are working feverishly to create AI capable of making complex decisions on its own, without the aid of a person. Because of this, a new set of business models will be required for a wide range of industries, from financial services and healthcare to energy and mining to industrial goods and the media and entertainment industry. As a result of AI, social media is on the rise. AI is used to personalize social media feeds by analyzing post history, determining interests, and making appropriate friend recommendations. Machine learning can also thwart misinformation and cyberbullying. Many machine-learning models are used in the diagnostic and recommendation domains (https://medium.com/dataseries/ai's-link-to-nature-and-how-it-is-tied-to-AI's-job). AI encompasses a wide range of technologies, from large-scale medical data processing to online shopping recommendation engines, including robotics, computer vision, and sensor fusion. New paradigms in computing are evolving as AI approaches permeate more and more elements of computing.

How AI may be designed to mimic nature's fundamental principles is Jin's primary research interest (Vinuesa et al., 2020). His work on evolutionary and developmental systems based on nature-inspired optimization will continue at the Bielefeld Institute for Systems Biology.

For the Alexander von Humboldt Professor of Artificial Intelligence at the University of Bielefeld, Professor Dr.-Ing. Yaochu Jin worked to understand and replicate intelligence in nature, notably how neuronal networks and body plans coevolve and expand. According to Jin, AI is built to be exceedingly precise. "However, when doubt or ambiguity enters the picture, there are difficulties." To avoid having to deal with too many questions at once, AI tries to concentrate on just one or two at a time. This is especially true if you're looking for connections or a solution to an issue that isn't well-defined.

Nature is well-equipped to deal with adversity. Jin asserted that we are born with a body that has undergone millions of years of evolution. To give you an example: For a long time now, nature has demonstrated the anatomy of the brain to be correct. To put it another way, neuroplasticity means our brains are always rearranging their internal structure. Thus, it is capable of evolving (Rothman, 2020). If you want to change your brain, you can learn a new language or take up a new sport. "If you allow identical twin cats to grow up in diverse environments, you can also see this. Despite the fact that their genes are identical, their neurological systems differ" Jin was a late addition to the conversation. As the healthcare system expands and infrastructure prepares to meet current needs, telehealth technology is also relevant in emerging countries.

12.2.21.1 HEALTHCARE AND ARTIFICIAL INTELLIGENCE

Improved human performance is the primary goal of AI, which has the potential to boost both doctors and patients (Figure 12.5).

FIGURE 12.5 An elderly man using a pulse oximeter to measure his blood oxygen levels.
Photo credit: Tunstall Telehealthcare. https://creativecommons.org/licenses/by/2.0/.

When technology was first introduced, it was primarily used to automate mundane, time-consuming processes and reduce the amount of paper that was being used in the healthcare industry. It also made it easier for healthcare providers and insurers to share information about their patients. Every aspect of healthcare–from inpatient care to clinical research to pharmaceutical development–is covered by insurance these days (Kalis et al., 2018; Carlton, 2019). For the sake of saving money and improving patient outcomes, AI applications are reshaping the workings of the healthcare sector.

As long as these jobs are being performed, AI has expanded its applicability beyond enhancing back-office productivity to being a facilitator for improving healthcare outcomes, especially in the current COVID age. COVID has played a significant role in putting AI technology into practice, even as it has taken a heavy toll on people's health. Thanks to the advancements in technology, new models, treatments, and vaccines can be developed.

AI has a wide range of applications outside of the pandemic. AI has the potential to enhance human decision-making and information processing. In the healthcare industry, machine learning can help with the discovery of new pharmaceuticals and the speed of diagnostics. CT images are being analyzed with AI to look for pneumonia in individuals on COVID-19 medication (https://mobidev.biz/blog/technology-trends-healthcare-digital-transformation). This AI technology for radiotherapy, Project Inner Eye, was created by Microsoft. As a result, instead of taking hours, the patient's 3D contouring procedure can now be done in minutes. The project can be downloaded from GitHub. Project Hanover, a collaboration between Microsoft and PubMed, organizes biomedical research papers using AI. It speeds up the process of detecting cancer and helps doctors decide on the best course of action for each individual patient (Paul et al., 2021).

A wide range of patient care and administrative tasks can benefit from the use of AI in healthcare. The use of AI in modern business and everyday life is becoming more and more ubiquitous. It's possible that a variety of AI and healthcare technologies may have a significant impact on healthcare, yet the methods employed to support these technologies can vary substantially. AI in healthcare has the potential to surpass humans when it comes to recognizing illness. AI will not replace people in the healthcare industry for many years to come (Marr, 2018; Kalis et al., 2018; Carlton, 2019).

Currently, there are a number of AI-based solutions that can improve patient care and perhaps save lives. Pattern recognition, robotics, and natural language processing, such as speech recognition and translation, are all examples of these applications. In addition, machine learning, a "technology that trains software algorithms to learn from and act upon fresh data to enhance performance continuously," is becoming more popular. Assistive technology: People with Alzheimer's disease may benefit from this AI-driven device, which can assist them with their daily routines. If you're like Brian Leblanc, 59, whose early-onset Alzheimer's disease was identified in 2014, you've probably started using Alexa on your Amazon Echo Dot. There are a number of ways in which AI and its subsets can be used to improve healthcare and the medical field.

For the evaluation of pigmented lesions, Mela Find makes use of infrared light. Skin malignancies such as melanoma can be diagnosed using algorithms by dermatologists. Even though a biopsy is still the best way to confirm a diagnosis, this technology can aid in the process.

Using computer algorithms, a robotic arm and hand can detect moves that patients are unable to perform during therapy and aid them. In stroke patients, robotic-assisted treatment may help improve motor recovery while minimizing labor costs, which are the most expensive part of providing physical therapy interventions.

Echocardiographic images of patients' hearts can now be captured without the need for specialist training thanks to an AI-powered program developed by the Food and Drug Administration. Using machine learning, the software can now recognize heart problems in high-quality 2D ultrasound images and even record video recordings.

12.2.22 ADVANCES IN TELEHEALTH AND TELEMEDICINE

Patients and healthcare professionals alike can benefit from telehealth services, which we define as the application of electronic information and communication technology to support long-distance clinical healthcare services, health education, and the improvement of overall health and well-being through better access to treatment. There are a variety of tools for collaborating, including videoconferencing, the Internet, store-and-forward imaging, streaming media, and terrestrial and wireless communications. Telemedicine refers to "telemedicine," whereas "telehealth" refers to "telehealth" services that can be offered remotely (https://aspe.hhs.gov/topics/health-health-care/advances-telehealth).

There are both challenges and opportunities for healthcare workers who care for patients with various health conditions as a result of the COVID-19 crisis, which has led to a rise in telemedicine. CoV-19, which is the virus that causes SARS-CoV-2, spreads less frequently when patients use telehealth, particularly virtual consultations rather than in-person ones. Using two-way, real-time audio and video connection between a patient and a healthcare provider, the practice is known as "telemedicine." Peripheral examination equipment that enhances video chats saves data and forwards it can be provided via telemedicine (Monaghesh & Hajizadeh, 2020; Kissi et al., 2022).

Net neutrality (to ensure stable, secure internet connectivity) and the desire of providers and patients to use tools may change depending on clinical needs, budget, storage capacity, and other factors. Video otoscopes, electronic stethoscopes, dermatoscopy, retinal imaging systems, and intraoral scopes are all examples of these devices. As long as there isn't a standard or broad use of these devices, adoption may be hampered. Clinics may also need to approve certain applications. Thomas Bird created the term "Telemedicine" in the 1970s. In the current sense, telemedicine was first used in 1906. As early as the 1800s, telemedicine was already being utilized to overcome geographical obstacles and resolve healthcare problems. In The Lancet (1879), a doctor diagnosed a youngster over the phone in the middle of the night; this was an early example of telemedicine.

Health services such as examinations and consultations can be provided over vast distances via telemedicine, which is also known as telehealth or e-medicine. There is no need for an in-person visit for healthcare providers to diagnose and treat patients thanks to the use of Telemedicine. Patients can now consult with their doctors from the comfort of their own homes, thanks to advances in modern technology (Gajarawala & Pelkowski, 2021). Symptoms, medical issues, and more can be discussed with a healthcare provider via video, web portals, and email via telemedicine. You can get a diagnosis, learn about your treatment options, and even get a prescription via telemedicine.

Despite the similarity in pronunciation, there is a substantial difference between the terms "telemedicine" and "telehealth." WHO describes telemedicine as "healing from a distance." Medical care can be obtained without having to book an appointment with your physician or go to their office. In the same way, telemedicine is an old idea in a new package. The use of telemedicine has the potential to save money, improve the quality of care, and increase the level of patient participation. Thanks to technological advances in telemedicine, which began to appear in the late 1950s, seniors are now able to age in place at home. Patients in rural areas who previously couldn't get in touch with a doctor due to a lack of options now have the option of doing so via a digital connection. Even medical devices in faraway locations can be viewed and captured by doctors and patients via a computer screen in real time. Telemedicine software enables virtual doctor appointments at any time of day or night. Many people find that in-home doctor visits are a handy choice.

For many patients who require treatment, telemedicine is an established method of care that can help them get it. Recent developments are largely responsible for its widespread acceptance and success. If telemedicine is done correctly, evidence-based feeding therapies can be disseminated over the world. Because so many individuals are turning to virtual healthcare, the telehealth market is predicted to expand to $55.6 billion in the next five years.

12.2.22.1 TELEMEDICINE HAS NUMEROUS ADVANTAGES

Telemedicine isn't simply good for the patient; it's also good for the medical community. It's also good for the service provider.

Specialists can be accessed more easily via telemedicine. Traveling between hospitals and out-of-the-way specialists' offices is eliminated. Additionally, telemedicine eliminates travel expenses. In terms of gas money, it saves the patient money.

Reduced interference with childcare is another advantage of telemedicine. Doctor's appointments do not need arranging a babysitter for patients' children, so they do not have to worry about it. Elderly in-home care is another example of this. In addition, medical expenses are significantly lower on the whole.

12.2.22.2 REASONS WHY THE SERVICE PROVIDER IS ADVANTAGEOUS

Because of the advantages of telemedicine (Litvak et al., 2020), doctors are now able to treat many patients at once. Allowing providers to see as many patients at once as they can handle, the "line" is effectively abolished. It also reduces the number of cancellations and no-shows. With one-on-one patients, providers also earn more money.

There are three main categories of telemedicine, which include store-and-forward, remote monitoring, and real-time interactive services.

Patients and doctors can communicate in real time using interactive telemedicine/telehealth. It is possible to perform these sessions in the patient's home or in a medical kiosk. HIPAA-compliant video conferencing software or phone calls can be used for interactions.

The term "telemonitoring" refers to the use of mobile devices to monitor vital indicators such as temperature, blood sugar, and blood pressure in the patient's home. One healthcare practitioner can store and transfer patient information, such as a patient's test results, to another healthcare provider using asynchronous telemedicine, or "store and forward."

12.2.22.3 TELEMEDICINE PROVIDES A LOT OF CONVENIENCE

Telemedicine in primary care is typically done via phone calls. Non-emergency medical concerns are those for which the patient does not need to see the doctor. Face-to-face consultation is still necessary while using telemedicine, but it is not a replacement for it. Telemedicine's primary function is to make medical consultations and treatments more accessible to patients and practitioners alike by eliminating the need for in-person meetings. As a result of this, it is less expensive than visiting a doctor or other healthcare practitioner on a regular basis.

Has an emergency, but the doctor's office is closed? Incorporate telemedicine into your practice (Lurie & Carr, 2018; Litvak et al., 2020). Those with diabetes, high cholesterol, or high blood pressure can greatly benefit from the use of the drug. If you don't have an urgent medical problem but still need help with things like medication tweaks, lifestyle regimens, prescription refills, or access to a support group, telemedicine may be a good option for you. Additional telemedicine consultations might be arranged to follow any necessary blood tests or vital sign monitoring. Telemedicine has been practiced by doctors for a long time, but not under the term of telemedicine. Broadband internet technology has made both audio and video calls more accessible to a greater range of people, making this an attractive alternative to the traditional phone system.

Medical services can be accessed and managed remotely via computers and mobile devices through the use of telehealth technology. Healthcare services may benefit from the usage of these technologies at home or by your physician (Perrone et al., 2020).

If you have diabetes, telehealth may be able to help you. All of the following options are available.

You can have a nurse look at your meal plans, medication dosages, and blood sugar levels online by uploading them from your phone or other device. Learn how to keep track of your carbohydrate intake by downloading an app and watching a video. An app can help you determine how much insulin you need based on your diet and exercise habits. An online patient portal allows you to access your lab results, schedule appointments, get refills for prescriptions, and communicate with your doctor.

12.2.22.4 MEDICATIONS AND TESTING SUPPLIES CAN BE ORDERED ONLINE

Instead of booking an appointment with a specialist, your doctor can perform mobile retinal photography in their office. There are a variety of ways to send reminders for Text, email, and phone calls that can be used to provide preventive healthcare.

As the coronavirus disease 2019 (COVID-19) pandemic continues to spread, telehealth providers face an increased danger of legal liability. Health and convenience are both enhanced by the use of telemedicine, which allows patients and their doctors to communicate by video or phone. Patients can now "see" their doctors through a computer or smartphone, which is becoming more common among medical professionals.

12.2.22.5 TELEMEDICINE: A QUICK GUIDE

- Medical care can be obtained in a variety of ways depending on the services provided by your physician. There are a number of ways in which telemedicine can be used, but two of the most prevalent are:
- A website for patients.
- As long as you have a login and password, a patient portal lets you communicate with your doctor or nurse, obtain refills for prescriptions, and schedule appointments. If you've had any lab work or imaging tests, you'll be able to get the findings from your doctor. Rather than dialing and waiting on hold for a response, you can send a message directly to them.
- Using a virtual assistant to schedule appointments.
- It is possible to schedule an appointment with your doctor through phone or video conference with some medical professionals. Meetings with mental health professionals and urgent care clinics are often available.
- Images, video, and text that are recorded and stored before being sent are known as "storage-and-forward." This method is referred to as asynchronous because it does not involve any real-time communication between the participants. When a patient texts their dentist a picture of an injured tooth, they're providing an example.
- The transfer of health or medical data such as blood pressure, glucose measurements, and blood oxygen levels via remote patient monitoring (RPM).

Telehealth has had a positive impact on both patients and providers (Burroughs et al., 2020). The use of technology in the treatment of chronic disease care in the United States has resulted in improved physiologic measures, enhanced satisfaction, and better medication adherence.

Telemedicine isn't used to replace face-to-face visits but rather to supplement them. It has some drawbacks, which we'll discuss in more detail later, but its advantages are undeniable.

Specialists are readily available. While many people have a regular doctor they can call, not everyone does. All kinds of specialists can be accessed 24 hours a day, seven days a week, without an appointment on many online medical networks.

It's less expensive: Even if you have decent health insurance, seeing a doctor or therapist can be pricey. Appointments for telemedicine are often less expensive than those for in-person consultations. It removes a barrier to care by reducing out-of-pocket expenses.

Uninsured patients may be unable to see a doctor because of the financial burden of medical bills. The majority of cash-pay telemedicine services do not require health insurance or a referral from a primary care physician.

Medical care isn't always as accessible as it could be for people who live in remote areas. Telemedicine makes it possible for patients who reside distant from a medical facility to see a doctor as quickly as possible. It saves time and keeps people off the road during inclement weather, such as a snowstorm or hailstorm.

Even if you have outstanding health insurance, seeing a doctor or therapist can be pricey, even if you're covered. Visiting a doctor in person is more expensive than using telemedicine. There is less of a financial barrier to receiving medical care as a result.

People who do not have appropriate health insurance may be unable to see a doctor because of this. Cash-pay telemedicine, which does not necessitate insurance or a referral, is offered by several internet businesses.

People who live in rural settings may not always have quick access to medical care, despite the many advantages of country living. Telemedicine allows those who live far from a medical center to see a doctor fast. During bad weather, such as a snowfall or hailstorm, it keeps people off the road and saves them time.

Babies and children often suffer fevers or become ill in the middle of the night, necessitating medical attention. A doctor can diagnose and prescribe medication via telehealth services rather than having parents rely on an internet search.

There's no need for a babysitter; parents get sick, too, and taking small children to the doctor's office can be a hassle. Telehealth does away with the requirement for childcare and the associated costs.

Due to the epidemic, several medical offices have either closed or reduced their working hours. Virtual private networks (VPNs) have made it possible for doctors and therapists to treat patients from the comfort of their own homes over the internet. As a result of this service, patients and doctors are able to communicate at all hours.

For doctors, telemedicine reduces medical overhead expenditures, such as the necessity for receptionists.

Telehealth services can now be billed to Medicare and many commercial health insurance organizations, regardless of where the patient or practitioner is located. However, because state regulations and insurance plans differ, be sure you check your policy's criteria and restrictions before seeing a physician remotely.

Monitoring devices that send data to doctors can alert patients to new symptoms, deteriorating health, and even crises in the home. It's a lifesaver for folks who need quick interventions.

For many people, the availability of online psychiatric treatment has made it hard to see therapists face-to-face. When it comes to mental health issues, telemedicine has made it easier for people to begin or continue receiving therapy.

Those in need of emergency psychiatric care, such as those who are at risk of self-harm, can get in touch with a doctor at any hour of the day or night.

Online therapist-led support groups can help people who are dealing with challenges such as grief or infertility connect with each other.

Reducing anxiety: Many people postpone going to the doctor because they are afraid. Those with agoraphobia may be reluctant to seek help because they are afraid that they will be alone when they do so. If you have a medical concern, you don't even have to leave your house to seek answers or get care.

For patients undergoing physical therapy, telemedicine allows physical therapists to watch and observe exercises and therapies that are carried out in the patient's home. It has the ability to expedite the healing process following accidents and surgeries.

12.2.23 ELDERLY PEOPLE'S IMMUNITY IS BOOSTED BY VITAMIN D SUPPLEMENTATION

Supplements can enhance the immune system and reduce inflammation in older persons, according to a new study. A recent study found that supplementing the vitamin D levels of elderly people who were vitamin D deficient considerably improved their immune system. According to the findings (https://www.ed.ac.uk/roslin/news-events/latest-news/vitamin-d-supplements-boost-immunity-in-elderly), vitamin D deficiency may contribute to the decline in immunity that occurs as people age. The data suggests that supplementing elderly people with vitamin D may enhance their overall health by raising their immunity to specific diseases and possibly increasing the efficiency of immunizations. There may be some correlation between severe cases of COVID-19 and vitamin D deficiency, according to the study's findings.

Varicella-zoster virus (shingles) causes a decrease in the elderly's immunological responses and an increase in their levels of inflammatory markers like vitamin D, according to the study. It sparked an investigation into whether vitamin D supplements could impair the virus's ability to infect people. There has been a huge rise in the number of people taking vitamin D supplements. Using skin samples from older volunteers, they examined the immunological responses. Increased immune response and decreased inflammation in the elderly can be achieved by taking vitamin D supplements. Vitamin D, which is a generally available, inexpensive, and safe supplement, appears to boost immunity in older persons, according to this intriguing study. A weakened immune system makes us more vulnerable to viral infections like COVID-19, influenza, and shingles as we get older. Vitamin D supplementation, according to our findings in this study, can improve immunity.

Vaccine efficacy decreases with age, and systemic inflammation rises as a result. Older people's skin is less responsive to VZV challenge because of these changes. The elderly are more likely to suffer from a deficiency in vitamin D. Taking vitamin D supplements can help the body maintain its own immune system, as well. Shakoor et al. (2021) studied the age-related changes in cutaneous antigen-specific immunity in 65-year-olds. An oral vitamin D3 dose of 6,400 IU/day was administered orally to 18 elderly participants ($n = 18$) who were vitamin D deficient for a period of 14 weeks. Antigen-specific immunity against VZV is determined by skin biopsies collected from injection sites before and after vitamin D3 supplementation.

According to Lanham-New and colleagues, there was a decrease in VZV cutaneous immune response and an increase in non-specific inflammation in older adults (2021). A rise in generalized skin irritation in the elderly has been linked to vitamin D deficiency. A study by Chambers et al. (2021) found that treatment with vitamin D3 improved the cutaneous VZV antigen challenge response in older adults.

An enhancement in antigen-specific immunity in older adults with low vitamin D levels has been shown to be possible by therapy with vitamin D3. According to a study published in the journal Pediatrics, patients with vitamin D deficiency during the current COVID-19 epidemic had a worse prognosis. To maintain good respiratory health, vitamin D can help limit the production of antimicrobial peptides (such as cathelicidin). Supplementing with vitamin D3 is an easy, low-cost, and risk-free way to improve protection against SARs-CoV-2 infection. Scientists believe that supplementing with vitamin D3 could be an inexpensive, straightforward, and readily available treatment for improving antigen-specific immunity in the elderly.

12.2.24 T-CELLS HAVE AN IMPORTANT ROLE IN THE SPREAD OF COVID-19

CD4+ T cells and CD8+ T cells help produce immunoglobulins. A cytokine produced by T cells, interferon-gamma, is among the most important cytokines. COVID-19 infection is associated with lymphopenia, which affects CD4+ T cells, CD8+ T cells, and B cells. This symptom is more apparent in sicker patients. The results of this research have been published in science (https://www.science.org/doi/10.1126/science.368.6493.809). T cell responses in critically ill patients may be faulty, hyperactive, or inappropriate; further research will help guide therapy alternatives if it can be found to be either Cross-reactivity with seasonal or endemic coronaviruses has been found. In most people with COVID-19, both CD4 and CD8 T cells respond strongly and broadly, and some of these T cells also show a memory phenotype, which bodes well for long-term immunity in those people.

T cell subsets' functions in protection and pathogenesis are critical to the therapy and prevention of COVID-19 disease. COVID-19 (Leslie, 2020) and maybe the immunological memory that forms following the recovery from SARS-CoV-2 infection appear to have a role for T cells (based on increasing data). Even though they appear to have both CD8+ and CD4+ T cell responses, patients with severe disease may have deficient, exaggerated, or otherwise inappropriate T cell responses. Reduced T-cell counts in COVID-19 patients and functional depletion of those that remain are concerning (Moise et al., 2020). T-cell response patterns may vary from patient to patient, indicating that different treatment modalities may be required depending on the immunotype of each patient.

SARS-CoV-2 infection displays a complicated pattern of T cell responses. It's possible to draw conclusions about the population's immune system, however. It seems unlikely that population-level T cell response tracking will be practicable in the near future, as opposed to monitoring antibody reactions. There needs to be more studies done on those who are asymptomatic or paucisymptomatic because there are significant methodological constraints and heterogeneity within studies. Priority should be given to evaluating vaccine recipients as a sub-group. T cell response demographic and clinical correlation studies were few in number and did not cover many potentially significant variables, such as ethnicity. T cell response and illness severity were studied in connection to one another, with the majority of the research being conducted in a hospital setting. There were no differences in the CD4/CD8 ratio between patients with asymptomatic and severe disease, while those with moderate or severe disease had dramatically reduced peripheral counts. There were some issues with the study, such as a small sample size and a lack of reporting of sample selection techniques, but overall, the quality was good.

More extensive studies conducted multivariate analyses and found independent associations for specific subsets; with persistent RNA-positivity, mainly in smaller studies with some risk of selection bias; and with older age, including in one large higher-quality study, lower peripheral T cell counts were linked to non-survival.

Lower peripheral T cell counts were associated with non-survival in two more comprehensive studies that performed multivariable analyses and discovered independent associations for specific subsets; with persistent RNA positivity, primarily in smaller studies with some selection bias; and with older age, including in one large, high-quality study.

Where may T cells be found in the body, and how quickly? Between the 10th and 21st day after infection, the majority of people infected with SARS-CoV-2 will develop antibodies. It does, however, describe the detection of T cells in SARS-CoV-2-infected people one week after the onset of symptoms. Detection of T cells at such an early stage may be crucial.

12.2.25 NANOTECHNOLOGY IS RESHAPING THE HEALTHCARE INDUSTRY IN THE YEARS AHEAD

Because of the potential applications of nanotechnology in healthcare, new possibilities are emerging across the entire life sciences sector. There are numerous ways in which nanotechnology can improve healthcare, including diagnostics, disease monitoring and monitoring, surgical equipment, regeneration, vaccine research, and drug delivery. It's also making it possible to find new drugs to treat a wide range of disorders thanks to improved drug discovery techniques.

Nanotechnology is the study and engineering of systems at the molecular and atomic levels. When working with nanoscale materials, engineers use molecular chemistry and physics as well as traditional engineering methods. The human body's failure to absorb the full amount of medicine is a major concern for the healthcare business. This is where nanotechnology comes into play. It is possible to target specific cells in the body with the drug thanks to nanotechnology, which increases the precision of treatment while simultaneously reducing failure or rejection. There is essentially no limit to the number of innovative diagnostics and treatments that can be developed because of nanotechnology's rapid developments. To sum it up, the ability to examine the human body, its drug treatments, and medical gadgets at the nano-level ensures significantly greater precision in medicine. Examples of how the healthcare industry makes use of this technology include diagnostics, medical equipment, and nanomedicine.

Nanotechnology is revolutionizing healthcare in four key ways. They are (1) nanomedicine; (2) nanobots; (3) nanofibers; and (4) nanotech-based wearables.

Nanomedicine (Hua & Wu, 2018) employs NPs in medical devices, nanoelectronic biosensors, and molecular nanotechnology to treat and diagnose a wide range of ailments, including cancer, heart disease, and diabetes. Nanomedicine is currently being used to develop smart tablets and cancer treatments (Martins et al., 2020).

12.2.25.1 SMART PILLS

- The term "smart pills" refers to nanoscale electronic devices that resemble pharmaceutical pills in form and design but perform more advanced functions like sensing, imaging, and medicine administration. Previously, nanotechnology-enabled the creation of the PillCam, a capsule equipped with a tiny video camera and dose-tracking pills.
- A new type of intelligent pill called an "Atmo Gas Capsule" analyzes the gases in the stomach and reports on any irregularities when taken orally. In addition to monitoring oxygen and carbon dioxide levels, it can also detect harmful substances like radon, which it can identify through its sensors. Diagnosis of the digestive system can be made using this method.
- Detection of cancerous digestive organs.
- Personalizing diet and nutrition recommendations by tracking dietary sensitivities.

The 'Smart Sensor Capsules' developed by MIT eliminate the need for stomach injections of medications. Tracking vital signs and monitoring treatment are two of the functions of this drug, which can be taken orally (Trafton, 2021). Compartments are also included for pre-loading drugs into certain parts of the body.

12.2.26 A NEW APPROACH TO THE IDENTIFICATION AND TREATMENT OF CANCER

A major downside of standard chemotherapy and radiation is that it damages the body's healthy cells. To treat skin cancer, new nanomedicine technologies are being used, which allow medications and other treatments to be delivered efficiently and with low hazardous side effects to specific tumor areas and target cells (Figure 12.6). To detect cancer cells in circulation, a novel technology called NanoFlares is being developed.

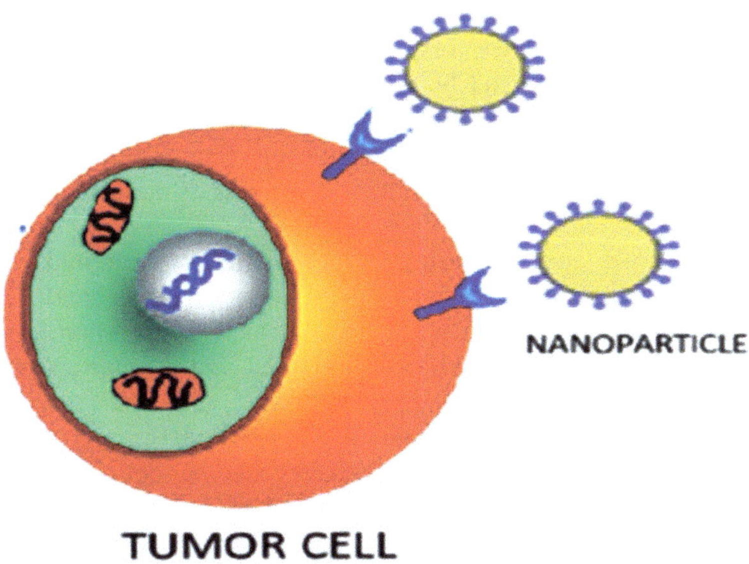

FIGURE 12.6 Monoclonal antibodies, peptides, and nucleic acids on the NPs surface are able to recognize receptors located on the plasma membranes of specific target cells and bind with high affinity to the receptor site.

Source: Reprinted with permission from Cassano et al., (2021), Molecules 2021, 26(4), 785; https://doi.org/10.3390/molecules26040785

For the identification of cancer cells, researchers have developed Nano-Flares, which can bind themselves to the cancer cells' genetic targets and generate light when that target is detected.

Researchers at the University of California, San Diego, are developing a method for collecting and analyzing nano-sized exosomes in search of biomarkers indicative of pancreatic cancer. Small surgical robots, known as nanobots, can be inserted into the body to repair or replace intracellular structures that have been destroyed. It's also possible for them to self-replicate so that they can fix genetic faults or even eradicate disease. Nanobots are currently being explored for eye surgery via a microneedle inserted into the eye. The construction on this property is not yet complete. Surgeons may be able to direct this needle with the use of a specialized magnetic field. Another use for nanobots is to drill through artery blockages. Scientists at Michigan State University and Stanford University collaborated to manufacture carbon nanotubes coated with a medicine that can eat away at arterial plaque. Cardiovascular disease can be prevented by using it.

Nanobots resembling unfolded cubes are being developed by Toronto scientists for rapid biopsies of suspected cancerous tumors. Upon reaching their target tissue, these cubes fold up and take a sample. A human surgeon's biopsies can take longer and be less accurate. In order to target specific disease cells, nanobots generate custom DNA. DNA has been used to create an "origami nanorobot" that can carry a molecule. Using the robot, they were able to demonstrate that they could deliver compounds that would cause the death of specific kinds of cancer cells.

Bioengineering and tissue-engineering applications all make use of nanofibers in various forms. 'Smart bandages' that scientists are developing will dissolve into the flesh when a lesion is healed. These smart bandages' nanofibers can include clotting agents, antibiotics, and even sensors to detect infection.

The Institute of Advanced Studies in Science and Technology in India has developed pH-responsive smart bandages that can deliver medicine to the wound at the proper pH level. Nanotechnology-enhanced cotton patches can be made more affordable and environmentally friendly by employing ordinary and readily available materials like cotton and jute. Nanofiber-based drug delivery systems can be used to deliver drugs to specified locations and at specific times in order to accomplish desired therapeutic effects. In order to accomplish quick drug release, nanofibers utilize an appropriate polymer, a high surface-to-volume ratio, improved loading efficiency, and mesh with improved permeability and retention.

When it comes to vaccination, Vaxxas is already working on commercializing their 'nano-patch' vaccines, which are essentially a nanofiber patch that injects the vaccine directly into skin cells via NPs. Vaccine refrigeration is no longer required, reducing the danger of infection and simplifying the operation. Using wearables based on cloth-based nanotechnology for RPM is an innovative yet extensively used method of healthcare. If there are any changes in the user's body that could be hazardous, the device notifies both the wearer and medical specialists. Wearables that have nanosensors embedded in the cloth can capture medical data such as heartbeat, sweat components, and blood pressure.

An American start-up invented this revolutionary technology. Nano-SENSE, a clinical study on heart failure care and diagnostic validation, was initiated in 2019. The creators of an AI platform called Simple SENSE are from the U.S.A. Nanosensors embedded in the fabric of this garment recorded phonocardiography, stroke volume, and cardiac output, all of which were patented and FDA-approved. A multi-channel ECG assessment system monitors heart rate variability, respiration, and thoracic impedance, as well as activity level, posture, and movement. For the COVID-19 remote diagnostic research, Nanowear has worked with Hackensack Meridian Health Systems. Nanowear nanosensors can detect physiological and biomarker changes that indicate deterioration and demand medical intervention in individuals with COVID-19.

12.2.26.1 NANOPARTICLES (NPS) IN MEDICINE

There is a possibility that nanotechnology will revolutionize medical practice (Utreja et al., 2020; Murthy, 2007). In medical nanotechnology, NPs are utilized. Using the company's nano-robots, they can address problems at the cellular level. In modern medicine, NPs serve a critical role, from medical imaging contrast agents to carriers for gene transfer into specific cells. NPs are distinct from bulk materials because of their chemical reactivity, energy absorption, and biological mobility. Using NPs in modern medicine has numerous advantages. NPs can aid in research and treatment that would otherwise be unattainable. The toxicity of NPs, on the other hand, raises a host of new concerns for human health and the environment.

The use of NPs in medicine is an aspect of nanotechnology (Lin, 2015). The use of NPs to deliver medications, heat, light, or other chemicals to certain cell types (such as cancer cells) is one use of nanotechnology currently being developed in the medical industry (Utreja et al., 2020). People can alter their surroundings at the molecular and atomic level thanks to the new scientific subject of Nanotechnology (Keskinbora & Jameel, 2018). Although it is aware of the possible dangers to human health, it continues to use it (Ray, 2019). Using NPs (Hamid & Manzoor, 2020) could lead to the development of tumor-targeted thermal scalpels for cancer treatment. Using gold nanoshells coated with polyethylene that absorbs near infrared light can slow the growth of a tumor. NPs made of gold or silver have antifungal, antibacterial, and anti-inflammatory properties, making them excellent for use in anti-wrinkle creams and burn creams. They are biocompatible, non-toxic to cells, and non-toxic to the environment. They're numb. Toxicity *in vitro* appears to be dose- and volume-dependent, with bigger AuNPs causing cytotoxicity and smaller AuNPs having no effect. Thus, ROS production is higher in cells with smaller diameters.

Designing nanotechnology solutions for global health must take ethical, environmental, social, cultural, and legal considerations into account (Salamanca-Buentello & Daar, 2021). It is possible that the developments in nanomedicine could save the lives of individuals in the poorest countries. Equality is a moral precept as well as a formally stated policy objective. Even though there are numerous examples of medical nanotechnology being used in low and middle-income countries (LMICs) to alleviate global health challenges, this translation from lab to village remains incomplete (Hull, 2019).

Pay attention to the social and cultural contexts in which nanomedical technologies are applied. Because of this, even the most cutting-edge technological solutions have limitations that are influenced by the context in which they are born. A comprehensive assessment of nanotechnology's impact on public health requires that both the risks and benefits of the technology be considered (Hull, 2019).

12.2.27 INTERNET OF MEDICAL THINGS

As the global population continues to get older, so does the demand for early biological diagnosis and proactive healthcare. The development of novel biomedical diagnostic gadgets, circuits, and microsystems can lead to new approaches to health management. Biosensors, bioelectronics, and integrated circuits are all part of these devices, circuits, and microsystems for biomedical use (Figure 12.7).

They are also utilized in lab-on-a-chip systems for the detection and monitoring of biophysical and biochemical parameters. With regard to healthcare-related technologies, professionals from every key field—electronics, chemistry, physics, biology, and materials science—have been drawn in by the multidisciplinary nature of healthcare-related technologies. An IoT system is a generic name for the billions of connected objects and equipment that communicate with each other and other systems through the internet.

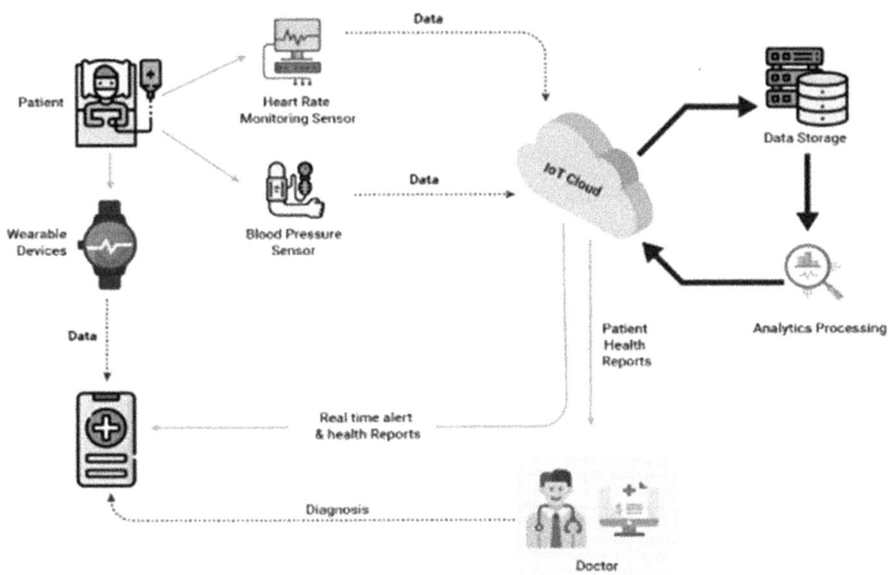

FIGURE 12.7 IoT is Transforming the Healthcare Industry.

Source: Reprinted with permission from Shrimali (2020).

The internet of medical things (IoMT) is a subset of internet of things (IoT) technologies that connects medical devices and applications (Steger, 2020; Hemanth et al., 2021). The Internet of Things (IoT) links patients, clinicians, and medical devices together. The phrase "Internet of Medical Things" (IoMT) is used to describe a wide range of medical devices that connect with each other through a secure network, including hospital equipment, diagnostic gear, and wearable technologies.

Although it encompasses a wide range of medical equipment, it can be categorized into four basic groups:

- IoMT for wearables;
- In-hospital monitoring;
- Home monitoring; and
- Community monitoring.

Physicians used to make house calls 60 years ago to see how their patients were doing. "House calls" via telemedicine have emerged as a result of advances in technology, allowing patients to communicate with healthcare professionals without having to travel to their offices. After the events of COVID-19, this method has become increasingly important. However, the IoT's healthcare possibilities go far beyond telemedicine. It also contains cloud-based systems that will bring together data from all of the devices above, which have hitherto been disconnected from one another in the healthcare industry. Healthcare can be improved as a result of the IoMT's efforts.

It is possible to think of an IoT healthcare facility as a collection of ubiquitous computing systems that are primarily concerned with the outside world. Internet-based medical devices collect patient data and expert feedback from medical professionals (https://www.researchgate.net/figure/Example-IoT-medical-devices-on-a-patients-body_fig2_326568227). The best example of this is continuous glucose monitoring for insulin pens. Every one of these gadgets is capable of interacting with one another and taking the necessary actions to save a life in a timely manner. IoT medical equipment would gather data and then transmit it to the cloud, where it could be acted upon by doctors. As a result, we can conclude that IoT in healthcare has the potential to benefit patients' health as well as the productivity of healthcare workers and the efficiency of hospital processes.

12.2.27.1 IoT IN HEALTHCARE: WHAT ARE THE POSSIBILITIES?

The Internet of Things (IoT) is "the linking via the internet of computing devices embedded in common things, allowing them to send and receive data." Connected devices and mobile apps are a key component of the IoMT, which aims to improve outcomes for patients and carers by bringing together patients, caregivers, physicians, and data.

IoMT progression can take many forms (Razdan & Sachin, 2021). It has already been implemented in emergency medical care, insurance policies, and pharmacies. Healthcare will be transformed by the Internet of Medical Things in 2021 (https://www.insight.com). IoT innovations have made emergency hospital wait times much lower. Bed availability is tracked using RFID tags, infrared sensors, and computer vision. It reduces the time it takes for patients to be identified in the emergency room. Emergency medication technicians (EMTs) can also benefit from this information when transporting patients to the hospital. EMTs can divert patients to a hospital with more beds if they are aware of a hospital's bed shortage.

Infrared sensors not only help with bed availability, but they also keep track of the hospital's blood supply and the types of blood that are available. EMTs can use this information to determine which hospital to send a patient to. Patients are given IoT-enabled ID wristbands when they arrive at an emergency hospital and register. Analyzing how much time patients spend in each step of the process and identifying areas for improvement is made easier with this tool. It is possible that an IoT-enabled badge that measures blood pressure and pulse, as well as body temperature and breathing rate, could be the future of emergency medical care. Patients' temperatures increase unexpectedly while they wait to be seen, and an alert goes out to the medical team.

12.2.27.2 IoT-BASED MEDICAL EQUIPMENT

A couple of years ago, it was impossible or difficult to follow medical components in a way that is now possible, thanks to the Internet of medical things devices. These gadgets can aid in the achievement of medical goals by offering a positive patient experience, regardless of whether they are used to provide inventories or monitor health issues. This can be seen, for example, in devices such as smartwatches and pillboxes. A biometrics checker and a medication reminder can be found on the same device. Smart IoMT in medical devices can improve diagnosis and treatment accuracy. Disabled individuals might considerably benefit from home-based gadgets that provide healthcare mobility solutions.

In healthcare, the Internet of Things (IoT) is vital. The Internet of Medical Things (IoMT) refers to a network of interconnected medical devices that may monitor biological signals and detect disease in patients without the need for human intervention due to advances in lightweight communication protocols and smart sensors (IoMT) (Alshehri & Muhammad, 2021). The Internet of Medical Things (IoMT) provides remote and isolated populations with easy access to life-saving medical treatment in densely populated areas. This part includes post-surgical care, virtual assistance, advanced real-time monitoring of patients, implantable sensors and cameras, and diagnostic and treatment planning. Additionally, security, IoMT machine learning, and medical data reduction are essential considerations that must be taken into account. In order to lower healthcare expenditures and improve overall system performance, IoMT is currently in the early stages of development.

12.2.27.3 THE PHARMACEUTICAL AND INSURANCE SECTORS ARE TWO SUCH EXAMPLES

The breadth and depth of IoMT's application in the pharmaceutical and insurance industries are illustrative. Healthcare commodities like prescriptions, medical supplies, medical equipment, and gadgets can be tracked and transported more efficiently using tracking technology and software. Temperature, vibration, and humidity may all be monitored and regulated with IoT sensors.

There are now insurance firms that are experimenting with measuring the appropriate behavior and health state of covered individuals and groups (https://kontakt.io/blog/how-internet-of-medical-things-iomt-is-revolutionizing-healthcare). When the data is reviewed, it will be able to lower insurance premiums and speed up the claims process. The benefits of collecting, accurately aggregating, and disseminating data in the form of several reports to numerous parties now necessary for insurance compliance are numerous. Records management can save a ton of money this way.

KEYWORDS

- **advances in healthcare**
- **artificial intelligence**
- **clinical applications**
- **coronavirus disease**
- **internet of medical things**
- **nanotechnology**
- **telehealth**
- **telemedicine**
- **vitamin D**

REFERENCES

Abbas, Z., & Jawaid, Ar. M. F., (2017). Biophilia and built environment: An implication for healthcare facilities. *Intern. J. Emerging Technologies*, *8*(1), 628–634.

Alshehri, F., & Muhammad, G., (2021). A comprehensive survey of the internet of things (IoT) and AI-based smart healthcare. *IEEE Access*, *9*, 3660–3678. doi: 10.1109/ACCESS.2020.3047960.

Andreucci, M.B., Loder, A., Brown, M., & Brajkovic, J. (2021). Exploring challenges and opportunities of biophilic urban design: evidence from research and experimentation. *Sustainability 13*(8), 4323; https://doi.org/10.3390/su13084323

Bio-Inspired Technologies Ingenia, (2008). Emerging technologies. *Ingenia 20*(37). https://www.ingenia.org.uk/getattachment/17547f90-3a2d-44f2-83da-50502deaa46b/Toumazou.pdf (accessed on 17 August 2023).

Bohr, A., & Memarzadeh, K., (2020). The rise of artificial intelligence in healthcare applications. *Artificial Intelligence in Healthcare*, 25–60. https://doi.org/10.1016/B978-0-12- 818438-7.00002-2.

Burroughs, M., Urits, I., Viswanath, O., Simopoulos, T., & Hasoon, J., (2020). Benefits and shortcomings of utilizing telemedicine during the COVID-19 pandemic. *Baylor University Medical Center Proceedings, 33*(4), 699, 700. doi: 10.1080/08998280.2020.1792728.

Carlton, S., (2019). *The Era of Exponential Improvement in Healthcare?* Blockchain in Healthcare Today. https://blockchainhealthcaretoday.com/index.php/journal/article/view/128 (accessed on 17 August 2023).

Cassano, R., Cuconato, M., Calviello, G., Serini, S., & Trombino, S. (2021). Recent advances in nanotechnology for the treatment of melanoma. *Molecules 26*(4), 785. https://doi.org/10.3390/molecules26040785

Chambers, E. S., Vukmanovic-Stejic, M., Turner, C. T., Shih, B. B., Trahair, H., Pollara, G., Tsaliki, E., et al., (2021). Vitamin D_3 replacement enhances antigen-specific immunity in older adults, *Immunotherapy Advances, 1*(1). https://doi.org/10.1093/immadv/ltaa008.

Gajarawala, S. N., & Pelkowski, J. N., (2021). Telehealth benefits and barriers. *J. Nurse Practitioners, 17*(2), 218–221. https://doi.org/10.1016/j.nurpra.2020.09.013.

Hamid, R., & Manzoor, I., (2020). Nanomedicines: Nano based drug delivery systems challenges and opportunities. In: Akram, M., (ed.), *Alternative Medicine.* IntechOpen. doi: http://dx.doi.org/10.5772/intechopen.94353.

Han, H.H., Jun, D., Moon, SH. et al. Fixation of split-thickness skin graft using fast-clotting fibrin glue containing undiluted high-concentration thrombin or sutures: a comparison study. SpringerPlus 5, 1902 (2016). https://doi.org/10.1186/s40064-016-3599-x

Hemanth, D. J., Anitha, J., & Tsihrintzis, G. A., (2021). *Internet of Medical Things: Remote Healthcare Systems and Applications.* Springer.

Hollander, J. E., & Carr, B. G., (2020). Virtually perfect? Telemedicine for COVID-19. *N. Engl. J. Med., 382*, 1679–1681. doi: 10.1056/NEJMp2003539.

http://www.cmmmagazine.com/cmm-articles/healthcare-products-and-materials-inspired-by-nature/ (accessed on 17 August 2023).

https://aspe.hhs.gov/topics/health-health-care/advances-telehealth (accessed on 17 August 2023).

https://evisit.com/resources/what-is-the-difference-between-telemedicine-telecare-and-telehealth (accessed on 17 August 2023).

https://healthtechmagazine.net/article/2020/01/how-internet-medical-things-impacting-healthcare-perfcon (accessed on 17 August 2023).

https://ied.eu/blog/importance-of-technology-in-healthcare/ (accessed on 17 August 2023).

https://nph.onlinelibrary.wiley.com/doi/full/10.1002/ppp3.10138 (accessed on 17 August 2023).

https://nutrition.bmj.com/content/bmjnph/3/1/106.full.pdf (accessed on 17 August 2023).

https://www.asme.org/topics-resources/content/9-bioinspired-medical-technologies (accessed on 17 August 2023).

https://www.ed.ac.uk/roslin/news-events/latest-news/vitamin-d-supplements-boost-immunity-in-elderly (accessed on 17 August 2023).

https://www.ingenia.org.uk/getattachment/17547f90-3a2d-44f2-83da-50502deaa46b/Toumazou.pdf (accessed on 17 August 2023).

https://www.insight.com/en_US/content-and-resources/2021/future-of-healthcare--internet-of-medical-things.html (accessed on 17 August 2023).

https://www.netscribes.com/nanotechnology-in-healthcare/ (accessed on 17 August 2023).

https://www.science.org/doi/10.1126/science.368.6493.809 (accessed on 17 August 2023).

https://www.tribuneindia.com/news/schools/peacock-feathers-and-butterfly-wings-inspire-material-for-next-gen-smart-sensors-92929 (accessed on 17 August 2023).

https://www.visualcapitalist.com/the-future-of-nanotechnology-in-medicine/ (accessed on 17 August 2023).

https://www.webmd.com/lung/how-does-telemedicine-work#1 (accessed on 17 August 2023).

Hua, S., & Wu, S. Y., (2018). Editorial: Advances and challenges in nanomedicine. *Front. Pharmacol,* https://www.frontiersin.org/articles/10.3389/fphar.2018.01397/full (accessed on 17 August 2023).

Hull, M. S., (2019). Chapter 3 – multidimensional impacts of nanotechnology on public health. In: Roig, B., Weiss, K., & Thireau, V., (eds.), *Management of Emerging Public Health Issues and Risks Multidisciplinary Approaches to the Changing Environment* (pp. 65–85). https://doi.org/10.1016/B978-0-12-813290-6.00003-2.

Jenkins, M. L., (2022). Health IT advances for the 21st century. *J. American Association of Nurse Practitioners, 34*(2), 405–409. doi: 10.1097/JXX.0000000000000613.

Kalis, B., Collier, M., & Fu, R., (2018). *10 Promising AI Applications in Healthcare.* https://hbr.org/2018/05/10-promising-ai-applications-in-health-care (accessed on 17 August 2023).

Keskinbora, K. H., & Jameel, M. A., (2018). Nanotechnology applications and approaches in medicine: A review. *J. Nanosci. Nanotechnol. Res*., *2*(2), 1–6. https://www.primescholars.com/articles/nanotechnology-applications-and-approaches-in-medicine-short-review.pdf (accessed on 27 September 2023).

Kissi, J., Quansah, D. K. K., Nutakor, J. A., Dankyi, A. B., & Adu-Gyamdi, Y., (2022). Telehealth during COVID-19 pandemic era: A systematic review. *AIMS Med. Sci, 9*(1), 81–97. doi: 10.3934/medsci.2022008.

Kong, X., Gao, P., Wang, J. et al. Advances of medical nanorobots for future cancer treatments. J Hematol Oncol 16, 74 (2023). https: //doi.org/10.1186/s13045-023-01463-z

Lanham-New, S. A., Webb, A. R., Cashman, K. D., Buttriss, J. L., Fallowfield, J. L., Masud, T., Hewison, M., et al., (2021). Vitamin D and SARS-CoV-2 virus/COVID-19 disease. *BMJ Nutrition, Prevention & Health, 3*(1), 106–110. https://doi.org/10.1136/bmjnph-2020-000089.

Leslie, M., (2020). T cells found in coronavirus patients 'bode well' for long-term immunity. *Science, 368*(6493), 809, 810. doi: 10.1126/science.368.6493.809.

Li, X., Hu, Z., Ma, J., Wang, X., Zhang, Y., Wang, W., & Yuan, Z., (2018). The systematic evaluation of size-dependent toxicity and multi-time biodistribution of gold nanoparticles. *Colloids Surf. B Biointerfaces, 167*, 260–266. doi: 10.1016/j.colsurfb.2018.04.005.

Lin, W., (2015). Introduction: Nanoparticles in medicine. *Chemical Reviews, 115*(19), 10407–10409. doi: 10.1021/acs.chemrev.5b00534.

Litvak, M., Miller, K., Boyle, T., Bedenbaugh, R., Smith, C., Meguerdichian, D., & Goralnick, E., (2021). Telemedicine use in disasters: A scoping review. *Disaster Medicine and Public Health Preparedness* (pp. 1–10). doi: 10.1017/dmp.2020.473.

Lurie, N., & Carr, B. G., (2018). The role of telehealth in the medical response to disasters. *JAMA Intern Med., 178*(6), 745, 746. doi: 10.1001/jamainternmed.2018.1314. PMID: 29710200.

Marr, B., (2018). *How is AI Used in Healthcare—5 Powerful Real-World Examples That Show the Latest Advances*? Forbes. https://www.forbes.com/sites/bernardmarr/2018/07/27/

how-is-ai-used-in-healthcare-5-powerful-real-world-examples-that-show-the-latest-advances/?sh=632cb7b5dfbe (accessed on 17 August 2023).
Martins, J. P., Das Neves, J., & De La Fuente, M., (2020). The solid progress of nanomedicine. *Drug Deliv. and Transl. Res., 10*, 726–729. https://doi.org/10.1007/s13346-020-00743-2.
Messersmith, P. (2016). Biological polyphenols as inspiration for biomedical adhesives and coatings. Front. Bioeng. Biotechnol. Conference Abstract: 10th World Biomaterials Congress. doi: 10.3389/conf.FBIOE.2016.01.02469.
Moise, L., Ross, T. M., Hoft, D. F., Martin, W. D., & De Groot, A. S., (2020). Exploit T cell immunity for rapid, safe and effective COVID-19 vaccines. *Expert Rev Vaccines, 19*(9), 781–784. doi: 10.1080/14760584.2020.1825946.
Monaghesh, E., & Hajizadeh, A., (2020). The role of telehealth during COVID-19 outbreak: A systematic review based on current evidence. BMC *Public Health, 20*, 1193. https://doi.org/10.1186/s12889-020-09301-4.
Murthy, S. K., (2007). Nanoparticles in modern medicine: State of the art and future challenges. *Intern. J. Nanomed.*, *2*(2), 129–141.
Pandey, N., Soto-Garcia, L. F., Liao, J., Philippe, Z., Nguyen, K. T., & Hong, Y., (2020). Mussel-inspired bioadhesives in healthcare: Design parameters, current trends, and future perspectives. *Biomater Sci.*, *8*(5), 1240–1255. doi: 10.1039/c9bm01848d.
Paul, D., Sanap, G., Shenoy, S., Kalyane, D., Kalia, K., & Tekade, R. K., (2021). Artificial intelligence in drug discovery and development. *Drug Discovery Today, 26*(1), 80–93. https://doi.org/10.1016/j.drudis.2020.10.010.
Perrone, G., Zerbo, S., Bilotta, C., Malta, G., & Argo, A., (2020). Telemedicine during COVID-19 pandemic: Advantage or critical issue? *Med. Leg. J.*, *88*(2), 76–77. doi: 10.1177/0025817220926926.
Qureshi, K. N., Ahmad, A., Piccialli, F., Casolla, G., & Jeon, G., (2021). Nature-inspired solution for coronavirus disease detection and its impact on existing healthcare systems. *Computers Electrical Engineering, 95*, 107411.
Ray, U., (2019). *The Health Impact of Nanotechnology.* https://www.azonano.com/article.aspx?ArticleID=5113 (accessed on 17 August 2023).
Razdan, S., & Sharma, S., (2021). Internet of medical things (IoMT): Overview, emerging technologies, and case studies. *IETE Technical Review*. doi: 10.1080/02564602.2021.1927863.
Rothman, D., (2020). *Artificial Intelligence by Example: Acquire Advanced AI, Machine Learning, and Deep Learning Design Skills* (2nd edn.). Packt Publishing.
Sakes, A., Van, D. S. I. A., De Kater, E. P., Posthoorn, P., Scali, M., & Breedveld, P., (2020). Development of a novel wasp-inspired friction-based tissue transportation device. *Frontiers in Bioengineering and Biotechnology*, *8*, 575007. https://doi.org/10.3389/fbioe.2020.575007.
Salamanca-Buentello, F., & Daar, A. S., (2021). Nanotechnology, equity, and global health. *Nat. Nanotechnol., 16*, 358–361. https://doi.org/10.1038/s41565-021-00899-z.
Sani, A., Cao, C., & Cui, D., (2021). Toxicity of gold nanoparticles (AuNPs): A review. *Biochemistry and Biophysics Reports, 26*, 100991. https://doi.org/10.1016/j.bbrep.2021.100991.
Shakoor, H., Feehan, J., Al Dhaheri, A. S., Cheikh, I. L., Ali, H. I., Alhebshi, S. H., Apostolopoulos, V., & Stojanovska, L., (2021). Role of vitamin D supplementation in aging patients with COVID-19. *Maturitas, 152*, 63–65. https://doi.org/10.1016/j.maturitas.2021.03.006.
Shrimali, R. How IoT is Transforming the Healthcare Industry. Embedded Computer Design (2020). https://embeddedcomputing.com/application/healthcare/telehealth-healthcare-iot/how-iot-is-transforming-the-healthcare-industry

Steger, A., (2020). *How the Internet of Medical Things is Impacting Healthcare*. https://healthtechmagazine.net/article/2020/01/how-internet-medical-things-impacting-healthcare-perfcon (accessed on 17 August 2023).

Trafton, A., (2021). MIT News Office. *Drug Delivery Capsule Could Replace Injections for Protein Drugs*. https://news.mit.edu/2021/monoclonal-drug-delivery-capsule-0830 (accessed on 17 August 2023).

Utreja, P., Verma, S., Rahman, M., & Lalit, K., (2020). Use of nanoparticles in medicine. *Current Biochemical Engineering, 6*(1), 7–24. doi: 10.2174/2212711906666190724145101.

Vinuesa, R., Azizpour, H., Leite, I., et al., (2020). The role of artificial intelligence in achieving the sustainable development goals. *Nat. Commun., 11*, 233 https://doi.org/10.1038/s41467-019-14108-y.

Xiao, X., Lan, Y., & Chen, J., (2021). Learning from nature for healthcare, energy, and environment. *The Innovation, 2*(3), 100135. https://www.cell.com/the-innovation/pdf/S2666-6758(21)00060-6.pdf (accessed on 17 August 2023).

Xu, P., Ni, Y., Lu, S., Liu, S., Zhou, X., & Fan, Y., (2021). The cushioning function of woodpecker's jaw apparatus during the pecking process. *Comput Methods Biomech Biomed Engin., 24*(5), 527–537. doi: 10.1080/10255842.2020.1838489.

Yang, J., Chen, J., Su, Y., Jing, Q., Li, Z., Yi, F., Wen, X., et al., (2015). Eardrum-inspired active sensors for self-powered cardiovascular system characterization and throat-attached anti-interference voice recognition. *Advanced Materials, 27*(8), 1316–1326. https://doi.org/10.1002/adma.201404794.

Yuk, K., Chakraborty,A., Cheng, S., Chung, C.-I., Jorgensen, A., Basu, S., Chamorro, L.P., & Jung, S., (2022). On the design of particle filters inspired by animal noses. J. R. Soc. Interface 19(188). http://doi.org/10.1098/rsif.2021.0849

Index

C

D

E

I

M

N

W

For Product Safety Concerns and Information please contact our EU representative GPSR@taylorandfrancis.com
Taylor & Francis Verlag GmbH, Kaufingerstraße 24, 80331 München, Germany

www.ingramcontent.com/pod-product-compliance
Ingram Content Group UK Ltd.
Pitfield, Milton Keynes, MK11 3LW, UK
UKHW022209190926
475702UK00003B/270
* 9 7 8 1 7 7 4 9 1 4 8 2 3 *